15K

ROOTSTOCKS
FOR FRUIT CROPS

ROOTSTOCKS FOR FRUIT CROPS

Edited by

Roy C. Rom
Department of Horticulture and Forestry
University of Arkansas
Fayetteville, Arkansas

Robert F. Carlson
Department of Horticulture
Michigan State University
East Lansing, Michigan

A WILEY-INTERSCIENCE PUBLICATION
JOHN WILEY & SONS
New York • Chichester • Brisbane • Toronto • Singapore

Library of Congress Cataloging in Publication Data:

Rootstocks for fruit crops

"A Wiley-Interscience publication."
Includes bibliographies and index.
1. Fruit trees—Rootstocks.   2. Fruit—Rootstocks.
I. Rom, Roy C. (Roy Curt)   II. Carlson, Robert F.
(Robert Fritz), 1909-

SB359.45.R66   1987   634'.0432   86-15730
ISBN 0-471-80551-3

Printed in the United States of America

10 9 8 7 6 5 4 3 2 1

CONTRIBUTORS

J. M. Audergon
Station de Recherches Fruitière Méditerranéennes
Montfavet-Avignon, France

Robert F. Carlson
Department of Horticulture
Michigan State University
East Lansing, Michigan

William S. Castle
Citrus Research and Education Center
University of Florida
Lake Alfred, Florida

Peter B. Catlin
Department of Pomology
University of California
Davis, California

P. Crossa-Raynaud
Station de Recherches Fruitière Méditerranéenes
Montfavet-Avignon, France

David C. Ferree
Department of Horticulture
Ohio Agricultural Research and Development Center
Ohio State University
Wooster, Ohio

Charles Grasselly
Station de Recherches Fruitière Méditerranéenes
Montfavet-Avignon, France

J. Dan Hanna
Department of Horticulture Science
Texas A&M University
College Station, Texas

v

Brian H. Howard
Plant Propagation Department
East Malling Research Station
East Malling, Maidstone, Kent, England

Gordon S. Howell
Department of Horticulture
Michigan State University
East Lansing, Michigan

Dale E. Kester
Department of Pomology
University of California
Davis, California

Richard E. C. Layne
Department of Agriculture
Research Station, Agriculture Canada
Harrow, Ontario, Canada

Porter B. Lombard
Department of Horticulture
Oregon State University
Corvallis, Oregon

Gale H. McGranahan
Department of Pomology
University of California
Davis, California

William R. Okie
USDA Fruit and Nut Research Station
Byron, Georgia

Ronald L. Perry
Department of Horticulture
Michigan State University
East Lansing, Michigan

Roy C. Rom
Department of Horticulture and Forestry
University of Arkansas
Fayetteville, Arkansas

Roy K. Simons
Department of Horticulture
University of Illinois
Urbana, Illinois

Melvin N. Westwood
Department of Horticulture
Oregon State University
Corvallis, Oregon

PREFACE

"A powerful root system, its wide and deep distribution in the soil, and a persistent and adequate annual growth of absorbing roots are the principal prerequisites of abundant fruit bearing."

—*Kolesnikov*

The propagation of vine and tree fruit scions on rootstocks has been a horticultural practice for over 20 centuries. The belief that any rootstock, as long as it was compatible with the scion cultivar, was satisfactory is no longer tenable. Pomologists have long had an awareness of a rootstock's importance and its utilization in horticulture. The realization that successful production is intimately related to stock—scion interaction and rootstock environmental adaption is now firmly established in the minds of growers and nurserymen as well as researchers. As modern tree fruit and vine-growing systems become more complex and intense, rootstock problems appear more acute. Therefore, it is important to have an understanding and knowledge of rootstocks: their characteristics, usefulness, and availability. This book fulfills that purpose.

Serious attempts to study rootsystems, to characterize them, and to develop improved rootstocks have only been attempted in the past 100 years. This is especially so for asexually propagated clonal rootstocks. This effort has increased internationally in recent years. While rootstock research had its beginnings in the eighteenth century, the principal emphasis was on scion growth control in apples by the rootstock. In the second decade of this century, the East Malling Research Station, England, assumed strong leadership in rootstock development when Sir Ronald G. Hatton selected and characterized the East Malling Apple Rootstock Series. From 1920 to 1950 the original E.M. Rootstocks (and others developed later) were extensively researched but were not widely used in grower orchards, particularly in the United States.

Commercial interest in fruit tree rootstocks became evident in 1958 when a small group of pomologists and growers met in Michigan to form the Dwarf Fruit Tree Association, which later expanded to the currently very active International Dwarf Fruit Tree Association (IDFTA). The association's purpose is to promote fruit tree rootstock research and to disseminate practical research information to the fruit industry through newsletters, annual meetings and proceedings (Compact Fruit Tree), and study tours. In 1975, this association established a research committee to coordinate and encourage submission of fruit tree research project proposals from universities and experiment stations. Support funds for this research are derived from members' annual dues (70%) and special grants from growers and fruit-tree-related industries. It is appropriate that this book's contributing authors agreed that royalties would go to the IDFTA rootstock research fund and be utilized to develop improved rootstocks for pome, stone, vine, and citrus crops for the future.

This book was conceived with the idea that bringing together the state of the rootstock art into a single volume would serve a worthy function. And this is precisely what it will do. Authors contributing chapters were selected from the international community of researchers. All have exceptional familiarity and proficiency with regard to the rootstocks they write about; as a result, this text contains a unique information record on the status and characteristics of the principal rootstocks for tree and vine fruit crops. As such, it will be valuable and stand as the only unified comprehensive rootstock resource text available for those engaged in teaching, research, and nursery tree and fruit production.

The editors wish to acknowledge with thanks the willing contribution and timeless effort of the authors who have assigned royalties from this book to the furtherance of rootstock development.

ROY C. ROM
ROBERT F. CARLSON

Fayetteville, Arkansas
East Lansing, Michigan
September, 1986

CONTENTS

ROOTSTOCKS
FOR FRUIT CROPS

INTRODUCTION

Roy C. Rom
University of Arkansas
Fayetteville, Arkansas

It has been stated that fruit growing is essentially photosynthesis management. This management must recognize the importance of the tree's roots, particularly when a specific rootstock is combined with a scion cultivar to form a multiple genetic system. Kolesnikov (2) states that

> The roots of fruit plants are as active as the leaves and the root system as a whole, interacting with the aboveground system, plays an important role in growth, development, and fruiting of the plant. It follows from this that the fruit grower must have a good knowledge of the structure and activity of the whole plant and be able to carry out daily control of the harmonious work of both leaves and roots.

Knowledge concerning the important function of roots, particularly specific rootstocks, becomes valuable and critical in understanding their application in fruit culture.

The origin of the word *root* dates back to 3000 B.C., when Proto-Indo-European was the common language. At that time the word w(*rad*) was used to describe the descending axis of a plant. Later, around 1500 B.C., the Proto-Germanic language formed the word *rot*, which the Old Norse language adapted about A.D. 500–1100. The word passed to the vocabulary of the Old English language with the Viking invasion, but was modified to *rote* around A.D. 1100–1500. The word *rote* was distinctly separate in meaning from an Old High German word *wurz*, which meant both "root" and "herb"; the English version of herb was *wort*. Modern English, which evolved after A.D. 1500, brough the word *root* into common usage. The Greek language derived from Proto-Indo-European contained the word *rhiza*, which yielded the term *rhizome* used in botany since the late eighteenth century. Similarly, the Latin word *radix* evolved to the botanical term *radical*, describing the plant's primary root. Thus roots have had a long history in the discussions and writing of mankind.

1

Dictionary definitions of the word root, that is, (1) the descending axis of a plant, (2) usually the underground portion of a plant, give little clue regarding the root's vital and complex role in the tree's physiology. Root systems serve the important functions of tree anchorage, absorption and transport of water and minerals, biosynthesis of growth regulators, and food storage, plus an ancillary function of contributing to the rhizosphere environment.

Natural graft unions, particularly between roots, are common in nature. The observation and study of this natural occurrence may have led ancient man to engage in the horticultural art of grafting - uniting parts of two plants so that they can grow and function as a single plant. The Chinese practiced grafting as early as 1000 B.C.. Aristotle (384−322 B.C.) discussed the art in his writings. Early Roman writers described it precisely. And St. Paul, in his epistle to the Romans 11:17−24, illustrated a point by describing a graft between good and wild olives (1).

Historically, seedlings were used as rootstocks and grafting was exploited as an asexual propagation method to perpetuate or multiply fruit scions (cultivars) selected on the basis of worthy horticultural characteristics. Little concern was attributed to the rootstock characteristic other than compatibility with the scion, with the possible exception of vigor control.

The use of rootstocks in Europe to dwarf apple and pear scions became a common practice beginning in the late fifteenth century, but applied primarily to garden production. Specialization in fruit growing, orchard plantings, occurred following the Industrial Revolution with increased interest in size control and stock scion relations occurring after the mid-eighteenth century. Yet in the United States there was no significant trend toward the use of dwarfed fruit trees until the mid-twentieth century. At this time their potential for increasing economic efficiency began to be realized (3).

In the philosophy of current production practices, it is recognized that, as a consequence of bud or graft propagation, a rootstock and a scion cultivar, each controlled by different genes, are brought together to form a compound genetic system—the tree or grape vine. This genetic system is sometimes called a "stion" (stock−scion combination). An accommodation or adjustment is made between these genetic systems and an equilibrium is reached, resulting in a characteristic metabolism for the unit. One of the most easily recognized results of the new equilibrium is found in scion vigor, manifested principally by tree size reduction, although some rootstocks are selected for increased scion vigor. Other physiological consequences that are more subtle but are increasing in importance relate to mineral uptake, precocity, hardiness, and yield efficiency. All these biochemical effects are utilized to grower advantage in modern high-density production schemes.

Today rootstocks are also used to meet a specific cultural need, aside from vigor management, with adaptability to climatic and edaphic conditions having a high priority requirement. Although for fruit tree species seed-propagated rootstocks are still used, there are many rootstocks propagated asexually as clones. Ideal rootstocks, by modern pomological criteria, provide options for solving problems that exist in the tree nursery as well as in the orchard. These criteria, as related to the production of trees, are: compatibility with scions, rootability if clonally propagated, high germination percent if seed propagated, freedom from nursery disease and for efficient and economic plant production. Those characteristics relating to production by the trees are: adaptability to the soil problems such as nematodes, soil pathogens, and insects; to soil characteristics pertaining to drainage, pH, depth, and nutrition; and to climatic excesses of heat and/or cold.

While there are at present numerous rootstock options for tree fruits, there are still major rootstock problems. Current rootstock breeding programs worldwide, are designed to develop rootstocks with strong attributes in many of the criteria listed, thus permitting the selection of rootstocks to endure one or more of the stress factors present in any given site or to expand their use into orchard conditions for which there are now no known acceptable rootstocks.

With the improved methods of root initiation, an interest in producing root systems directly on scions (own-rooting) has developed. The future to this approach lies in the philosophy that a scion's genetic makeup is such that its own roots can tolerate or resist the stresses found in a root zone environment or that the economic production system cycle of the orchard is shortened to the extent that root system stress is neither debilitating nor limiting during that period. A future answer may lie in combining scion and specific rootstock characteristics into one genetic system through breeding or genetic engineering.

The fruit tree's or vine's root system, whether seedling, clonal, or own-rooted, is an integral part of the plant, interacting in significant ways with the aboveground portion of the system. As such, root systems, previously unheralded in orcharding or vine culture, now command an understanding and attention equal to that given to scion cultivars in modern intensive orchard production systems.

REFERENCES

1. Hartman H. T., and D. E. Kester (1975). *Plant Propagation Principles and Practices*, Prentice-Hall, Englewood Cliffs, NJ, 372 pp.
2. Kolesnikov, V. A. (1971). *The Root Systems of Fruit Plants*, MIR Publications, Moscow.
3. Tukey, H. B. (1964). *Dwarfed Fruit Trees*. Macmillan, New York, pp. 11−27.

1

ROOTS

Roy C. Rom
University of Arkansas
Fayetteville, Arkansas

1. INTRODUCTION

The mature seed (embryo) has at the apex of its hypocotyl a well-defined root primorida. Upon germination, the first structure to emerge is the radicle, an embryonic or primary root, which ultimately develops into a well-defined and specialized anchoring and absorbing organ. In time, this primary root gives rise to lateral roots. When the primary root and its laterals develop more or less equally, the growth is described as a fibrous root system, a type typical of most fruit tree roots. Roots are described as vegetative parts of a plant system with distinct form and function which contribute to the total activity of the root–scion combination.

Roots of a species have their own distinct characteristics, yet, in the main, roots of all fruit species are quite similar in morphology, structure, and function (3). Ultimately each root system is a manifestation of its heredity and environment (31). The fundamental difference between roots and stems is found in their external and internal structural adaptation.

2. ROOT STRUCTURE

Since actively growing roots undergo extensive anatomical development, which relates to their function, it is worthwhile to examine, in some detail, their basic partitions: (a) the root cap, (b) the meristematic regions, (c) the area of elongation and differentiation, and (d) the region of maturation.

2.1. The Root Cap

The root's apex consists of a root cap formed during embryology from peripheral meristem cells. It is several cells thick at its tip and decreases in

5

thickness laterally (Fig. 1). The cap's conical shape is the result of its being pushed through the soil by the cell elongation occurring behind it. During growth this cap is reduced by abrasion with the soil, causing outer cells to disintegrate. It is continually renewed from within by cells of the meristem directly behind it. The cap persists during the entire life of a root tip.

The root cap's function is to protect the meristematic region, to perceive a gravitational stimulus (6), and to respond to light and tactile stimulus as well. The root cap exudes a slime, or mucigel, whose function in growth is not completely understood. It is thought to improve root contact with soil particles and provide a favorable medium for soil microorganisms. In the growth process, root caps exert a pressure of 15 or 25 bars, thus opening the pore spaces. This action is to some extent dependent upon soil particle mobility (3).

2.2. Meristematic Region

Directly behind the root cap is a mass of mitotically inactive cells known as the quiescent center (17), characterized by a very low rate of protein synthesis. All rapidly growing root systems exhibit this quiescent center, whose function is to serve as a template for cell dimension and pattern changes. The quiescent center is surrounded by mitotically active cells, forming a functional promeristem. This meristem represents a complex yet organized system, responding sensitively to physiological changes in the system, such as inhibition or stimulation of growth activity (52). There is a pattern to the orientation of cell division and direction of future expansion. The destiny of the cells produced, with regard to the kind of tissue they will eventually form, is determined by the meristem (11) and not by influences from mature root tissue.

2.3. Region of Elongation

Behind the meristematic region, at a distance of 1–2 mm, cell enlargement and, later, differentiation are the dominant processes. Since the meristematic region gives rise to cells, characteristic of the root tissue, differing in size and shape, a gradual adjustment in their cell wall growth is needed if the root is to develop as a unit. Walls of adjacent cells elongate together as the result of differential growth rates in differing parts of each cell's wall. Cell growth in this region follows a typically sigmoidal pattern beginning with slow initial expansion, then a short period of rapid expansion, and finally a decline to no expansion. Concomitant with this growth are complex changes in enzyme patterns, protein composition, respiratory activity, and cell wall structure formation (11). This region is relatively short and merges into the zone of differentiation and maturation.

2.4. Region of Differentiation and Maturation

When the newly formed thin-walled cells at the region of elongation attain a maximum size, they begin to differentiate into specific characteristic tissues

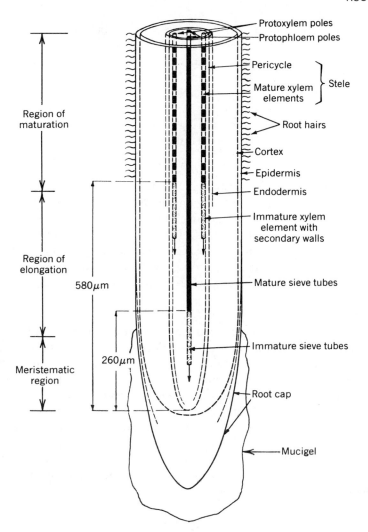

Figure 1. Diagram of a tobacco root tip showing order of maturation of various tissues. The distance from the tip at which various tissues differentiate and mature depends on the kind of root and the rate of growth. [Modified from Kramer (30) and Esau (16); reprinted with permission of the Division of Agriculture and Natural Resources, University of California.

which form the primary root axis. Differentiation and maturation of cells in the tissues being formed do not proceed with equal rapidity. This is particularly true of xylem and phloem formation. Three well-defined primary tissues are delineated: epidermis, cortex, and stele (Fig. 1).

2.4.1. The Epidermis The epidermis is composed of the outer layer of cells some of which contain external cell wall protrusions known as root hairs. They are unicellular, with thin cellulose walls, and are found in a definite zone behind the elongation region. Root hairs are common in fruit tree root systems; pecan and avocado roots are exceptions (30). Their formation is inhibited by extremes of soil moisture. They play an important function in the absorption of water (Section 5.2), particularly by greatly increasing cell surface. The tip of the root hair is permeated by a fuzz of cellulose strands which assist in establishing close contact with soil particles (11). Root hairs function for a short time, either dying or being sloughed off during root maturation. Neither the epidermis nor the hypodermis cell layer has intercellular spaces.

2.4.2. The Cortex   The cortex tissue, consisting mainly of parenchyma cells, is that portion of a young root axis between the epidermis and stele. Intercellular spaces exist, and cortex cells also have a greater frequency of plasmodesmata in their walls than do endodermal cells. Their radial and transverse walls are lignified in plants belonging to the Rosaceae. The cortex's function is primarily as a protective sheath for the stele. As growth occurs, the primary cortex is sloughed off and a cork cambium forms in the pericycle (Section 3.3).

2.4.2.1. The Endodermis.   The one-cell inner layer of the cortex, uninterrupted by intercellular spaces, is termed the endodermis, and its cells are characterized by Casparian thickening. The radial and end walls become suberized with cellulose and pectic substances, making this tissue relatively impervious to water diffusion. Thus the endodermal layer, with its Casparian strip, may act as a control over diffusion of substances between the outer cortex and the stele. By not permitting free movement of water, diffusion is through semipermeable membranes to the protoplast. The endodermis may also function to maintain pressure relations and prevent air accumulation in the conductive tissues. In woody plants, the endodermis is pierced by branch root formation, thus providing openings for lateral movement of solutions into the stele (14).

2.4.3. The Stele   The primary root of dicotyledons has a stele comprised of a mass of primary xylem tissue with two to several arms extending out to the pericycle, which is the stele's outermost cell layer. The pericycle is commonly a single layer of cells but frequently becomes multilayered, forming a cylinder just inside the endodermis. In some instances, this cylinder is interrupted by protoxylem cells which contact the endodermal cells and which may be a source of lateral rooting (Section 3.5). Between the xylem arms are primary phloem cells. Parenchymatous cells in the zone between xylem and phloem may differentiate into a cambium.

The stele's function is to provide longitudinal transport of water and solutes in the root system. The regions described are not clearly delineated, as some cell division and expansion is found in all regions.

3. ROOT DEVELOPMENT AND GROWTH

A tree accumulates in its life cycle a root system, more or less permanent, that is well developed and distributed in the soil.

3.1. Root System Development

Rooting patterns are genetically controlled but modified to such an extent by the environment that species' rooting peculiarities may be partly obscured.

   Overall growth is an expression of cell division and cell enlargement. The primary (zero order, or seed root) soon gives rise to roots of the first order (laterals), which in time produce laterals of the second order, etc. In apple and pear root systems up to eight orders are developed in one season (28). Small roots, under 10 mm long, can make up more than 80% of the total root number. Average root length decreases from the first order to higher orders, but to a lesser degree in the shallow portions of a root system. Average root length is practically constant within a cultivar (28). The relative activities of the root apical meristem and the cambial meristem, issuing lateral roots, determines the rooting pattern. If the level and duration of the apical meristem activity is low and short, a much-branched root system forms (11).

3.2. Classifications

Roots may be classified as skeletal (scaffold) or fibrous. Skeletal roots are primary roots, but they may include roots of the first, second, or third order. Skeletal roots may elongate to several meters in length and, by secondary growth (Section 3.4), thicken to several centimeters. Fibrous roots form, as laterals, on the third to higher orders of branching. They are short (a fraction of a millimeter to a few centimeters long) and thin (seldom over 3 mm in diameter).

   Fibrous roots may be categorized by relating their morphological development stage to their function:

   a. Axial roots whose primary function is rapid extension growth, thus providing a base for further orders of branching. They are white and, at their meristematic zone, are thicker than the central cylinder, before secondary thickening occurs.

   b. Absorbing roots are also white but distinguished from axial roots by a higher physiological activity; their main function is absorption of water and minerals (Section 5.2, 5.3). At peak root growth periods they comprise the most numerous group, constituting up to 90% of the total rootlet number, which in an apple root system may exceed 1 million. Absorbing roots are short (less than 5 mm) and thin. Their life span is short, a few weeks, and secondary thickening does not occur.

c. Intermediate roots develop from a few former absorbing roots. Their color is a light gray and they appear to be a transition stage of classification between absorbing and conducting roots.

d. Conducting roots are light to dark brown and arise from intermediate roots whose primary cortices have sloughed away. In time, they undergo secondary thickening and may eventually become skeletal roots.

The use of the terms *absorbing* (feeding) and *conducting* to describe or classify roots by function is questionable because, at this stage, function is not properly identified by root length or diameter. However, thinner roots are mostly ephemeral and do not transpose to conducting roots.

As a tree's root system invades an increasing volume of soil with both skeletal and fibrous roots, an additional type of classification may be made based on the growth direction. Root systems may thus be thought of as horizontal or vertical.

Horizontal roots are found paralleling the soil surface at depths of 0–100 cm, where the soil environment is generally most favorable. Most roots are found in the 0–80 cm layer, and in the apple as much as 70% of the root weight may be in the 0–30 cm layer. Roots may spread horizontally, occupying an area of 2–100 m², commonly 10–20 m², or to 1.5–2.0 times the crown diameter (34). This spread is often limited by mechanical resistance in the soil or by natural avoidance of soil occupied by neighboring tree roots of the same species (28), peaches as an example. This avoidance characteristic by no means holds true for all species.

Vertical roots, or "sinkers", may descent into the soil as much as 12 m. The depth of rooting is determined by soil conditions, scion cultivar, and rootstock. Seedling roots tend to have more vertical roots while those generated from adventitious roots of stem origin are strongly horizontal (29). Active root growth proceeds longer in vertical roots than in horizontal roots and these roots have larger secondary xylem.

Root distribution changes with tree age, differs among species and clones, and is affected by soil type and cultural practice. At wider tree spacing, root systems tend to be more horizontal with few vertical roots (4). As planting density increases, the trend is to more vertical growth, although intermingling may occur. The weight, length, volume, and surface area of a root system all decrease as planting density increases.

3.3. Secondary Thickening

Secondary growth is found in tree roots but not in the higher orders of branching (16). Cambial activity increases the stele's diameter. This deep-seated meristematic activity occurs in the outer pericycle cell layer. Successive cork cambiums form and, in the process, the cortex and epidermis are sloughed off and the endodermis layer is lost. The root bark that forms from the cork cambium is more permeable to water than is stem bark (31). Along with the expansion growth external to the pericycle, secondary xylem and phloem develop. They arise from vascular cambium at the inner side of the phloem.

Secondary growth of this nature, whether in root or stem, is utilized in the budding and grafting of trees.

As a result of secondary growth, root diameter increases and roots become woody, frequently resulting in a change in root color from white to brown. Root browning may result from other causes such as changes in soil temperature, root desiccation, or root feeding by soil flora and fauna all of which may cause death to the epidermal layer or result in cortical breakdown or total degeneration of the axis. Browning may occur in patches on a root (Fig. 2), and

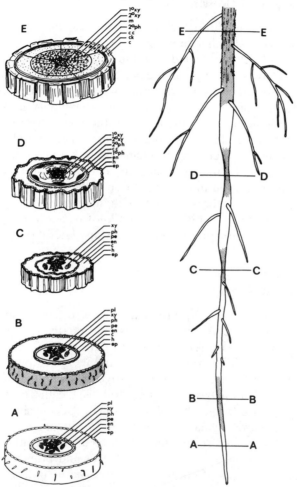

Figure 2. Diagrammatic representation of a grapevine root showing anatomical and structural changes along its length. (*A*) White primary root hairs. (*B*) Epidermis collapsed, giving a brown appearance, hypodermis and endodermis suberized. (*C*) Cortical collapse (browning) and degeneration of all tissues. (*D*) Cortical collapse with expansion of the stele due to development of secondary growth. (*E*) Woody, secondary growth with remnants of the cortex still attached.

Abbreviations: ep, epidermis; h, hypodermis; c, cortex; en, endodermis; pe, pericycle; ph and 1°ph, primary phloem; xy and 1°xy, primary xylem; pi, pith; cc, cork cambium; ck, cork; m, medullary ray; 2° ph, secondary phloem; 2° xy, secondary xylem. [After Richard and Considine (38), reprinted with permission of the publisher.]

this type of browning (result of phenol oxidation) is associated with a decrease in root diameter. Phenols in these primary root axis tissues may function to give young roots resistance to infection. The suberization of hypodermal and endodermal cells, to give the young root a barrier to desiccation, which occurs without a color change in grapes, (38) is not to be confused with root browning caused by epidermis death, cortical breakdown or root axis degeneration.

3.4. Secondary Root Growth

The primary root forms during early embryonic development and emerges at seed germination. All other roots are the result of lateral root initiation, a complex process closely associated with the plant growth regulator complex. Secondary roots have their origin in the pericycle, which may be considered as a potential deep-seated meristematic tissue (45). They form opposite protoxylem points, and their organization occurs prior to secondary growth in the root axis. They arise in acropetal sequence, with the youngest tissue nearest the apical meristem. A conical growing point is formed. This lateral root primordia enlarges and forces itself, both by mechanical pressure and digestive activity (4), through the endodermis and remaining cortex. Lateral root xylem and phloem connections are continuous with those of the main root axis. Development may also occur on suberized roots with much age and secondary thickening. The stimulus for this initiation is not known, but it may be associated with a change in root direction or pressure. Many lateral or secondary roots form, but few persist.

3.5. Adventitious Roots

The term *adventitious roots* by broad definition refers to roots that arise from aerial plant parts, underground stems, and relatively old roots (23). In another sense, it refers to roots that do not arise from root pericycle tissue. The organization of adventitious root structure and its development sequence is similar in all features to that of true roots. Adventitious roots are distinct from wound roots formed at the base of cuttings following callus proliferation (21).

Root primordia are formed by stem cambia at specific sites where branch and leaf trachea, parenchyma, or primary or secondary rays intersect with cambium tissue (48, 8, 23). This juxtaposition occurs frequently at stem nodes. Nodal rooting is associated with starch-rich leaf and branch gap areas (15) (Fig.3). Under conditions of etiolation, such as occur in layer propagation, there is a decrease in stem sclerification which is negatively correlated with adventitious rooting propensity (8). Once initiated, the root primorida forms a cap and enlarges until it reaches the outer bark surface. There it may project a bit as a node or burrknot initial. At this stage, it remains dormant until special conditions of low light, high humidity, and warm temperature stimulate further development (41). Adventitious rooting is not restricted to the nodal zone and may develop at any point along the internode or through lenticels as long as there is a retention of meristematic capacity in certain cells (28, 29).

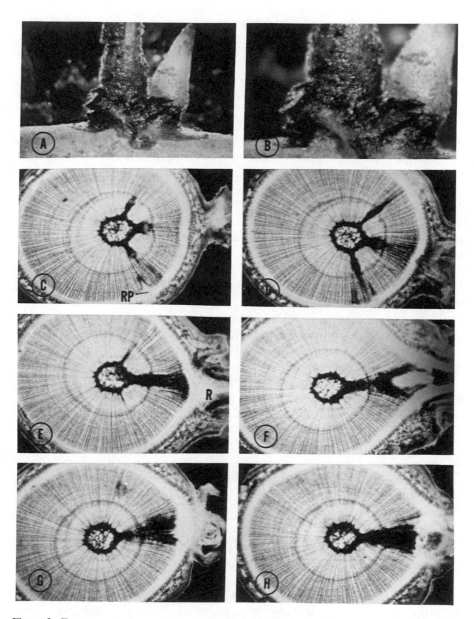

Figure 3. Emergence of two roots through a bud (*A*, *B*). Sequential transverse section of M-2 clonal apple rootstock stem showing: (*C*, *D*) branch gaps and root primordium (RP); (*E*, *F*) new root with cambial continuity, initiated from leaf gap; (*G*, *H*) position of bud with starch deposits extending into pith of cutting. [From Doud and Carlson (15).]

3.6. Root Cyclic Growth and Longevity

Whereas shoot growth duration in bearing fruit trees is short (1.5−2.5 months), root growth may last 9 months or more depending on such factors as species, culture, and environment. This growth occurs in one to three waves or cycles rather than evenly or in one cycle. The growth pattern may differ with soil depth and proceeds mainly at night (60%), although it also takes place by day (34). Apple roots have been observed to have some growth at all times of the year.

As a rule, root growth precedes shoot growth following soil warming, and reaches a maximum prior to vigorous vegetative growth. A second growth peak occurs in late summer after vegetative growth ceases and the crop is matured. Minor fluctuations in the pattern, which may be detected during the growing season, relate to internal tree processes such as flowering, seed development, flower initiation, or to such external causes as drought and variations in soil temperature. Cyclic fluctuations may be due to competition between roots and shoots for carbohydrate reserves (37). Shoot growth intensity is a factor which suppresses root growth, creating a cyclic pattern (24). If shoot growth is strong, white root production is suppressed in early summer; if it is weak, production of new roots proceeds into midsummer. Not all fruit trees follow the same seasonal cycle. Thus growth patterns vary among species. Patterns are also strongly influenced by environment and cultural practices which affect vegetative vigor, for example, pruning degree, crop level, and fertilizer use.

Root death, self-thinning, as well as new root growth are part of a natural cycle which returns minerals to the soil, feeds soil flora and fauna, and contributes to soil structure (4). Three types of root death are distinguished (11):

a. A systematic dying of short laterals on active roots occur after a brief period, 1−2 weeks, in apple (12), following suberization of roots from which they arose (27). These roots are often of the second order of branching (23).
b. Tips of main roots and those of a higher branching order frequently die back while the lateral roots continue the main growth course (52).
c. Old (more than five years) fibrous roots may also die completely, only to be replaced by new fibrous roots.

Skeletal or semiskeletal roots die, particularly after age five in the apple, and the proportion of dead to live roots increases with distance from the trunk and in the shallowest and deepest soil horizons (28). There is little information available on maximum root longevity and their capacity to remain functional. Root shedding or self-thinning is observed at bud break with a maximum loss early in the growing season. Any environmental stress conditions increases root shedding (4).

Fruit tree root systems are in constant need of new root formation. This is necessary to maintain a balance with the shoot's demand for water and nutrition (3). It is important to have new root growth in the fall so that an absorbing surface is in place to replenish stored nutrients utilized in the flush of spring growth, characterized by vigorous shoot development.

4. THE ROOT ENVIRONMENT

The environment in which the root system grows influences all phases of its development and function. The soil environment is more stable than the air; however, roots are more sensitive to changes occurring in the soil and more responsive to these changes than aerial plant parts. The failure of many fruit tree rootstocks or the problems associated with them can be attributed to their being affected in some unfavorable or limiting way by the soil environment.

4.1. Soil Texture

The physical and chemical nature of the soil directly affects root growth rate and mass. Extension growth is limited by soil porosity, pore size, and mobility of soil particles, each of which affects mechanical resistance (3). Root tips can exert a pressure of 15 to 25 bars and deform the soil. A greater force is exerted laterally by roots developing secondary thickening. There is a close relationship between soil water and mechanical resistance. Root soil contact is important for anchorage and absorption. Growing roots utilize existing soil channels, but observations indicate that up to 40% of new roots do not have good soil contact after the cortex is lost and root diameter decreases. Contact improves again for those roots undergoing secondary thickening (4). Roots respond to pressures in the soil by a modification in the size of vascular tissue, size of epidermal cells, diameter, and branching pattern. The response is not explained entirely on the basis of physical forces; growth substances may mediate the response (18).

Rooting depth increases as soil texture changes from sand to clay if both are deep, well drained, and not compacted. In sand and loam soil, scaffolds are shallow but with vertical sinkers. In clay soils, scaffold roots descend into the subsoil at an angle (4).

4.2. Soil Oxygen

Root system oxygen requirements vary with species; most fruit trees, especially those of the Rosaceae family, have high root oxygen requirements and function best at above 10% oxygen concentration. Growth is much reduced at a 3–5% level, and yet roots survive or have survived at 0.1–3.0% oxygen (10).

New root growth requires more oxygen than sustaining roots. Virtually no oxygen is required if the tree is dormant. On the other hand, active roots are highly sensitive to oxygen stress, with root hair activity suppressed within 30 min (28) after oxygen becomes limiting.

Soil oxygen deficiency can result from inadequate aeration due to insufficient pore space for gas exchange as in compacted or fine-textured soil and poor aeration due to soil flooding. In nonflooded soils, oxygen used by roots and soil organisms in respiration may deplete the oxygen supply to a point where stress occurs if the diffusion rate of oxygen into the soil is low (30).

4.3. Soil Moisture

Root growth is random yet severely affected by excess moisture when an oxygen deficiency results. Fruit tree root systems have varying degrees of adaptability to excess soil moisture. Pecans are most tolerant, followed by quince, pear, apple, citrus, plum, cherry, apricot, peach, almond, and olive. The latter four fruit species are very susceptible to excess soil moisture (wet feet), especially during active growth. There is an important relationship between soil moisture, soil air, and mechanical resistance in a soil with respect to root growth. Hypotropism, if it exists at all, is weak, with no evidence that root seek water (30).

It is generally claimed that soil moisture in a range between field capacity and permanent wilting percent is available for growth, although some parameters of growth show a reduction at soil moisture levels above permanent wilting percent in an apple study (26). Furthermore, permanent wilting percent cannot be thought of as a soil constant, as wilting is influenced by transpiration rate and cell osmotic concentration. Available water in part of the rooting zone may permit growth to continue, masking the water-deficient status in adjacent soil (39). Compensatory growth occurs in roots in a more favorable environment, that is, below emitters in a trickle-irrigated orchard, while restricted growth occurs in the unfavorable dry root zone section. Roots respond to their own local water potential and soils are known to have different water potentials through their horizons. A root system's capacity to endure developing water stress is found in its ability to extend rapidly into soil with a lower root density or more moisture. In fruit trees, strong vegetative growth which suppresses root growth occurs at a season when a soil moisture stress commonly occurs. Under a moisture deficiency, extension growth ceases and roots tend to suberize to their tips. Roots do not, after suberization, regain their capacity to absorb water easily when the soil is wetted.

4.4. Soil Temperature

Root systems are very responsive to changes in soil temperature. At any temperature, they are the most sensitive plant part. The optimum for root growth is in the 20–25°C range. Above 35°C, new root formation in Malus and Prunus rootstocks ceases (35); established fibrous roots can withstand this temperature but not without some internal injury. Above 24°C, roots lack succulence, their percentage of dry matter increases, more lignification develops, and early cortex death occurs. The active root number in the top 30 cm of soil may be limited in areas with high soil temperatures. As a rule, roots of plants adapted to warmer climates cease growth at higher temperatures than those of cooler climates (30). Root growth starts when the soil reaches a threshold temperature. This varies with fruit type—for example, apples and pears, 7−8°C; peaches and plums, 4−5°C; and citrus 16−18°C. Axial roots grow at lower ranges than absorbing roots. Nitrogen may be absorbed at 0.6°C in the apple root system but does not reach a maximum until 7.2°C (5).

Root freezing is characteristically associated with the lighter-textured and drier soils, especially in the absence of insulating snow cover. However, there is considerable difference among species and among clones. Susceptibility to low temperatures varies 3−4°C within a season, with maximum hardness at the winter season's end. Tenderness increases with rooting depth. At any time, the farther the roots are from the crown, in verticle or horizontal distribution, the less they are able to endure low temperature. This may be a factor of root diameter, as smaller roots freeze first. The critical low temperature for apple roots is −3 to −12°C. The most hardy apple roots will will survive −10 to −15°C. For pear, the critical minimum temperature is −9°C; for peach, −10°C; and for cherry, −15°C (20).

Root damage from freezing is limited to surface roots, as critical low temperatures do not normally penetrate deep into soils. Factors influencing frost as well as high-temperature penetration into soils are important considerations, as there is less temperature tolerance in roots than in tops of fruit trees. Freezing damage to tree root systems frequently occurs in shipping and during planting.

4.5. Soil Flora and Fauna

Numerous types of flora (bacteria and fungi) and fauna (insects, earthworms, and nematodes) are found in the soil. Some forms are beneficial to root growth while others are not.

The flora in the soil is greater near root systems than in the bulk of the soil because root systems excrete into the soil such exudates as polysaccharides, enzymes, vitamins, and organic acids. These exudates facilitate dissolution of minerals but primarily aid in the development of microorganisms which in turn contribute to the production of adequate soil conditions for continued root growth (28, 37). Bacteria on the root cap surface or peripheral cells dispose of root material sloughed off in the growth process or in senescence. Bacterial invasion in root hairs and epidermal and cortical cells is common, but the stele is generally not affected (41). Reduction of the cortex by bacteria leads to loss of root contact with the soil. Generally, however, soil microorganisms contribute greatly to availability and utilization of nutrients by tree roots.

Absorbing roots, under optimal moisture conditions, form a symbiotic mycorrhizal relationship with certain soil fungi. Two types of associations are found (28). Ectotropic mycorrhizal development modifies root structure by stimulating hypertrophy and branching. The fungal cells are found on the root surfaces or between cortical cells. They form on roots growing under moderate mineral nutrition deficiencies, a condition which enhances root carbohydrate accumulation. This mycorrhizal formation is most common on trees which produce both long and short roots. Endotrophic mycorrhizae, much less common on tree root systems, citrus excepted, penetrate the host root cells and form vesicles within them as well as extending into the soil. There is no modification of root anatomy by endotrophic mycorrhizae.

The presence of mycorrhizae, depending on the degree of their penetration into the soil from the root, increases mineral absorption, particularly in low-phosphorous soils. It also reduces the resistance to water entry in roots and increases resistance to root attack by organisms that injure or infect roots. Mycorrhizal associations are generally mutually beneficial, but occasionally the fungus may be parasitic.

4.6. Soil Nutrient Status

The soil is the principal source or medium for plant nutrients. Soil acidity or alkalinity (pH) is of primary importance in determining the availability of nutrients through its effect on solubility. Aluminum and manganese toxicity problems are likely to exist in acid soils while iron chlorosis, a deficiency problem, can occur in alkaline soils. Orchard trees grow best at pH 5.5 to 6.5. Certain tree species used as specific rootstocks have a wider or narrower adaptable pH range.

The element concentration of the soil solution in the immediate vicinity of the root affects plant growth. Calcium and boron have a direct effect on root growth. Although cation exchange occurs between the soil solid phase and the root, there are no grounds for assuming that the soil solution is bypassed (42). Since living roots occupy only a fraction of the soil volume, diffusion of elements through the soil becomes important in terms of replenishing the soil solution at the root's surface. Chloride and nitrate ions move more slowly in a soil solution than in a free solution, but they do not interact with the solid phase as do potassium and phosphate, which have very low diffusion rates. Ions which interact with the solid phase have variable and slow diffusion depending upon the concentration, solid-phase composition, and soil moisture status (43). High concentration of salts can occur in arid regions or through excessive fertilizer use. The direct effects of such concentrations usually relate to high osmotic effects and decreased water absorption, but they can also hasten root maturation (22).

4.7. Alleopathy

Root growth is adversely affected by alleochems released by plants. These substances include plant exudates, leachates, and decomposition products. The term *alleopathy* is used to describe this negative interaction between roots and plant products. The replant problem common in orchards and vineyards is associated, to a degree, with toxic compounds released by decaying fruit tree root systems (25).

4.8. Root Injury

Direct injury to roots by some insects, pathogenic fungi, and nematodes is a common experience and an important part of the replant problem (54). Mechanical injury incurred during orchard operation is frequently overlooked as a limitation to normal root development.

5. ROOT FUNCTION

The root is a basic plant organ which has the following specific functions in the plant's life cycle: (a) anchorage, (b) absorption and transport of water, (c) absorption and transport of nutrients, (d) conversion or synthesis point for growth regulators, and (e) storage of food reserves. A sixth and ancillary function is the root's contributions to the rhizosphere. Roots exist in a complex interrelationship with other plant organs.

5.1. Anchorage

Rootstocks are variable in their anchorage characteristics, and success in anchorage may be related to soil texture and depth. Rooting depth increases as soil texture changes from and to loam to clay. As a rule, most roots, on a total weight basis, are formed in a loam soil. Effectiveness of the anchorage is related to genetic characteristics which control spatial distribution, root density, root strength, and rooting depth. Factors that affect the total growth of the tree also affect anchorage. Deep-planted apple trees developed a new root system near the soil surface but initially maintain enough roots at a depth to give good anchorage. Recent work by the author has shown that in apple, scion cultivar can influence root development. A firm attachment to the soil is dependent upon adventitious root development and root branching as in a fibrous root system. Root hairs are important in this binding, but the development of sclerenchyma in older roots is the source of root strength. Size-controlling clonal apple rootstocks have a high bark−xylem ratio. As this proportion of living phloem to xylem tissue increases, the roots become more brittle—M-9 for example,— anchorage decreases, and staking is required. It is difficult to study root systems in situ and to obtain accurate information as to spatial distribution. Most studies relate to quantitative analysis, not anchorage (28).

5.2. Absorption and Transport of Water

A principal function of a root system is to supply water to satisfy the tree's transpiration demand. A tree's xylem is a continuous extending from the leaves almost to the root tip, yet it does not open directly into the soil. Water movement into the root is not a simple process, since it must pass sequentially

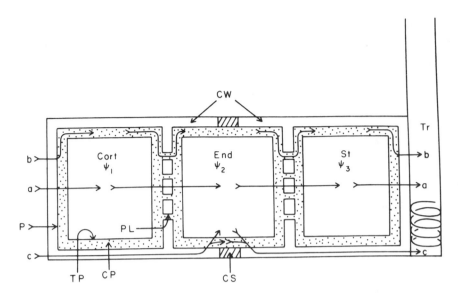

Figure 4. Diagram showing three pathways for water movement across the root cortex: (*a*) vacuole to vacuole, (*b*) movement in the symplast, (*c*) movement in the cell walls. Cort = cortical cell; End = endodermal cell; St = stelar cell; Tr = trachea; CP = cytoplasm; CA = Casparian strip; CW = cell wall; P = plasmalemma; PL = plasmodesma; TP = tonoplast. [From Weatherly (53), with permission from the publisher.]

through cells of the root cortex, endodermis and stele before it enters into the xylem tracheae. Water movement into and through a root system occurs when there is a lower water potential, as a result of transpiration, in the xylem tracheae than in the soil. There are three possible modes of entry (Fig. 4):

a. Water may pass from cell vacuole to cell vacuole, down a gradient of water potential, passing repeatedly through wall structures. There is little flow by this means.

b. Water may also pass from cell to cell through the symplast, which is composed of plasmodesmata strands connecting and binding cells together. It exits from the plasmalemma cells adjacent to the xylem tracheae. Suberization is not a hindrance to symplastic movement.

c. Water may move through cell walls (apoplastic space) in response to hydrostatic pressure. However, at the endodermal layer, which is impermeable to water, it must enter the cell and move through vacuoles or plasmalemma (55).

The latter two methods described are most probably the most important systems of water entry.

Water entry is sensitive to environmental conditions such as aeration, temperature and water supply. As soil dries, there is a strong drop in the water potential at the root surface, and roots may contract. This creates a gap in the root/soil interface, further increasing resistance to water entry (52, 4). Since water movement in soils is slow, root extension into unoccupied soil is important for continued absorption. Thus at the time of greatest water need by a tree, when shoot growth is at a maximum and transpiration rates are likely to be at their highest level, new root growth is at its lowest level. Factors which reduce photosynthesis in the tree reduce new root growth before reducing shoot growth (31). Furthermore, since fruit trees have a lower relative root length per soil area than herbaceous root systems, older roots must meet the transpiration demand (2).

Water absorption is highest at the root apex, a 4–6 mm zone. At this location, plant cells have the lowest permeability, but water absorption occurs at a relatively similar rate in the rest of the system. In secondary thickening (Section 3.3), the endodermis fractures or sloughs off and the xylem is enclosed in a layer of phloem cambium and suberized cells. The suberin is a fat like substance that impregnates the cell wall, but cracks in it allow water to enter directly into parenchyma tissue. Endodermal cells also rupture during branch root formation.

In all probability, a large portion of water absorption occurs through the older roots that have undergone secondary thickening. Although absorption is best in unsuberized roots, in fruit trees these roots constitute only a small percentage of the total root surface area and account for only a fraction of water absorbed (4, 30).

The existence of active water uptake against a concentration gradient has not been satisfactorily demonstrated. If it exists, its importance is doubtful (9, 52) and of questionable value (30, 4). Root pressure in grapes resulting from water being secreted into the xylem system by osmotic forces occurs but is not a major factor in water absorption (31).

5.3. Nutrient Absorption

The root is the chief organ for absorption of nutrients. The extent to which absorption occurs is dependent upon root length, distribution in depth, and the degree to which root contact with the soil is affected by temperature, oxygen, moisture, and nutrient concentration.

Nutrient ions move into the root system in solution by way of the apoplast, through simple diffusion, or into the symplast. This latter type of entry and movement occurs from root hairs, root epidermal, and cortex cells (2). At some point, in roots with intact Casparian strips or those with secondary thickening, nutrient ions must pass through plasmalemma membranes which maintain continuity between cortex and stele. Thus the properties of plasmalemma cells in roots are important in the exchange of ions between the soil solution and plant aerial parts. Differential membrane permeability is an important characteristic of these cells.

Ions move through membranes by active and passive means. Inorganic salts, being water soluble, move through aqueous protein channels in the cell membrane. Certain ions move through these channels faster than others. The movement rate and distribution at equilibrium is controlled by the membrane's specificity, the ion concentration on each side, and the electrical charge balance. Plant cells have an internal negative charge; thus they attract cations along electrochemical gradients.

At some membrane sites, energy transfers occur, providing electrogenic transport (nutrient pumping) which directs an ion or cation movement through membranes, into or out of cells, against electrochemical gradients. Both electrochemical and electrogenic processes operate in a fashion to keep the cell interior negatively charged with respect to its exterior (19). Active ion uptake, particularly divalent ions, is dependent on respiration. Studies with sunflowers indicate that an intermediate controller may function in the root cell. This controller appears to regulate the application of energy required for active entry of K^+, Cl^-, and NO_3^- (9). Ion absorption decreases with distance from younger to older root parts, and selectivity changes also. Calcium is readily absorbed in the youngest portions.

Once a nutrient element enters the xylem, backward diffusion is prevented by cell membrane selectivity. Nutrients then flow in the transpiration stream. Some nutrients may, however, enter phloem cells and move to "phloem sinks" in the root tips or to regions of food storage (19). At this stage, all ions are mobile.

The leakage of low-molecular-weight organic molecules from root cells results in the mobilization of absorbed micronutrients (minor metal cations), which then become available for movement into roots (32).

A functional equilibrium continues between shoots and roots for water uptake in the tree; a similar relationship exists for nutrient uptake. Root cytokinins and other hormones play a role in maintaining this growth balance, but no hypotheses to explain this root function have been proposed (39).

5.4. Conversion or Synthesis of Growth Regulators

An important root function is that of the biosynthesis and transport of plant hormones, that is, auxins, gibberellins, cytokinins, abscisic acid, and ethylene. These natural hormones markedly influence tree growth.

5.4.1. Auxins   Indoleacetic acid (IAA) if found in root tips, the stele, and the cortex. It is transported acropetally, from root base to root tip, through the vascular system or in living cells on the central cylinder. The synthesis source is thought to be in the more mature root tissues rather than in the root's apices (7). Auxin in all but the lowest concentration (10^{-8}) is inhibitory to root elongation, and supraoptimal levels occur in the root system. Growth stimulation is probably mediated by the presence of ethylene (38). Auxin transported to the root from shoot sources has a role in root growth by reducing cytokinin synthesis. Auxin applied to shoots moves rapidly, by polar transport, into the roots. The roots may actually serve as a sink for excess auxin synthesized in shoots or as a site for oxidative inactivation (49). It was thought that an asymmetric auxin distribution within a root was responsible for the geotropic response; however, three factors work against this conclusion: limited production in the root cap, transport in the wrong direction, and the inability to stimulate adequate ethylene production.

5.4.2. Gibberellins Roots serve a a source of gibberellins which move in the xylem sap. Evidence of gibberellin synthesis in roots is indirect (50). Gibberellinlike substances have been found in root xylem exudates, with the amount varying with root environment conditions. While the synthesis site may be the root tip, the root tip may serve, more importantly, as a conversion site for leaf-synthesized gibberellin to a more active form (13). The full role of gibberellin in root growth is vague. It has no function in lateral root initiation and its role in root elongation is doubtful (50).

5.4.3. Cytokinin Cytokinins are produced in root apical meristems. The levels vary with growth stage, development, and root stress. Reduced synthesis levels of cytokinins and gibberellins are found in plants exposed to drought, flooding, poor nutrition, and high soil temperatures (50, 49). This is strong evidence that cytokinins are synthesized in root tips, although there has been no direct demonstration of this. The role of cytokinins in root growth, rather than that of direct elongation, is found in its interaction with other hormones affecting cell division. A role of inhibiting lateral root initiation is possible (50). Root-produced cytokinins, through xylem transport, influence shoot physiological activity by controlling protein and CO_2 metabolism in leaves, enzyme formation in leaves, lateral shoot development, release of buds from dormancy and, fruit set (50, 36).

5.4.4. Abscisic Acid Abscisic acid (ABA) is an inhibitor of cell division and elongation. It can be stimulatory to other processes depending upon the endogenous level and its interaction with other plant hormones (50). ABA is found in root xylem extracts and synthesis occurs in root cap cells. It is also synthesized in other plant parts. ABA concentrations affect tree physiological responses in many ways: inhibiting root growth and stem elongation, inhibiting K and P ion absorption and accumulation, increasing cell wall permeability to water, accentuating leaf senescence, increasing leaf abscission, inducing rest in buds, and countering the activity of gibberellins (49). Conditions causing plant stress lead to increased ABA levels, yet root synthesis is not affected by water stress.

   ABA may have a possible role in geotropic root response, as it is asymmetrically distributed on the lower side, within the root. Gibberellins are also asymmetrically distributed, with concentration on the upper side. ABA is quite mobile in the trees. Transport in young roots is basipetal in the xylem and phloem. Some lateral transport occurs.

5.4.5. Ethylene Ethylene is not, in the strict sense, a true plant hormone, but its production throughout the tree has specific regulatory effects such as breaking of dormancy, control of flower induction, ripening and senescence, and the induction of adventitious roots. Directed transport is not a feature of ehtylene action. Its detection as a gas in various tree parts probably occurs after it has asserted its influence (1). Roots are highly sensitive to elongation growth inhibition by ethylene above endogenous levels. Endogenous produced ethylene, particularly in soils where the gas may diffuse quickly, remains low and root elongation is promoted, although the mechanism is not known (18). It is theorized that normal ethylene levels are below the inhibition threshold, but that auxin at physiologic or higher levels stimulates evolution of ethylene above this threshold (49). Ethylene interferes with auxin transport. Thus the auxin – ethylene ratio in roots controls the action of each. Auxin stimulates ethylene root growth inhibition. Ethylene, by hindering auxin action, reduces auxin stimulation of ethylene production.

Wounding elucidates ethylene evolution. Lateral branching of root systems is thought to be an ethylene response. Natural wounding of roots during root primordia emergence or in the cortical or pericycle cell of roots resulting from bending results in ethylene levels which stimulate lateral root extension growth rather than root initiation (52, 18). The latter may be an auxin-induced ethylene-production response.

6. STORAGE

A tree's roots serve as an important storage organ. Primary roots store food, starch primarily, in their cortex. Root undergoing secondary thickening store starch in parenchyma and sclerenchyma cells of the xylem and phloem. Roots, particularly those of dwarfing clonal apple stocks, have a higher proportion of parenchyma cells than stems. Some roots have special adaptations for storage usually manifested as fleshy growths, for example, carrot, and beet. This adaptation is not characteristic of tree fruit root systems, however (16).

7. CONTRIBUTION TO RHIZOSPHERE

An ancillary function of roots is their contribution to the rhizosphere. The rhizosphere is the soil zone, immediately adjacent to the roots, whose environment is influenced by roots of all tree species. The boundary between the rhizosphere and nonrhizosphere soil is not rigid. In general, it includes only the area of root hair occupancy. Organic substances move into the rhizosphere as the result of solubilization of root tip muligel, droplets secreted from root hairs, exudates from living cells, and releases from the disintegration of cortex cells (43). The combined effect is to supply an energy substrate for heterotrophic rhizosphere organisms, whose populations are greatly increased in the rhizosphere over that in the soil bulk. The amount of exudate is significant and the stimulation to flora activity could be inhibitory to root growth through competition for soil minerals (30). In young apple roots, during the first growing season, almost half of the dry weight accumulated can be lost to the rhizosphere (40). Root exudates facilitate the symbiotic relationship between soil flora and root systems (Section 4.5), but exudates may include compounds with alleopathic effect on other root systems. Rapid root growth favors increased root exudation, while nitrogen and especially phosphorous deficiencies decrease exudation. Some herbicides may also increase root exudates (44).

CONCLUSION

The root systems on which fruit species are propagated have been classified by Kolesnikov (28) as: (1) those of generative origin, seedlings; (2) those derived vegetatively by means of root cuttings or layers; and (3) those of vegetative origin from shoots sprouting from the root or stem of parent plants, with roots at their base. Cultivars grown on their own roots, for the most part, would fall in the second category.

The prerequisites of a rootstock include scion compatibility, adaptability to the edaphic and climatic environment in which it is utilized, a substantial vertical and horizontal root distribution, plus a strong, persistent, and adequate annual growth pattern.

REFERENCES

1. Abeles, F. B. (1972). Biosynthesis and mechanisms of action of ethylene, *Ann. Rev. Plant Phys.*, **23**, 259–292.
2. Atkinson, D. (1982). The growth and activity of the fruit tree root system and its mycorrhiza, ISHS 21st Int. Hort. Cong., Hamburg, 1131 (Abstr.).
3. Atkinson, D. (1980). "The growth and activity of fruit tree root systems under simulated orchard conditions," in *Environment and Root Behavior*, D. Sen, Ed., Geobios Inst., Jodhpur, India, pp. 171–183.
4. Atkinson, D. (1980). The distribution and effectiveness of roots of tree crops, *Hort. Reviews* **2**, 462–489.
5. Bajter, L. P., J. P. Magness, and L. O. Regeimbal (1939). The effect of root temperature on growth and nitrogen intake of apples, *Proc. Amer. Soc. Hort. Sci.*, **37**, 11–18.
6. Barlow, P. W. (1975). "The root cap," in *The Development and Function of Roots*, J. G. Torrey and D. T. Clarkson, Eds., Academic, London, pp. 21–55.
7. Batra, M. W., K. L. Edwards, and T. K. Scott (1975). "Auxin transport in roots: Its characteristics and relation to growth," in *The Development and Function of Roots*, J. G. Torrey and D. R. Clarkson, Academic, London, pp. 300–322.
8. Beakbane, A. B. (1961). Structure of the plant stem in relation to adventitious rooting, *Nature*, **192**, 954–955.
9. Bowling, D. J. F. (1981). "Evidence for an ion uptake controller in *Helianthus*

annus.," In *Structure and Function of Plant Roots*, R. Brower, O. Gasparikova, J. Kolek, and B. C. Loughman, Eds., Martinus Nijshoff, Dr. W. Junk, The Hague, pp. 179–185.

10. Boynton, D., and W. Reuther. (1938). Seasonal variation of oxygen and carbon dioxide in three different orchard soils during 1938 and its possible significance, *Proc. Amer. Soc. Hort. Sci.*, **36**, 1–6.

11. Carthy, J. D., and C. L. Duddington (1962). "Physiology of roots," in *View Points in Biology*, H. E. Street, Ed., Butterworths, London, pp. 1–49.

12. Childers, N. F., and D. G. White (1942). Influence of submersion of roots on transpiration, apparent photosynthesis and respiration of young apple trees, *Plant Phys*, **17**, 603–618.

13. Cozier, A., and D. M. Reid (1971). Do roots synthesize gibberellings? *Can. Jour. Bot.*, **46**(6), 967–975.

14. Dumbroff, E. B., and D. R. Peirson (1971). Probable sites for massive movement of ions across the endodermis, *Can. Jour. Bot.*, **49**(1), 35–38.

15. Doud, S. L., and R. F. Carlson (1977). Effect of etiolation, stem anatomy and starch reserves on root initiation of layered Malus clones, *J. Amer. Soc. Hort. Sci.*, **102**(4), 487–491.

16. Esau, K. (1965). *Plant Anatomy*, 2nd ed., Wiley, New York, pp. 481–531.

17. Feldman, L. J. (1984). The development and dynamics of the root apical meristem, *Amer. Jour. Bot.*, **71**(9), 1308–1314.

18. Feldman, L. J. (1984). Regulation of root development, *Ann. Rev. Plant Phys.*, 35, 223–242.

19. Galston, A. W., P. J. Davies, and R. L. Satter (1980). *Life of the Green Plant*, 3rd ed. Prentice-Hall, Englewood Cliffs, NJ, pp. 170–197.

20. Gardner, V. R., F. C. Bradford, and H. D. Hooker (1952). *Fundamentals of Fruit Production*, 3rd ed., McGraw-Hill, New York, pp. 391–408.

21. Garner, R. J. (1944). *Propagation by Cuttings and Layers, Recent Work and its Application with Special Reference to Pome and Stone Fruits*, Imp. Bureau of Hort. and Plant Crops. Tech Comm. No. 14.

22. Hayward, J. E., and W. M. Blair (1942). Some responses of 'Valencia' orange seedlings to varying concentrations of chloride and hydrogen ions, *Amer. Jour. Bot.*, **29**, 148–155.

23. Hayward, H. E. (1983). *The Structure of Economic Plants*, Mcmillan, New York, pp. 39–56.

24. Head, G. C. (1967). Effects of seasonable changes in shoot growth on the amount of unsuberized roots on apple and plum trees, *Jour. Hort. Sci.*, **42**, 169–180.

25. Israel, D. W., J. E. Giddens, and W. W. Powell (1973). The toxicity of peach tree roots, *Plant and Soil*, **39**, 103–112.

26. Kenworthy, A. L. (1949). Soil moisture and growth of apple trees, *Proc. Amer. Soc. Hort. Sci.*, **54**, 29–39.

27. Kinman, C. F. (1932). A preliminary report on root growth studies with some orchard trees, *Proc. Amer. Soc. Hort. Sci.*, **29**, 220–224.

28. Kolesnikov, V. A. (1971). *The Root System of Fruit Plants*, MIR Publications, Moscow, pp. 11–79.

29. Kolesnikov, V.A. (1930). The root system of fruit tree seedlings, *Jour. Pomology and Hort. Sci.*, pp. 903–913.

30. Kramer, P. J. (1983). *Water Relations of Plants*, Academic, New York, pp. 146–156.

31. Kramer, P. J., and T. T. Koslowski (1960). *Physiology of Trees*, McGraw-Hill, New York, pp. 12–58.

32. Lineham, D. J. (1984). Micronutrient cation sorption by roots and uptake by plants, *Jour. Exp. Bot.*, **35**(160), 1571–1574.

33. Lockhard, R. G., and G. E. Schneider (1981). Stock and scion growth relationships and the dwarfing mechanism in apple, *Horticultural Reviews*, **3**, 316–372.

34. Horst, L. Y. R., and G. Hoffman (1968). Growth rates and periodicity of tree roots, *Int. Rev. of Forestry Res.*, **2**, 181–236.

35. Nightengale, G. T. (1935). Effects of temperature on growth, anatomy and metabolism of apple and peach roots, *Bot. Gaz.*, **96**, 581–639.

36. Phillips, D. J. (1964). Root shoot hormone relation, II: Changes in endogenous auxin concentrations produced by flooding the rootsystem of *Helianthus annus.*, *Ann Bot.*, **28**, 37–45.

37. Quinlan, J. (1965). The pattern of distribution of 14 carbon in a potted apple rootstock following assimilation of 14 carbon dioxide by a single leaf, *Rpt. E. Malling Res. Sta.* (1964), pp. 117–118.

38. Richard, S., and J. Considine (1981). "Suberization and browning of grape roots," in *Structure and Function of Plant Roots*, R. Browwer, O. Gasparikova, J. Kolek, and B. C. Loughman, Eds., Martinus Nijhoff, Dr. W. Junk, The Hague, pp. 111–115.

39. Richards, D., and R. L. Rowe (1977). Root shoot interaction in the peach, the function of the root, *Ann. Bot.*, **41**, 1211–1216.

40. Rogers, W. S., and G. C. Head (1969). "Factors affecting the distribution and growth of roots of perennial plants of woody species," in *Root Growth*, E. Whittinton, Ed., Butterworth, London, pp. 280–295.

41. Rom, R. C., and S. A. Brown (1979). Factors affecting burrknot formation on clonal Malus rootstocks. *HortScience*, **14**(3), 231–232.

42. Rouira, A. D., and C. B. Davey (1974). "Biology of the rhizosphere," in *The Plant Root and Its Environment*, E. W. Carlson, Ed., University of Virginia Press, Charlottesville, pp. 153–204.

43. Russell, R. S. (1977). *Plant Rootsystems and Their Function and Interaction with the Soil*, McGraw-Hill, New York, pp. 113–135.

44. Russell, R. S. (1977). *Plant Rootsystems and Their Function and Interaction with the Soil*, McGraw-Hill, New York, pp. 143–165.

45. Siegler, A.E,. and J. J. Bowman (1939). Anatomical studies of root and shoot primordia in 1 year apple trees, *Jour. Agri. Res.*, **58**, 795–803.

46. Skene, K. G. M. (1975). "Cytokinin production by roots as a factor in control of plant growth," in *The Development and Function of Roots*, J. G. Torre and D. T. Clarkson, Eds., Academic, London, pp. 365–396.

47. Spurway, R. A. (1980). "Root patterns and water supply," in *Environment and Root Behavior*, D. Sen, Ed., Geobios Inst. Jodhpur, India, pp. 118–122.

48. Swingle, C. F. (1927). Burrknot formation in relation to the vascular system of apple stems, *Jour. Ag. Res.*, **34**, 533−544.

49. Thimann, K. V. (1977). *Hormone Action in the Whole Plant*, University of Massachusetts Press, Cambridge, pp. 263−288.

50. Torrey, J. G. (1976). Root hormones and plant growth, *Ann. Rev. Plant Phys.*, **27**, 435−459.

51. Vakhmitrov, D. D. (1981). "Specialization of root tissue in ion transport," in *Structure and Function of Plant Roots*, Ed Brower, R. Gasparkova, O. Kolek, J. Loughman, and B. Loughman, Eds., Martinus Nijhoff, Dr. W. Junk, The Hague, pp. 203−208.

52. Wardlow, C. W. (1986). *Morphogenesis in Plants*, Methuen, London, pp. 246−285.

53. Weatherly, P. E. (1975). "Water relations of root systems," in *The Development and Function of Roots*, J. G. Torrey and D. T. Clarkson, Eds., Academic, London, pp. 297−403.

54. Yadava, U. L., and S. L. Doud (1980). The short life and replant problem of decidious fruit trees, *Hort. Review* **2**, 1−116.

2

PROPAGATION

Brian H. Howard
East Malling Research Station.
East Malling, Maidstone, Kent, England

INTRODUCTION

For today's critical market requirements scion varieties are selected from among the products of conventional breeding programs or, following natural or induced mutations, on the basis of detailed characteristics related to cropping potential and fruit quality. Little attention was given in the past to their capacity for producing adventitious roots from stems and hence their capability of being grown as self-rooted trees.

In the strict context of propagation, the use of rootstocks overcomes this deficiency by equipping the difficult-to-root scion with a root system of another plant using various grafting methods (29).

Historically, the most commonly available rootstock was a seedling of the same species as the scion. Seedlings are still widely used today for the production of ornamental and amenity trees, where the visual characteristics of the scion are of prime importance. The management of fruit trees in the orchard is increasingly detailed and demanding, with uniformity, size control, precocity, and cropping efficiency paramount. Modern intensive orchards are designed to optimize light interception (57) and produce early crops (86), so the often large and variable trees produced on seedling rootstocks are unacceptable and they remain in use only where a suitable clonal alternative is not available.

Rootstock production by vegetative propagation ensures uniform trees with similar cropping characteristics, and for apple in particular provides a range of size-controlling clonal rootstocks (68).

There is considerable activity worldwide by research and development services to produce clonal rootstocks for species where they are not currently available. This is increasingly paralleled by attempts to propagate directly scion varieties on their own roots.

The importance to growers of planting correctly identified material to meet specific market requirements, and high-quality planting stock to yield early crops of marketable fruit, has influenced the attitudes of nurserymen and the development of nursery industries, especially in Northern Europe, North America, and South Africa. Elite planting stock, such as that designated in the United Kingdom as EMLA (19), is produced by research institutes as being true to name and free from known virus and mycoplasma diseases. It is supplied to nurserymen who are able to satisfy government plant health requirements, which include freedom from soil nematodes and minimum isolation from similar fruit or related species. This reduces the risk of virus spread through the soil and by insects or pollen, as well as limiting mistakes through nursery activities such as grafting. Particular attention is given to avoiding contact with outbreaks of notifiable diseases, which in the United Kingdom include fireblight and plum pox.

Standards are maintained through regular nursery inspections by government officers, and the distribution of plant material may be organized through associations of nurserymen such as the Nuclear Stock Association (Tree Fruits) in the United Kingdom.

Increasing awareness by growers of the importance of planting the highest-quality trees to obtain early returns on their investment (86) has resulted in greatly increased effort by nurserymen in many countries to produce larger, more branched and uniform trees with balanced root systems. Speculative production has been reduced, with an increasing proportion of the nursery crop being contracted forward by the purchaser. During the nursery production stage the grower may inspect his trees in the nursery and request special manipulative or spray treatments to alter the normal pattern of lateral development to suit his intended orchard system.

It is inevitable that the increased costs of improved tree production result in higher prices being paid by the grower. Such is the emphasis placed on starting the new orchard on the best possible footing that the cost of trees and stakes amounts to more than half the total expense incurred in planting a new orchard (12).

1. VEGETATIVE PROPAGATION

Vegetative or asexual propagation invariably involves the induction of adventitious roots in stems. The alternative process of inducing the development of adventitious shoots on roots used as root cuttings (79) is rarely used commercially for fruit and for nut propagation. The term *adventitious* indicates that such organs arise in tissues unrelated to the normal hypocotyl- and radicle-derived meristems of the seedling plant. Adventitious roots arise through processes of mitotic cell division from meristematic centers in stem tissue such as the cambium, and adjacent cortex, especially the phloem parenchyma, or from callus developed in response to wounding the stem. Favored sites for root development are adjacent to the nodal buds. Roots emerge through the cortex via the gaps in the sclerenchyma sheath, which are associated with the entry of leaf and bud vascular traces (22). The relationship of adventitious roots to the vegetative bud at the node are sometimes characteristic of the clone or species.

Vegetative propagation systems fall into two basic classes of either division or cuttings. Systems employing the principle of division are those where a potential complete plant develops while attached to the parent stockplant. At harvest the shoots are divided from the stockplant complete with roots. Stooling, layering, and marcotting are methods of division.

The general characteristics of division methods for appropriate rootstocks selected for their ability to develop adventitious roots are that they offer a reliable and constant production system with the stockplants and crop occupying the same site. Root development in each successive crop of annual shoots is induced by mounding soil or an artificial substrate around their bases. It is achieved in a high proportion of the crop under the beneficial support, in terms of water, nutrients, and endogenous growth regulators, of the roots of the stockplant, which in turn ensures vigorous growth and associated beneficial auxin production in the shoot.

In contrast, true cuttings are isolated from the parent plant before root development occurs and must be held in stress-free environments and treated with synthetic auxin to encourage rooting.

It follows that the majority of production costs are incurred in division systems, whether or not there is a market for the product, but that in a steady market situation they have the advantage of reliability., Cuttings need not be collected when there is reduced demand, thus wasting only the cost incurred in maintaining the stockplants; but this increased flexibility is offset by a greater risk of failure. In practice, rootstocks are propagated by cuttings when the combination of genotype and propagation technique results in the same consistency and reliability provided by division methods, the latter being used for less predictable subjects.

1.1. Stooling

Sometimes referred to as mound or stool layering, the stooling technique is commonly used for the Malling and Malling Merton series of apple rootstocks which were selected for their ability to propagate in this way.

A stool is formed by planting a rootstock and allowing it to establish for one growing season before cutting off most of the aboveground part to encourage shoots to develop from the stump. In practice, these are likely to be a mixture of normal shoots from dormant axillary buds at the base of the stem and adventitious shoots from the root system. In early summer, friable soil is mounded up to and between the new shoots, and this operation is repeated at intervals until about 15 cm of the proximal shoot length is covered. To avoid weakening shoots, soil is never allowed to cover more than half the number of leaves present. After leaf fall the soil is cleared away and the rooted shoots removed for transfer to the budding nursery, leaving more stumps to continue the process in subsequent years.

The mounding and removal of soil was physically demanding work and harvesting unpleasant in winter months when done by hand. Today all aspects of the method are mechanized, with soil being mounded by ridge plough (Fig. 1), but it is essential to pull the soil between the shoots on the first occasion with a suitable hand tool to avoid air spaces and poor contact in the center of the ridge. Thoroughly irrigated sawdust is used in timber-producing areas such as Oregon (Fig. 2).

A high standard of pest and disease control is maintained through the growing season with special emphasis on diseases, such as powdery mildew and scab, that might severely restrict growth and carryover to the one-year maiden tree stage. At harvest the soil is ploughed away from the sides of the ridges and the rootstocks are cut by tractor-mounted rotary saw blade (Fig. 3) or pneumatic hand pruners.

Production per hectare depends on row spacing arranged to accommodate particular machinery, but from a meter of stoolbed comprising three plants about 40 rooted shoots of the rootstock MM.106 should be harvested. In light soils almost half of these may be too small to use and a proportion of the remainder will need growing elsewhere for another season to become usable.

The important components of the stooling process that are associated with

Figure 1. Soil being placed by ridge plough around apple stools in the United Kindgom (East Malling Research Station).

Figure 2. Sawdust being deposited between apple stools prior to placing around shoots in Oregon (Oregon Rootstock Inc.).

high levels of adventitious rooting (Fig.4) are the stimuli caused by ultrasevere pruning and the exclusion of light from the shoot base, resulting in blanched tissue (53). The physiological mechanisms of these processes are the subject of investigation since they have wide application in propagation systems.

Research to improve the stooling method has been aimed at increasing the number and size of shoots produced so that a greater proportion of the harvested crop is suitable for budding the next season, sometimes at a height of 45 to 60 cm above ground for high-worked trees in northern Europe.

For the dwarfing rootstock M.9, production through successive harvests was reduced consistently by initially establishing stoolbeds with other than the highest-grade plants (Table 1).

The root systems of stools established from large and small plants retained their original proportional differences when excavated after five cropping seasons, which may be attributable to the annual removal of almost all above-ground growth, with little opportunity to accumulate reserves in this weak-growing clone. Such effects from establishing beds with different grade plants were not obtained in the more vigorous MM.111 (44).

Continual cropping of stoolbeds leads to the production of a high proportion of small unusable shoots, which to a large extent may be overcome by adequate fertilizer application and irrigation in areas where the climate is conducive to rapid and sustained growth, as in Oregon. In the cooler, growth-limiting conditions of Northern Europe harvesting costs may be reduced by removing

Figure 3. Tractor-mounted rotary saw used to harvest apple rootstocks in the United Kingdom (A. F. Todd, Collier Street).

the small waste shoots of MM.106 rootstock during the growing season, especially when demand is not excessive. This is done by exploiting the fact that the potentially weak shoots grow first in spring and may extend to 10 cm before the potentially more vigorous shoots start to grow. During this period in early May the beds are sprayed with a 4% aqueous solution of the sucker eradicant Tipoff, which contains the auxin naphthalene acetic acid (NAA), killing the developing weak shoots. In soils of adequate fertility and moisture the loss of many small shoots is compensated by a small gain in large shoots (Table 2) in addition to the saving in harvesting and grading costs (48).

Shoot size may also be increased by selectively harvesting only those which reach a required size each year, allowing the remainder to grow for a further season *in situ*, or harvesting only biennially. Selective harvesting of only large shoots increased the accumulated crop of useful M.27 and MM.106 rootstocks over a two-year cycle compared to annual harvests (Table 3).

Biennial harvests tended to produce excessively large shoots. Rooting was

Figure 4. Rooted apple stool shoots after grading (East Malling Research Station).

TABLE 1. Annual Production of M.9a Rootstocks Suitable for Budding[a]

Establishment Quality	Cropping Years					
	1969	1970	1971	1972	1973	Totals
Good	6.7	15.5	19.7	25.7	6.2	73.8
Poor	1.5	7.5	8.3	15.3	6.8	39.4

[a]Related to the quality of original planting material used to establish stoolbeds (no. per nonirrigated experimental plot, originally of eight plants). Note that the 1973 crop was subjected to drought.

improved by retaining shoots on the stool for a second season, but there was a small increase in the number of misshapen plants and in those that produced laterals and that were infected with powdery apple mildew. A benefical resid-ual effect of increased production over all shoot grades was measured for two

TABLE 2. Production of MM.106 Stoolshoots following Treatment of Stoolbeds during April with 4% Tipoff (no. per m)

	Rootstock Grade			
	Small Waste	Replanting	Normal Budding	High Budding
Irrigated				
Tipoff	4.2	8.6	10.2	7.8
Control	27.2	21.0	13.1	4.3
Nonirrigated				
Tipoff	0.5	1.3	3.4	4.1
Control	23.8	15.4	9.2	4.5

TABLE 3. Stoolshoot Production following Annual or Selective Harvesting (no. per m, accumulated over two years)

		Rootstock Grade		
Rootstock	Harvests	Unrooted and Waste	Replanting	Budding
M.27	Annual	13.9	14.4	11.3
	Selective > 60 cm	5.1	7.1	14.2
MM.106	Annual	20.6	14.6	11.6
	Selective > 75 cm	7.6	7.1	20.7

subsequent harvests, suggesting that when the demand for a particular crop is not excessive, beds can be rested in this way with benefit (100). Where mechanized harvesting is practiced, only the biennial system will be possible.

Attempts to improve rooting in stooled plants include girdling the bases of stems by causing new shoots of peach rootstocks to grow through wire netting (27). Experience at East Malling suggests that some clones such as M.9 may, for a reason not understood, produce the majority of new roots *below* the girdle, and that only a proportion of shoots are suitably girdled when using a mesh of constant size.

1.2. Layering

This technique involves retaining a proportion of the crop of new shoots at harvest when the majority are removed. Those retained are bent down the following spring before bud-burst and held against the soil with pegs to enable a thin layer of friable soil or other suitable substrate to be spread over them. If only 3–4 cm of covering soil is used, the axillary buds of the layered shoot will push through with the result that their bases will be truly etiolated and root relatively easily. Among the effects noted in the etiolated zone is an increase in starch and a decrease in phloem sclerenchyma (22). The remaining soil-mounding sequence is done as for stooling.

The preparation of layer beds is tedious and, because a proportion of the shoots needs to be retained each year, the method is not suitable for mechanized harvesting. The first shallow dressing of soil should also be applied by hand to ensure even covering of the layered shoots which do not lie in a uniform horizontal plane.

Because shoot bases are etiolated in the layering method, rather than blanched as in the stooling method, layering was practiced for the most recalcitrant subjects such as *Prunus avium* F12/1 sweet cherry rootstock in Northern Europe. Production fell off after a number of years and many small shoots were produced, an effect attributed to the *in situ* development of a soil-borne "replanting" disease caused by the continually reused soil being mounded to the bases of each annual crop of cherry shoots. The soil-borne fungus *Thielaviopsis basicola*, which causes root rot and reduced growth in *Prunus* spp., was implicated, and layer bed production was improved by spring drenches of benomyl (84). It is possible that a similar association exists between the falloff in cropping of some apple stoolbeds and the development of the apple replant disease organism *Pythium sylvestris*, but existing control methods of fumigation with chloropicrin (tear gas) or drenching with formalin (formaldehyde) is suitable only for unplanted soils.

The use of the layering method has declined both for the management reasons described and, in the case of cherry, because the difficult-to-propagate *Prunus avium* F12/1 was replaced with the semidwarfing *Prunus avium* × *P. pseudocerasus* hybrid cherry rootstock 'Colt,' selected for its ability to root readily from cuttings.

Perhaps the most frequent current use of the layering technique is to economize on planting material by starting apple stoolbeds as layer beds for one season, enabling the original planting material to be reduced by about 60%, but obtaining frequent intermediate plants from the layered shoot, all of which are then managed by stooling in subsequent years.

1.3. Winter Leafless Cuttings

Cuttings propagated without leaves between autumn and spring are often referred to as "hardwood cuttings." As foresters look increasingly towards horticultural propagation techniques to clone desirable source trees, as a step beyond propagation by improved seed from specific crosses, the terms *hardwood* and *softwood* used by horticulturalists to describe woody and succulent shoot cuttings are likely to be confused with the foresters' use of the same terms for deciduous broad-leaved and coniferous species, respectively. For this reason, the more descriptive term *winter leafless cuttings* is preferred.

The use of such cuttings has long been a method of propagating easy-to-root subjects, and especially those capable of developing preformed root initials in shoots during growing season, such as *Ribes*, *Salix* and *Populus*.

When alternatives to stooling and layering for fruit tree rootstocks were considered, the winter cutting was seen as a suitable alternative because cuttings of a size similar to stoolshoots could be harvested from stockplants at the end of the growing season. The flexibility that this offers by management deciding each year whether certain clones should be propagated (in contrast to their routine production annually or biennially on stoolbeds) is offset by the need to root and establish cuttings after removal from the stockplant, and the need usually to grow them for a season before they are sufficiently well rooted to be lined out for grafting.

Because of the wide variety of approaches to both stooling and cutting methods it has not been possible to make economic comparisons. The method has been developed particularly at East Malling Research Station, partly in direct support of rootstock producers, but also because the technique has wide application for the propagation of woody plants generally.

This appreciation of the potential for propagating winter cuttings has been endorsed by their use at other centers, such as in Oregon and in South Africa, for the propagation of pear rootstock selections and by commercial nurserymen in the United Kingdom, who propagate all cherry ('Colt') and plum rootstocks by this method together with an increasing proportion of the quince rootstock production for pears and a small proportion of the easier-to-root apple rootstocks such as M.26, MM.106, MM.111, and M.27.

Special uses of the winter cutting technique include its application to unrooted shoots from the stoolbed, especially when these are relatively numerous in the first year of stoolbed production (77).

Cuttings are also used in the early buildup of a new or reissued clone by budding each available bud into a nurse rootstock (45). These must be virus-free and preferably with distinct morphological characteristics such as purple leaves and stems, to avoid early contamination of the desired clone with off-types. The rootstock buds are grown into stockplants to provide cuttings, giving a more economical use of limited plant material whose use directly for cuttings, some of which are likely to fail, puts at risk about 30 buds in a normal 60-cm long winter apple cutting.

The winter cutting technique in its current state of development requires the establishment of stockplants in hedges at a spacing of between 30 and 100 cm in the row, depending on the relative availability of plant material and land area (Fig. 5). Interrow spacing is of the order of 2 to 3 m for tractor access. Row orientation should be for management convenience, the lack of differences in rooting of cuttings from the north and south sides of hedges suggest that under normal field conditions shade is not detrimental.

During the winter removal of cuttings and the subsequent pruning of framework shoots to three to four buds, the hedges are severely winter pruned. This stimulates shoot growth the following season, which has a high rooting potential compared to shoots from nonpruned or lightly pruned bushes.

Figure 5. Hedge for the production of winter plum cuttings (East Malling Research Station).

Within conventional hedges severity of pruning can influence subsequent rooting, rooting being reduced by pruning the annual framework to one-third its length, rather than to a few basal buds, and increased by deheading the bush (Table 4).

The fact that markedly enhanced rooting results from physically reducing the height of the hedges, and that shoots arising from the trunk of normal stockplants root more readily than those from the top of the bush, suggests that proximity of shoots to the root system of the stockplant may have a causal effect, possibly through the production of a root-derived adventitious root-inducing factor. Attempts to explore this hypothesis further by raising the root system of conventional hedge plants by enclosing their trunks in bags of potting compost have so far given inconsistent results.

TABLE 4. Rooting (%) Response to Severity of Pruning Apple Hedges

Pruning	M.26	MM.106	MM.111	Means
Ultrasevere (23-cm framework)	88	48	63	66
Severe (45-cm framework)	75	48	60	61
Normal (100-cm framework)	42	25	58	42
Light (100-cm framework and one-third annual shoot length retained)	23	25	33	27

Severely pruned hedges, stools, and layer beds have three characteristics in common with juvenile phase material, namely, enhanced rooting ability, increased production of short axillary laterals (spines), and prolonged leaf retention in the autumn, in contrast to more lightly pruned trees.

Rooting potential may fall off above the proximal part of the shoot (54, 76) and cuttings are ideally removed at the base of the annual shoot (Fig. 6) where the rooting potential in apple is highest, influenced by the presence of a rosette of buds with their associated nodal rooting sites (43).

Cuttings are shortened to a convenient length. In countries where annual growth rate is slow it is usual to prepare cuttings of sufficient length (60 cm) to enable the scion bud to be inserted into it subsequently.

Roots are initiated from or near the cambium (Fig. 7) and rooting is improved markedly in many subjects by wounding, especially when the cutting base comprises less than ideal internode tissue. Splitting the shoot base through its diameter for a length of 2 cm is particularly effective (55). Ranks of roots emerge (Fig. 8) from the callus that develops from the longitudinally exposed cambium, and appear to be associated with the development of cambial outgrowths formed in the new callus as the cambium fails to rejoin across the split. In less severe incision wounds a modified cambial outgrowth occurs in the parenchymatous tissue of the callus in which the cambium is able to bridge the wound (66).

Figure 6. Collecting basal cuttings (East Malling Research Station).

Figure 7. Root primordium developing adjacent to the cambium in a 'Myrobalan B' plum winter cutting (East Malling Research Station).

Normal and wounded cuttings require auxin treatment. The synthetic auxin indolylbutyric acid (IBA) is widely used (54, 76, 13) due to its ability to induce a fibrous root system and not to translocate in the stem and inhibit bud growth to the same extent as naphthalene acetic acid (NAA).

IBA may be applied to the bases of cuttings as the acid dissolved in an organic solvent such as ethanol or acetone. Alternatively, the potassium salt may be dissolved in water or a powder formulation may be used. Of the liquid treatments, the 50% aqueous organic solvent "quick dip" is preferred in the United Kingdom, possibly penetrating the tissues more effectively than water and evaporating relatively quickly.

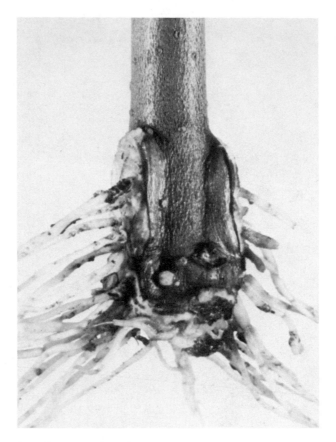

Figure 8. Rooting of IBA-treated nonbasal, internodal M.26 winter cuttings after splitting (Howard, Harrison-Murray, and MacKenzie [55]).

Because IBA treatment is essential to the success of many rootstocks considerable research has been done to understand the factors influencing its uptake by the cutting and to standardize treatments to give consistent results.

IBA is mostly taken up through the exposed vascular system at the cut base of the cutting. That auxins achieve their effects in the cambium and secondary phloem indicates that they must move laterally in the tissues (13). The epidermis is a less efficient uptake pathway for liquid-applied IBA, but it may supplement that taken in through the base (70, 49). The depth to which cuttings are dipped in the solution influences uptake. The main reason is that solution applied to the epidermis runs off the waxy surface and collects at the cut base, where it is absorbed (49). It also appears that IBA applied to the epidermis at a position not adjacent to the cut base partially negates the effect of the auxin taken in through the cut end (49). The effect on rooting of IBA solution applied to the epidermis and running to the cut base was investigated by drying cuttings in the vertical position with their bases above the shoot tips or vice versa (49).

The normal method of allowing the solvent to evaporate with the bases projecting over the edge of a bench gives closely similar results to those of cuttings in the "base down" position.

The time during which cuttings are isolated from the stockplant before receiving IBA determines the extent of the partial suction that develops as water evaporates from the cut base, and hence the suction that determines the quantity of IBA absorbed (71). This and the time that cuttings are held in the IBA solution (71), and dipping depth interact to vary the rooting response to IBA concentration (54) and are standardized in practice. Nurserymen are recommended to treat cuttings routinely 24 hr after collection (or some other standard time), dipping them for 5 sec to a depth of 1 cm (Fig. 9). Under these conditions plum and apple are usually treated at a concentration of 2500 ppm

Figure 9. Shallow dipping cuttings for IBA treatment (East Malling Research Station).

in a 50% aqueous solution of acetone, and cherry and quince at 1000 ppm for shoots not already bearing preformed roots.

IBA applied in power formulation using talc as the carrier can be made almost as effective as the liquid treatment by wetting the base of the cutting in a 50% aqueous organic solvent such as ethanol or acetone before treating with the powder (Table 5). The powder should be retained on the cutting for as long as possible and not removed when handling or planting, and it is not critical to minimize treatment to the epidermis, which is a more effective uptake pathway than for liquid treatments, presumably because powder is retained in considerable amounts on the edpidermis (50).

The timing of propagation is influenced by the readiness with which the subject roots and changes in its rooting physiology. Cuttings of the cherry rootstock 'Colt' that have developed preformed roots while growing on the hedge (Fig. 10) may be harvested and cold-stored or planted directly into their new location at any time between leaf fall and bud burst.

Those rootstocks such as plum and quince which leaf out early in spring are best propagated in the autumn, when their buds are fully dormant. Many of these are relatively free rooting, and autumn planting directly into the soil provides sufficient residual soil warmth to enable rooting to proceed slowly, so that cuttings are established by the following spring in countries with relatively open winters (Table 6).

Some subjects such as apple are not sufficiently stimulated by direct soil planting and must be propagated with the benefit of base heat. The time when this is done in apple is determined by physiological changes in the shoot related to changes in the activity of phenolic rooting cofactors that synergize with the applied IBA to induce rooting (5, 6), and possibly of rooting inhibitors also (88), although their exact role in the sequence of root initiation and development is not understood.

Typically, rooting potential in apple cuttings is at a peak in late winter and early spring (40), but this does not always appear to be related in various species to the influence of vegetative buds released from dormancy (30).

TABLE 5. Rooting (%) of MM.111 Winter Cuttings in Response to IBA[a]

Experiment	Carrier	Predip	Rooting %
1	Talc	Nil	35
	Talc	Water	48
	Talc[b]	50% Ethanol	57
2	Talc[b]	50% Ethanol	74
	50% Ethanol	Nil	79

[a]Mean of 5 concentrations from 625 to 10,000 ppm; applied either in talc preceded or not by different 5-sec predips, or dissolved in 50% ethanol.
[b]Data combined from two experiments with a common treatment.

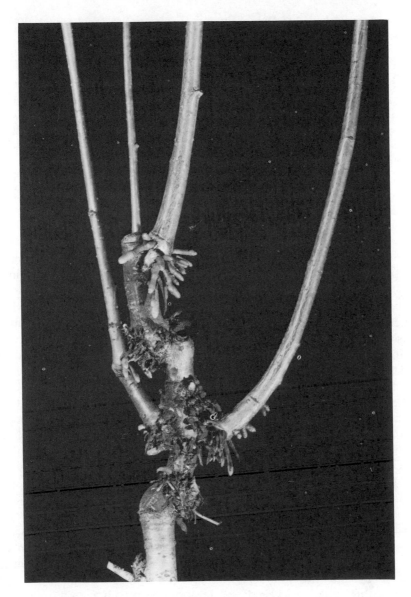

Figure 10. Preformed roots at the base of 'Colt' shoots (Howard [43]).

Studies with apple and plum suggest that disbudding treatments used to investigate this possible relationship, and which at first sight suggest a causal relationship (26), may have incurred an artifact whereby desiccation or water absorption through the disbudding wounds was the factor most affecting rooting.

TABLE 6. Establishment of Quince Cuttings Related to Season of Direct Field Planting or Prior Stimulation with Basal Heat

	Collection Date		
	Mid–Nov.	Mid–Jan.	Mid–Feb.
Prior treatment for 2 weeks at 21°C	76	85	12
Directed planted into field	61	8	2

Among results which support this conclusion was the similar response to disbudding from making the same-size wounds in the internodes of cuttings (41) and the modifying effects from applying antidesiccants to the disbudding wounds in relatively dry and humid environments (47). Environmental influences on winter cuttings are not fully understood, the requirement for research into this aspect being less obvious than for relatively stress-sensitive leafy summer cuttings. Nevertheless, recent studies show that the pattern of changes in seasonal rooting of apple cuttings, conventionally declining to a low level in midwinter and rising to a peak in late winter and spring, can be reversed to give low autumn and spring responses if the aerial humidity is reduced from 85% to 55% (Table 7).

Whether or not the adverse effect of low relative humidity is directly associated with a shoot more sensitive to water stress at the beginning and end of dormancy or whether the increased water loss from cuttings at the lower humidity interferes with the production or movement of the water-soluble phenolic cofactor remains to be resolved.

The need to apply basal heat (42) to those cuttings not able to exploit relatively long exposure to low soil temperatures necessitates using an artificial substrate. Leafless winter cuttings do not have the ability to transpire excessive water taken up from the rooting medium, and rotting of cutting bases due to excessive water and inadequate air is the most common cause of failure. The use of sphagnum peat as a major component of peat/sand composts contributes largely to this problem because of its great water-holding capacity, its increased water-holding and reduced aeration when compressed during handling, and its reluctance to drain. Detailed studies of the water-release curves of a range of substrates including sand, grit, and perlite (Table 8) have shown the advantages of coarse-particle materials with the correct balance of air and water (31) to provide approximately the fivefold increase in air content required by these cuttings compared to that present in compost used for growing plants in containers.

The system currently being developed at East Malling is to use granulated

**TABLE 7. Seasonal Rooting (%) of M.26 Apple
Winter Cuttings in Different Environments**

Dates (1983/84)	RH 80−90%	RH 50−60%
November 9	71	32
December 14	30	69
January 18	15	69
February 22	61	78
March 28	90	51

**TABLE 8. Rooting (%) and Basal Rotting of M.26 Winter Cuttings
in Relation to Air and Water Contents (% v/v) of Different Grades
of Perlite at Two Suctions**[a]

	Coarse Perlite		Fine Perlite	
Suction (cm H_2O)	5	35	5	35
% Rooting	68	85	28	53
% Basal rotting	5	3	40	28
% Air	67	70	17	33
% Water	18	14	66	45

[a]After Harrison-Murray.

pine bark (34), which does not compress in the way that peat does, drained over
a bed of approximately 20 cm of fine sand. The thermostatically controlled
heating elements are placed at the interface of bark and sand, with sand
brushed through the heating mats to ensure continuity of drainage. A shallow
layer of bark, approximately 7.5 cm deep, is used and bundles of cuttings with
their bases level are worked into the medium with a twisting action until their
bases are judged to be 3 cm above the heating elements. A temperature of 20°C
at the cutting base is suitable for rootstocks and the bundles are wrapped in
polythene to minimize desiccation (Fig. 11), although the extent of the wrapping is determined by the humidity of the propagating room. With an aerial
relative humidity of 90%, good results have been obtained by wrapping the
proximal half of the bundle only or not wrapping at all. The main advantage of
bark over peat-containing composts is that is can be rewetted regularly and will
drain rapidly. It is very difficult to rewet peat composts during use and not
produce adverse air/water conditions. Rooting peach cuttings in polythene
bags (4) also avoids these problems.

Cuttings stimulated in heated bins must be transferred bare root to the field

(Fig 12) or planted in beds of compost (Fig. 13) to establish. In free-rooting subjects, extended periods of bottom heat are associated with subsequent reduced establishment (4, 46).

Establishment utilizes carbohydrate reserves in pecan (93) and excessive root production exhausts carbohydrate resources in leafless apple cuttings, leaving inadequate reserves to maintain the cutting during early root and shoot development in the field (15). Premature bud development similarly utilizes carbohydrate and suppresses establishment (16), for which reason modern propagation houses are insulated, with minimal or no natural light, and fitted with air coolers. Arrangements should be made to prevent the coolers progressively drying the atmosphere in the same way that cold stores need water replaced at intervals. At East Malling a small timer-controlled "fogging" unit operates at intervals, but regular wetting of the floor and of bins containing bark is adequate.

Cuttings of easy-to-root subjects may be planted in the field after a period of only two to three weeks' basal heat and with only a small percentage of cuttings showing a few roots, the remainder being healthily callused, which is indicative of successful stimulation.

Slower-to-root subjects require the presence of a few roots on all cuttings to assist with the early stages of establishment. In most rootstocks a proportion of the basal-heat-induced roots survive transplanting and further roots emerge from the rootstock stem during the growing season.

Figure 11. Winter cuttings in polythene-wrapped bundles on granulated pine bark at 20°C base heat (East Malling Research Station).

Figure 12. Winter apple cuttings established in the field (East Malling Research Station).

Figure 13. Winter apple cuttings established in raised beds of prepared compost (East Malling Research Station).

1.4. Leafy Summer Cuttings

Leafy cuttings must be relatively small to minimize water stress and consequently are not suited to the general production of rootstocks, which, being only part of the process of tree production, must not extend the production cycle to a marked degree.

Leafy cuttings are sometimes used to facilitate the rapid multiplication stage when introducing a new clone. The cherry rootstock 'Colt,' with its readiness to root from both winter and summer cuttings, was multiplied effectively by the latter method.

An analysis of the root-inducing components of the stoolbed method (53) (which is effective at inducing roots in M.9, whereas cuttings of the same clone from normal hedges generally fail to root) highlighted opportunities for pre-conditioning shoots during their growth by severe pruning and etiolation so that their subsequent rooting by both summer and winter cuttings was enhanced. Covering M.9 hedges for a month from bud burst with black polythene-clad structures raised the rooting (Fig. 14) of summer cuttings of M.9 from 14 to 84% (32).

Studies with M.26 bushes grown at different light levels in growth rooms

Figure 14. M.9 cutting rooted after pre-etiolation (right) compared to a cutting grown normally (East Malling Research Station).

showed that much heavier rooting was obtained in cuttings from plants grown at the lowest irradiance if treated with IBA (17). Current studies at East Malling indicate that complete darkness is not essential for an effective "etiolation" response (33). Elongated internodes, thin, small leaves, and anthocyanin development in the stem, characteristic of juvenile material, often accompany etiolation treatments. The difficult-to-root black walnut was successfully propagated by inducing adventitious shoots from seedlings or the lower parts of older trees and treating them with high levels (5000−8000 ppm) of IBA (89). Etiolation, blanching, and severe pruning treatments further extend the opportunity to use leafy cuttings for the rapid bulking up of limited material.

1.5. Micropropagation in vitro

The general relationship whereby progressively smaller cuttings offer increasing opportunity for rapid multiplication extends to micropropagation *in vitro*, but not because shoots from the stockplant are divided into still smaller propagules.

Techniques which enable plant organs or tissues to grow in chemical media are generally described as tissue culture techniques and were first demonstrated before 1935. Their exploitation for propagation comes from the ability under laboratory conditions to encourage axillary buds in the meristem of a shoot tip to produce small lateral shoots. These are repeatedly separated and cultured themselves at intervals of a few weeks to give very large potential shoot multiplication rates without major seasonal constraints and without the need for numerous large stockplants. This type of regeneration *in vitro* is often described as axillary shoot proliferation and is more productive than simply propagating very small shoot tips (meristem tip culture). It minimizes the chances of introducing off-types through mutation that occur with greater frequency when a stage requiring the development of adventitious shoots from organized stem tissue or from disorganized callus tissue (adventitious shoot initiation and organogenesis) is involved.

The axillary shoot production method starts with an "explant" in the form of the distal 5 mm or so of newly growing shoots, which is removed from the stockplant and taken to the laboratory, often in a flask of water. Rapidly growing shoots provide explants which multiply well, and so dormant plants are often containerized from cold store as required and forced into rapid growth in the glasshouse. They are not watered overhead, so as to minimize the microflora population on the leaves and stems. Because of the potentially high multiplication rates normal requirements for the stockplant to be confirmed as true to type and free from systemic diseases are paramount when employing *in vitro* techniques.

The cultures are eventually supported in growing media containing sugar (69) as a source of carbon, which also encourages and supports growth of fungi and bacteria that may be present superficially on the explant. For this reason the explant is washed in running tap water, surface sterilized by immersion for prescribed times in ethanol, a chlorine-containing product such as bleach, or hydrogen peroxide, and rinsed in sterile water. Typically, the apical meristem, 5–10 mm in length, will be dissected from the shoot tip, treated with 10% calcium hypochlorite for between 10 and 30 min depending on experience, rinsed, and placed into the culture vessel.

Because fungal and bacterial spores are present also in the laboratory air the sterilized explant is handled in a stream of filtered air in a laminar flow transfer cabinet (Fig. 15) equipped with a high-efficiency particulate air (HEPA) filter. All instruments and the necks of vessels are heat-sterilized by flaming, and the media into which the explants are put will have been sterilized by autoclaving or, in the case of some heat-unstable growth regulators, by passing through very fine filters.

During the multiplication stages (Fig. 16) the objective is to produce as

Figure 15. Commercial micropropagation laboratory laminar flow sterile transfer cabinets (Oregon Rootstock Inc.).

Figure 16. Flasks containing cultures held under growth room conditions (Oregon Rootstock Inc.).

many shoots as possible; this is the equivalent stage to planting and growing a hedge for normal cuttings.

The medium, often employing an agar base to give it jellylike consistency, is supplied with sucrose, inorganic salts, and growth hormones of which a cytokinin must be present in relatively large amounts (e.g., 1 mg/L) with auxin present at about 10% of that level.

Shoots multiply through successive subcultures until sufficient are available to root as cuttings (Fig 17) by transferring them onto a medium with increased auxin and with cytokinin and gibberellic acid omitted. Alternatively, the shoots may be briefly exposed to a high-auxin medium (58) and then rooted in a nonsterile compost as for normal cuttings, or treated as cuttings by dipping in auxin solution or powder before sticking in the propagating bench. Small preformed plugs of peat-based compost held together with resin are useful when rooting directly, and transfer from the high humidity of the culture vessel to the greenhouse is best done by weaning in a fine "fog" system.

Figure 17. Rooting cultures of GF677 peach rootstock (Laboratorio di Micropropagazione Vitro Coop, Cesena).

Individual laboratories will practice variations in the general method described, especially with respect to the concentrations of inorganic salts and growth regulators used (36, 91, 60, 73), but within the range of approaches employed the following scientific features emerge.

Cultures, while retaining their genetic integrity, may undergo "epigenetic" changes which are maintained under favorable conditions but are not permanent. The most obvious is that rooting ability increases with increasing numbers of transfers from an old culture to a new one in subjects difficult to propagate, such as apple scion varieties (96). This is accompanied by apparent phase change from the adult to juvenile state, which may be preserved in the plants eventually produced if they are prevented from growing too large by restricting pot size (94) or if they are severely pruned as hedges. The plum rootstock 'Pixy,' propagated *in vitro* and then grown as a hedge, retained juvenile characters of excessive spininess and ready rooting from winter cuttings five years after being micropropagated (Table 9). The opportunity exists to exploit this feature in conventional nurseries without it necessarily adversely affecting tree performance subsequently in the orchard.

The preconditioning of cultures to root may be enhanced by growing them in agitated liquid culture (95), or by including phloroglucinol, a derivative of the phenolic glycoside phloridzin, in the agar culture medium (59).

The transplanting stage is critical as propagules are weaned to a normal

TABLE 9. Shoot "Spininess" and Rooting of 'Pixy' Plum Winter Cuttings from Hedges Raised Either from Normal Cuttings or *in Vitro*

Derivation of Cutting Hedge	No. of "Spines" per Hedge Plant	% Rooting[a]	% with Lateral Roots[a]
Conventional winter cuttings	16	18	2
Micropropagated cuttings	124	61	23

[a]Indicative of speed of rooting. Most cuttings were healthy and would be expected to root.

environment (Fig. 18) and encouraged to develop cuticular wax (8) and a photosynthetic capability when removed from the sugar-containing medium.

Checks to growth may occur as the propagules are removed from the agar to the compost, which causes 'Pixy' plum propagules to become dormant, necessitating a cold treatment or spray with gibberellic acid (51) to encourage active growth. These problems are minimized if transplanting is done in early spring at the beginning of the normal growth cycle.

The advantages of rapid multiplication in a sterile environment which prevents reinfection with systemic diseases of initially healthy material, and facilitates international movement of plant material, are offset by the expensive facilities required for micropropagation and particularly the fact that very small plants, less than 5 cm tall when transplanted, need to increase in size between ten and twenty fold before being usable. This problem is particularly relevant in rootstock production where the rootstock is only the first stage in the production of the final tree, which attracts increasing costs the longer it remains in the nursery.

It is perhaps significant that two highly successful micropropagation laboratories producing fruit tree rootstocks are in areas with relatively favorable climates for rapid growth. Oregon Rootstock Inc., situated in the Willamette Valley near the West Coast of the United States (latitude 45°N), enjoys a relatively long, hot summer with adequate irrigation water available. Its joint venture, Microplants Nurseries Inc., is responsible for the production and release of East Malling—derived rootstocks under licence in the United States, incorporating the *in vitro* stage into a comprehensive nursery organization.

At Cesena, near the Adriatic coast of Italy (latitude 44°N), four micropropagation laboratories produce in excess of 2.5 million rootstocks each year, primarily of the difficult-to-propagate peach × almond hybrid GF677 to replace local peach orchards damaged by heavy water-retentive soil.

Typical is the Laboratorio di Micropropagazione Vitro Co-op, whose 1984 micropropagation production included 800,000 peach × almond hybrid GF677, and 250,000 plum 655/2 rootstocks, both as rootstocks for peaches, and a total of 60,000 rootstocks of three clones for sweet cherry.

Figure 18. Growing-on plantlets of GF677 peach rootstock (Laboratorio di Micropropagazione Vitro Coop, Cesena).

Peach rootstock production is programmed so that most plants are weaned and ready for field planting in the autumn, to be budded the following August. The rootstock is smaller when planted than conventionally raised stocks, but growth is made up during the budding year and in the following maiden tree year to give impressive stands of clean-stemmed well-shaped trees. In contrast, rootstocks micropropagated in south east England (latitude 51°N), where there is a temperature disadvantage in every month compared to Cesena, must be grown initially in heated glasshouses before being transferred via unheated glass to the field.

It is likely that the rapid multiplication and health advantages of micropropagation will be exploited universally, with major commercial production areas determined by favorable climate for growing-on the small plants. There is a precedent for large-scale propagation of chrysanthemums and carnations in Malta and the Canary Islands (latitudes 35 and 30°N respectively) where climate and labor advantages exist, returning the small plants to the United Kingdom to be grown-on for cropping. The small size and sterile condition of micropropagules lends itself readily to their production in northern European laboratories, their growing-on farther south, and their distribution to European fruit-growing areas much in the same way as seedling rootstocks for ornamental trees are distributed throughout Europe from production areas in Germany and Denmark. Where macropropagation methods meet commercial requirements, and where other climatic features such as long open winters characteristic of the United Kingdom can be exploited for planting and lifting field-grown plants, it is unlikely that micropropagation will be used to produce the finished rootstock, but rather to produce plants for stoolbeds and conventional hedges, incorporating juvenility effects where relevant.

2. SEEDLINGS

Root systems can be provided for difficult-to-root scion cultivars by budding or grafting onto seedling rootstocks, which are used to a greater or lesser extent in the commercial production of all fruit and nut crops with the possible exception of grape (Table 10). Seedlings are the most common commercial source of rootstock for almond, apricot, citrus, peach, pecan, and walnut.

Capability for vegetative propagation exists in most crops, but may be developed only locally for many reasons, including specific scion–rootstock incompatibility, which is highlighted by the use of clonal rootstocks, as exemplified among plums.

There are few reasons to choose seedlings for any intrinsic value other than cold resistance and their tolerance of dry soils compared to some shallow-rooting clonal rootstocks.

Generally, the use of seedlings reflects ready availability of seed and the lack of a clonal alternative. The fact that seedlings of most crops are, by and large, less likely to be virus infected than clonal material was in the past a strong point in favor of their use, but reports of virus transmission through seed in apple, pear, plum, peach, cherry, citrus, and walnut suggest that seedlings offer only the best available source of healthy rootstocks in the absence of a comprehensive plant-health program for clonal rootstocks.

There are many technical requirements for successful seedling raising, including the need to sterilize seedbeds against soil-borne diseases and nematodes, pretreating seeds to overcome dormancy and/or hard seed coats, and relating sowing rates to expected germination ability to obtain even stands of uniform rootstocks. Each species has its own technical requirements.

On a small scale the need for cold stratification under moist conditions can be met by mixing seed with moist sphagnum peat in a polythene bag placed in a refrigerator. Generally, the cold requirement is less for species grown at lower latitudes. For example, apple, pear, and walnut are stratified at about 4°C for 90 days, whereas almond and citrus require only 30 days. Species with a wide geographic range such as cherry and peach are normally given the longer period to ensure rapid uniform germination.

Physiological and technical studies aimed at improved seed germination and seedling production are as numerous as those in support of vegetative propagation techniques (35). The improved rooting of walnut and pecan cuttings by nurse-grafting them into germinating pecan nuts suggests that some common endogenous factors are involved in these distinct processes (2). Germination capacity is lost with storage, which in citrus is associated with dehydration.

TABLE 10. Commercial Use of Seedling, Clonal Rootstocks, and Self-Rooted Scions

Crop	Seedlings	Clonal
Almond (*Prunus dulcis*)	Almond, peach, almond × peach	'Marianna 2624' plum
Apple (*Malus pumila*)	Various *Malus* spp. including apomicts	Many clones selected for ease of propagation, size control, cold hardiness
Apricot (*Prunus armeniaca*)	Apricot, peach, 'Myrobalan' plum	—
Cherry (*Prunus* spp.)	*P. avium* (sweet) *P. mahaleb* (sour)	*P. avium* F12/1 *P. avium* × *P. pseudocerasus* 'Colt'
Citrus (*Citrus* spp.)	Various *Citrus* spp.	—
Grape (*Vitis* spp.)	—	*Vitis* spp. as rootstocks and self-rooted varieties
Peach, nectarine (*Prunus persica*)	Peach, apricot	'Brompton' plum (*P. domestica*) 'St. Julien A' plum (*P. insititia*) Self-rooted scions
Pear (*Pyrus communis*)	*P. communis* *P. calleryana* *P. betulaefolia*	Quince clones (*Cydonia oblonga*), selected clones of *P. communis*
Pecan (*Carya illinoensis*)	Pecan	—
Plum (*Prunus* spp.)	*Prunus* spp., having regard for compatibility	Clones of *Prunus* spp., having regard for compatibility
Walnut (*Juglans regia*)	*J. regia, J. hindsii, J. regia* × *J. hindsii* ('Paradox')	—

Chemical changes involving proteins and proteolitic enzymes have been described in walnut during stratification (61) and gibberellic acid is involved, for example, in citrus and pecan germination processes (10, 20).

Pregerminated seeds may be sown to optimize conditions for this process and to reduce the production period (67).

Typical commercial problems include irregular supplies of seed due to unfavorable weather, losses to birds, and inadequate labor for collection. For these reasons fresh cherry seeds are not always available in northern Europe, which partly may explain the long use and continued development of clonal rootstocks for cherry. Irregular seed supplies result in surpluses being stored for more than one year, with little information supplied to nurserymen at the time of purchase and consequent poor germination. Apple seeds, on the other hand, are readily available from processing factories where it is usually possible to stipulate seed from preferred varieties. Citrus seeds are similarly obtained from crushed fruit, separated from the flesh with aid of pectolytic enzymes (3).

2.1. Apomixis

Attempts to obtain the beneficial characteristics of clonal rootstocks in rootstocks of seedling origin have been made by exploiting the phenomenon of apomixis. Apomicts are produced when an asexual reproduction process replaces the sexual process that normally leads to embryo formation. A cell in the embryo sac or nucellus fails to undergo meiosis but forms a zygote genetically similar to the seed-bearing plant. Some plants may produce a mixture of both sexual and vegetative embryos, and others may be totally apomictic. Although vegetative in origin, the resulting plants are initially juvenile.

Various forms of apomixis occur in fruit plants. In citrus a mixture of sexual and apomictic embryos may develop in the same fruit. Pollination and fertilization are essential (35), the apomictic embryos arising in the nucellus by a process described as nucellar embryony, the multiple embryo production being called polyembryony. In apple a number of species produce a high proportion of apomicts in their seedlings, including *Malus hupehensis*, *M. sieboldii*, *M. sikkimensis*, and *M. toringoides*. Because apple was thought not to pass virus from an infected mother plant to its seedlings, the possibility was investigated of using apomicts as a source of virus-free clonal rootstocks (11, 80). During the ensuing 25 years, the use of apomicts has not been advocated commercially and it is possible to speculate on the reasons. Apomicts are no longer seen as a preferred source of virus-free rootstocks following reports of virus transmission through normal apple seedlings (99, 1). Furthermore, many apomictic selections were highly sensitive to latent virus infection in scions budded onto them, leading to incompatibility. The success of the EMLA scheme (19), and its counterparts in many fruit-producing countries, in maintaining a very low frequency of virus infection in clonal fruit rootstocks and scions, together with improved virus detection methods such as enzyme-linked immunosorbent assay ELISA (18), and the maintenance of healthy clones by tissue culture will have reduced the advantage once seen for apomicts in terms of a source of healthy material.

Orchard performance of apomicts in tests so far conducted has failed to match that of dwarf and semidwarfing clonal rootstocks. There appears to be a carryover of juvenility in that 'Cox', 'Worcester Pearmain', and 'Lord Lambourne' all cropped later on apomictic seedlings of *M. toringoides* than on the medium vigorous clonal rootstock M.2 (11). The majority of apomictic selections, especially of *M. hupehensis* and *M. sieboldii*, were of intermediate vigor in the orchard. Very few selections induced improved crop efficiency based on trunk cross-sectional area, compared to trees on M.9. Only apomictic selections of *M. sargentii* were more efficient than would be expected from a yield response determined by tree size alone (80).

All workers have reported inconsistent responses due to specific behavior of particular scion varieties on certain apomictic rootstocks. For example, 'Cox' grew more vigorously on *M. toringoides* than on *M. sikkimensis*, whereas the reverse was true for 'Lord Lambourne' (11). Some apomictic rootstocks improved crops in 'Cox' while others were particularly suitable for 'Golden Delicious' (80).

3. TREE RAISING

Producing rootstocks is the first step in tree raising, after which they are transplanted to a grafting nursery where the rootstock and scion are combined and where the developing tree may be manipulated, particularly with respect to the number and position of lateral branches.

Successful tree growth in the nursery requires attention to four factors:

a. Planting well-graded virus-free rootstocks and using healthy scion wood.
b. Avoiding root damage from nematodes and replanting diseases present in soil previously planted to a similar crop.
c. The use of an effective grafting technique.
d. Good horticultural practice, including planting rootstocks sufficiently early to ensure good growth when grafted, with benefits of irrigation and effective weed control.

Replanting diseases are avoided by planting into virgin soil or soil not having previously carried the same crop. In practice, nurserymen often do not have sufficient land available to avoid these problems and previously used soils are fumigated or sterilized with appropriate chemicals.

Causal agents of replanting disease have been identified in *Prunus spp.* as the fungus *Thielaviopsis basicola* (83) and in apple possibly as *Pythium sylvestris* (81). Soil treatment with a broad-spectrum fumigant such as chloropicrin (tear gas) is widely used, but a code of practice is enforced aimed at preventing its escape into the air with risks to health of those living nearby. Drenching with formalin is recommended as a less unpleasant way of treating old apple soils (82). In the United States a mixture of dichloropropane—dichloropropene (DD) for nematode control and methylbromide for fungus control is often used, which requires similarly careful application as for chloropicrin (Fig. 19). Treatment of citrus nurseries with methylbromide may result in poor plant growth due to adverse effects on beneficial mycorrhiza whose primary function is to assist in the uptake of phosphorus (64). Dressing seedbeds before sowing with phosphates largely overcomes the problem (98).

Figure 19. Fumigated soil covered by polythene sheets (Oregon Rootstock Inc.).

3.1. Production Systems

The graft can be made at any time, but seasonal constraints often limit opportunities in practice.

Bench grafting, usually by the whip-and-tongue method (29), is done, as the name implies, at the bench before the rootstock is planted. The need for the rootstock to establish in the soil and the union to form at the same time, with root and shoot growth consequently retarded, results in the production of weak trees by this method unless the summer is relatively long so that early deficiencies can be made good later.

The potential management advantage offered by bench grafting is that the graft can be made mechanically; a variety of vine-grafting machines have been tested, but the relative lack of uniformity in size and shape between most top fruit scions and rootstocks limits their application. Machine grafting is widely used for grape, which produces cylindrical, evenly sized stock and scion stems that callus together readily with the aid of warmth supplied in callus boxes, leading to vigorous growth when planted in the field.

Grafting by using a single bud (budding) is the most commonly practiced hand method, being relatively quick to do and economical of scion material. Budding the new established rootstock in early summer (June budding in the Northern Hemisphere) suffers from some of the disadvantages of bench grafting in that the scion bud, often overwintered in the cold store from the previous season, begins to grow soon after budding. This necessitates the removal or breaking of the rootstock shoots above the bud to avoid forcing the scion into horizontal growth, which in turn causes a general loss of vigor in the rootstock, requiring an extended growing season to produce adequate-size maiden (one-year) trees. At East Malling only weak apple trees with few or no lateral branches (unfeathered) of the normally free-feathering variety Cox were produced in one growing season from bench grafting or spring budding; but if left in the nursery for a second growing season, particularly high-quality, large, and well-branched trees resulted (52).

In temperate climates, and where the emphasis is placed on high-quality trees, it is normal to practice summer-dormant bud budding, inserting the scion bud into the rootstock stem during middle to late summer of the first rootstock growing season, where it remains dormant under the inhibitory influence of the rootstock shoots. The rootstock is cut off to the scion bud in the following winter and the scion grows vigorously in the second season, after which it is harvested for sale.

Attempts to reduce the time taken to produce nursery trees by budding the scion into the rootstock while it grows as a shoot on the stoolbed or hedge can be seen in principle as advancing the time of bench grafting, with the apparent advantage that the stock–scion union is already formed at transplanting. This method is not recommended, however, because it results in the parent rootstock accumulating the viruses present in all the scions budded onto it. This malpractice was responsible for the generally high level of virus infection in many apple clonal rootstocks, and even with today's more healthy scion varieties there is never a total guarantee of freedom from all viruses. For the same reasons it is not recommended to propagate cuttings from the tops of lined-out rootstocks after budding in the summer and when pruned back the following winter.

Normally, a scion bud is grafted directly into the rootstock stem at a height which takes into consideration the risk from soil-borne diseases and the distance above ground that branches are required. The fact that the soil-borne fungus *Phytophthora cactorum* infects by soil splash, and that infection was assisted by the moist environment induced by weeds at the base of the stem before the wide use of herbicides, resulted in the recommendation that relatively susceptible apple scion varieties should be budded at least 30 cm high onto relatively resistant rootstocks in the United Kingdom (92). Differences in varietal sensitivity, fungal strain virulence, and the propensity of rootstocks such as M.26 apple to produce burrknots leading to misshapen stems below the union (75) have lead to reports of varying success with this approach. In northern Europe *Prunus spp.* are usually budded at 15cm, with apple budding approaching 30 cm for rootstocks not producing burrknots. Some growers require unions up to 60 cm above ground on specially selected rootstocks to provide further size control and to enable the lowest scion laterals to be retained as the early bearing branches, with reduced risk of the crop being damaged by resting on the ground. Alternatively, high budding allows deeper-than-normal planting in the orchard to avoid the cost of staking (74). In cherry and plum a framework consisting of the trunk and primary branches arising from the main crotch was grown of rootstocks such as *Prunus avium* F12/1, considered to be less susceptable to cankers caused by *Pseudomonas spp.* than many scions. There is evidence from plum inoculation experiments that the extent of the framework's sensitivity to infection may be influenced by the scion worked above it (85), which could be further affected in the orchard by scions of different susceptibility, and hence spore load, altering the potential for infection. Trees with these "resistance" frameworks were expensive to produce because of the time needed to train the rootstock stem and the expense of high grafting, sometimes with four or so scions. Such trees are no longer produced in the United Kingdom and are produced to a much lesser extent than previously on the West Coast of America (R. L. Perry, private communication), with major interest now in obtaining resistant scions.

Sometimes framework trees were grafted with their scion after they were planted in the final orchard position. In the same way trees can be "repropagated" by converting from one scion to another by topwork or framework grafting in the orchard. The old variety is cut back to selected two- or three-year-old branches, and various types of graft to suit the thickness and position of the branch are applied in late spring (29). The main reason for orchard grafting is to convert trees to currently economic varieties. Grafts positioned vertically, particularly on upright branches, will grow strongly, and those inserted horizontally will produce fruit buds and crop the following year, which provides a flexible system for renewing and reshaping trees without prolonged loss of crop (Fig. 20).

The disadvantages of frameworking are that the new scion may be sensitive to latent viruses present in the tree, and considerable aftercare is required to cut the grafting ties, remove unwanted growth from the original scion, and balance the tree by selective pruning.

Multiple component trees with an interstem between the rootstock and scion may be produced to overcome incompatibility between stock and scion, to improve anchorage, to introduce a dwarfing component in the absence of a dwarfing rootstock, and to minimize the risk of winter cold injury to the main trunk above the snow line.

Where a choice of rootstock exists, as among the different species of *Prunus* available for plums, incompatibility is avoided by selecting suitable rootstocks for a particular scion. For example, the variety 'Victoria' is generally compatible, but 'Oullins Golden Gage' is not grafted onto 'Myrobalan B', 'Common Mussel' or 'Marianna' stocks and these are little used in modern nurseries. Some varieties of pear, including clones of 'Williams' ('Bartlett'), 'Bristol Cross', 'Packham's Triumph', and 'Dr. Jules Guyot', are incompatible with quince rootstocks used in northern Europe and a compatible bridge of a suitable variety such as 'Beurré Hardy' or 'Comice' is inserted in the tree. This is usually done by a double operation in which either the interstem is grown first to receive the scion bud later, or the scion is budded into the plant providing the

Figure 20. Frameworking young apple tree (right) to change variety (Howard-Le Fruit Belge **396** (4), 224–234).

interstem and the interstem with dormant scion bud is winter-grafted onto the rootstock. Methods of inserting the compatible bridge in a single budding operation (72) have been developed, but do not give consistently good results and are not used widely in commerce.

When a supply of rooted dwarfing rootstocks is not available a dwarfing interstem such as 'Clark Dwarf,' which is probably the rootstock M.8, is inserted between a seedling root system and the scion variety. A further refinement is the provision of a winter-cold-resistant lower trunk, often of 'Virginia Crab' or 'Hibernal' (38). Systems for raising multiple-component trees were developed primarily in the United States and named after the place (e.g. Geneva) or person (e.g. Clark) responsible (9). Research indicated the need for interstems of different lengths to control vigor of different scion varieties (14), and to some extent this approach appeared to be a way of overcoming the shortage of clonal rootstocks and adapting trees to the different hardiness zones and soil types in which they were required to grow. Multiple-component trees are expensive to produce, however, and could not be grown to a uniformly high standard in the nursery, nor perform uniformly in the orchard. The increased likelihood of virus infection being introduced in multicomponent trees may partly explain their variable growth and cropping in the past. The effect of multiple unions, which may be of varying degrees of completeness and efficiency in the young tree, has not been throughly investigated in this context.

Currently, interstem apple trees are being developed in a few major nurseries using virus-free material. 'Clark Dwarf' stems are being worked onto the cold-resistant 'Antonovka' rootstock, and the dwarfing M.27 introduced for additional size control, in four-component trees (28).

3.2. Budding Methods

The widely used T- or shield-budding technique exploits the convenience of slipping the small piece of rootstock stem carrying the scion bud into a "pocket" on the rootstock made by lifting the corners of "bark," at the intersection of a "T" cut. The scion shield is securely held awaiting tying and is partly protected from desiccation. For decades the variable growth common in T-budded nurseries, especially the tendency of some buds to develop a rosette of leaves in spring before resuming growth some weeks later, was accepted as normal. A study of the anatomy of graft union formation, however, showed that the T-bud method fails to place together the cambium of rootstock and scion whose fusion is essential for union formation (56).

Previously it was erroneously assumed that when lifting the "bark" the tissues of the rootstock stem separated at the cambium, leaving cambial cells on the surface of the xylem to make contact with the cambium at the cut surface of the scion. In fact, the tissues separate in the secondary xylem and the rootstock cambium is exposed only at the cut edges of the bark which overlie the scion shield. Cambial continuity is achieved only if the intervening voids are infilled with callus, and it is because this may not be complete when plants become dormant at the end of the budding season that a check to scion growth occurs the following spring. The chip-bud method, which substitutes a piece of scion stem carrying the bud for a similar piece of rootstock stem (Fig. 21), naturally places the cambium of stock and scion together, resulting in early union formation. This minimizes the risk of winter cold damage and leads to uniform vigorous growth the following season of trees with strong unions (Fig. 22) and reduced likelihood of canker infection because bud shields carrying spores are not inserted beneath the rootstock bark as in T-budding.

The benefits of chip budding compared to T-budding have been shown in all

Figure 21. Apple chip bud in place (Howard-J. R. Hort. Soc. **94** (5), 201–2).

plants tested including ornamental tree species (90). The advantage of chip budding is likely to be greatest in areas which experience relatively cool summers such as northern Europe, where the opportunity for union formation is limited by cool weather after budding. Despite this, the technique has been adopted for apple and cherry in favorable growing areas such as Oregon, where the increased growth and uniformity of trees compared to those T-budded is a sales advantage.

Patch budding, although more difficult to carry out than chip budding, offers similar advantages of good scion-rootstock contact of the cambium, but in this

Figure 22. Union in one-year-old chip-budded apple tree (Howard-J. R. Hort. Soc. **94** (5), 201–2).

case around the cut edges of the patch. Both chip and patch budding are used to propagate the Persian (English) walnut *Juglans regia* and pecan (37).

Buds are tied to the rootstock with polyethylene tape. Self-degrading rubber ties must only be used for chip buds in circumstances when rapid union formation can be expected (87). In view of the importance not to carry diseases from the nursery to the new orchard, bud sticks may be treated to disinfect them from canker spores (7) or disinfect them from pests (97).

3.3. Tree Shaping

Vigorous scion growth obtained by chip budding results in tall maiden trees with early, and hence extensive, lateral production in those varieties which are weakly apical dominant, such as 'Cox' and 'Golden Delicious' apple. For strongly apical dominant varieties, such as 'Spartan' and 'Red Delicious', chemical feathering agents may be sprayed during midsummer to induce lateral production (78). Chemicals such as M & B 25-105 (*n*-propyl 3-*t*-butylphenoxy-acetate) achieve their effect by temporarily preventing the natural auxin indo-lylacetic acid (IAA) from being exported from the scion shoot tip and thus release the lower axillary buds from imposed dormancy. This group of chemicals is most effective when applied to vigorously growing trees and, unlike earlier materials such as the long-chain fatty acit "Offshoot O," do not kill the apical bud, which resumes growth after a few weeks. Other chemical treatments sometimes increase lateral production by a direct stimulatory effect on axillary buds, one example being Promalin, a mixture of benzyladenine and gibberellin 4/7 (23) which is relatively effective with 'McIntosh' and 'Red Delicious' varieties, but not with 'Bramley', 'Spartan', and 'Discovery' (J. D. Quinlan, personal communication).

The growth inhibitor abscisic acid (ABA) is also implicated in the control of axillary bud growth (62). Higher-than-normal levels of ABA were detected in axillary buds of cherry prevented from growing by the presence of shoots in addition to the scion shoot. Shoots of the rootstock retained for longer than the normal period in spring effectively inhibit early scion lateral production, and laterals emerge close behind the shoot tip when the additional shoots are removed. Delaying rootstock shoot removal is used at East Malling as a method of raising the height of cherry (Fig. 23) and pear laterals (63).

Maiden trees are harvested once growth has stopped and the majority of leaves fallen, defoliation sometimes being advanced by chemical sprays (65). Trees are lifted mechanically (Fig. 24) and stored at between 1 and 5°C at high relative humidity to minimize desiccation before shipment.

4. CURRENT RESEARCH AND FUTURE PROSPECTS

There is a worldwide interest in extending the range of dwarfing clonal rootstocks currently based on the East Malling (M) and Malling Merton (MM) series, as indicated by the three international apple rootstock trials planted in 1976, 1980, and 1983. Currently 27 states in the United States, four Canadian provinces, and Mexico are assessing rootstocks from Poland, the United Kingdom, United States, and the Soviet Union. Many new rootstocks propagate well by conventional stoolbeds. Those, such as 'Budagovski 490', which root

Figure 23. Progressive increase in height of cherry laterals (center and right) caused by increasing the number of rootstock shoots retained in early summer (East Malling Research Station).

particularly freely on the stool and also by winter cuttings show a tendency towards preformed root production at aboveground nodes, resulting in burrknot production. This may be detrimental in trees budded above 15 cm by causing misshapen trunks and sites for canker infection.

It is unlikely that all genotypes selected in the future with potential as rootstocks will also have the capacity to produce adventitious roots readily. The increasing interest in the possibility of reducing the high cost of intensive

Figure 24. Over-row mechanized lifting of one-year trees (Oregon Rootstock Inc.).

orchard establishment by producing self-rooted scion varieties controlled in the orchard by growth-regulating chemicals (101) will add to the propagation challenge. The introduction of rooting ability as a selection criterion in current scion variety breeding programs will not be of benefit in this respect for some decades.

Examples of currently difficult-to-propagate clonal rootstocks can be found in pear, where dwarfing pear stocks are sought to replace quince rootstocks to give improved compatibility and wider tolerance of calcareous soils. These include the 'Old Home × Farmingdale' pear rootstocks of which 12 of the original 500 selections are propagated by the Carlton Nursery in Oregon, some with low success rates (J. K. Ballard, personal communication). A similar situation exists with a series of dwarfing pear stocks produced at Anger in France and others at the University of Stellenbosch in South Africa. As the potential worth of such rootstocks becomes apparent, they justify increasing propagation research effort, with every indication that previous difficulties can be overcome.

Research at East Malling in 1984 showed almost all Anger pear clones to root in high percentages from leafy cuttings with either the benefit of a four-week period of etiolation or the propagation of rapidly growing shoots under the stress-free conditions of a ventilated high-humidity (fogging) unit.

Research at Stellenbosch on the propagation of the semidwarfing BP1 and the vigorous BP3 pear rootstocks showed that it was essential to avoid excessive soil moisture when planting into the nursery in order not to depress establishment rates, a problem similar to that affecting survival of unrooted cuttings in heated bins generally.

A precise recommendation to South African nurserymen involves specifying May as the month to collect cuttings, their treatment with antidesiccant, auxin, and fungicides, the application of four weeks' cold treatment at 5°C, and their planting with or without the protection of plastic tunnels depending on the winter rainfall (D. K. Strydom, personal communication).

It is reasonable to expect that intensified research currently taking place into the vegetative propagation of recalcitrant fruit, ornamental and forest trees will further increase propagation possibilities in fruit tree rootstocks and scions. Currently this includes the realization that shoots of difficult-to-propagate subjects can be preconditioned during growth to respond to auxin treatment when propagated subsequently as cuttings. The role of severe stockplant pruning and etiolation/blanching has been discussed in this respect, and it is likely that the benefits of preconditioning together with technical advances during the cutting treatment stage will be effectively combined by producing initially unrooted cuttings on stoolbeds followed by their treatment with auxin and basal heat. Current successful experience in rooting the apple scion variety 'Cox' at 30°C suggests that higher-than-normal basal temperatures may need to be employed (54).

The use of micropropagation to precondition source plants by a process of rejuvenation offers further opportunity for improved propagation, with either macrocuttings propagated by conventional summer and winter techniques from rejuvenated hedges or microcuttings propagated directly *in vitro*. The latter offers the first prospects of clonal rootstocks for walnut (21).

Longer-term research aims to change the genetic basis of adventitious rooting by processes of gene transformation, whereby the auxin-regulating capacity of rhizogenic strains of the bacterium *Agrobacterium tumefaciens*, responsible for hairy root disease, is incorporated into the genome of difficult-to-root plants, where it is hoped that its physiological effect of increasing natural auxin production will enhance rooting potential (39).

Whether or not self-rooted scion varieties will replace trees normally produced on rootstocks will depend on how effectively growth control in the orchard can be achieved through applied chemicals, the extent to which the scion meets other requirements such as good anchorage and cold resistance, and the potential cost saving balanced against consideration of tree quality.

For rapidly growing, short-cycle subjects such as peach, self-rooted orchards of responsive varieties raised by winter cuttings (24) or micropropagation (G. D. Paoli, private communication) may prove economic. However, with slower-growing subjects, such as apple, especially in northern latitudes, it may require the same time to form a 1.5- to 2-m self-rooted tree with adequate branches, as is required for raising a tree by budding onto a rootstock. Much of the high quality reflected in a tall, straight stem with frequent laterals derives from the direction of the growth potential from a large, well-established rootstock through one scion bud. By the same principle, ultraseverely winter-pruned hedges produce relatively few large, feathered shoots per stockplant, which, if successfully rooted and established, may produce "instant trees," thereby reducing the period required for forming well-shaped self-rooted scion trees in the nursery. Whether or not there is a net cost benefit from this approach is likely to depend on the extent to which stockplant maintenance can be mechanized. Current progress with stoolbeds suggests that prospects are good.

REFERENCES

1. Allen, W. R. (1969). Occurrence and seed transmission of tomato bushy stunt virus in apple, *Can. J. Plant Science*, **49**, 797–799.

2. Assal, R. (1977). Nouvelle méthode de multiplication végétative de rameo physiologiquement adultes, de pistachier, noyer et pacanier, *Fruits*, **32**(5), 309–319.

3. Barmore, C. R., and W. S., Castle (1979). Separation of citrus seed from fruit pulp for rootstock propagation using a pectolytic enzyme, *HortScience*, **14**(4), 526–527.

4. Bartolini, G., and C. Briccoli-Bati (1975). Rooting of peach hardwood cuttings in polyethylene bags, *Acta Hort.*, **54**, 59–61.

5. Bassuk, N. L., and B. H. Howard (1981). A positive correlation between endogenous root-inducing cofactor activity in vacuum-extracted sap and seasonal changes in rooting of M.26 winter apple cuttings, *J. Hort. Sci.*, **56**, 301–312.

6. Bassuk, N. L., L. D. Hunter, and B. H. Howard (1981). The apparent involvement of polyphenol oxidase and phloridzin in the production of apple rooting cofactors, *J. Hort. Sci.*, **56**, 313–322.

7. Bennett, M. (1973). Apple canker (*Nectria galligena*): Infection in the nursery, *Rep. E. Malling Res. Stn. for 1972*, pp. 155–156.

8. Brainerd, K. E., and L. H. Fuchigami (1981). Acclimatization of aseptically cultured apple plants to low relative humidity, *J. Amer. Soc. Hort. Sci.*, **106**(4), 515–518.

9. Brase, K. D., and R. D. Way (1959). Rootstocks and methods used for dwarfing fruit trees, Bull. No. 783 New York State Agricultural Experiment Station, Cornell University, Geneva, NY, pp. 16–18.

10. Burns, R. M., and C. W. Coggins, Jr. (1969). Sweet orange germination and growth aided by water and gibberellin seed soak, *Calif. Agric.*, **23**(12), 18–19.

11. Campbell, A. I. (1959). Apomictic seedling rootstocks for apples: Progress Report II, *Rep. Long Ashton Research Stn. for 1959*, pp. 50–56.

12. Cannell, M. G. R., and J. E. Jackson, Eds. (1985). "Future fruit orchard design: Economics and biology," in *Attributes of Trees as Crop Plants*, Institute of Terrestrial Ecology (Natural Environment Research Council), Monkswood Experimental Station, Abbots Rippon, Huntingdon PE17 2LS, UK.

13. Cappellini, P., and A. Nicotra (1973). Translocazione di alcune sostanze rizogene in talee lignose, *Annal dell' Instituto Sperimentale per la Frutticoltura*, **4**, 109–117.

14. Carlson, R. F., and D.Oh. Sung (1975). Influence of interstem lengths of M.8 clone *Malus sylvestris* Mill. Growth, precocity, yield and spacing of 2 apple cultivars, *J. Amer. Soc. Hort. Sci.*, **100**(5), 450−452.

15. Cheffins, N., and B. H. Howard (1982). Carbohydrate changes in leafless winter cuttings. I. The influence of level and duration of bottom heat, *J. Hort. Sci.*, **57**, 1−8.

16. Cheffins, N., and B. H. Howard (1982). Carbohydrate changes in leafless winter cuttings. II. Effects of ambient air temperature during rooting, *J. Hort. Sci.*, **57**, 9−15.

17. Christensen, M. V., E. N. Eriksen, and A. S. Andersen (1980). Interaction of stock plant irradiance and auxin in the propagation of apple rootstocks by cuttings, *Scientia Hort.* **12**(1), 11−17.

18. Clark, M. F., A. N. Adams, and D. J. Barbara (1976). The detection of plant viruses by enzyme-linked immunosorbent assay (ELISA), *Acta Hort.*, **67**, 43−49.

19. Cutting, C. V., and H. B. S. Montgomery, Eds. (1973). *More and Better Fruit with EMLA*, East Malling Research Station and Long Ashton Research Station, 29 pp.

20. Dimalla, G. G., and J. Van Staden (1978). Pecan nut germination—a review for the nursery industry, *Scientia Hort.*, **8**(1), 1−9.

21. Driver, J. A., and A. H. Kuniyuki (1984). *In vitro* propagation of Paradox Walnut rootstock, *HortScience*, **19**, 507−509.

22. Doud, S. L., and R. F. Carlson (1977). Effects of etiolation, stem anatomy, and starch reserves on root initiation of layered *Malus* clones, *J. Amer. Soc. Hort. Sci.*, **102**(4), 487−491.

23. Edgerton, L. J. (1979). Effects of some growth regulators on shoot elongation and branching of apple, *Proc. North East Weed Science Soc.*, **33**, 150−153.

24. Erez, A., and Z. Yablowitz (1981). Rooting of peach hardwood cuttings for the meadow orchard, *Scientia Hort.* **15**, 137−144.

25. Esen, A., and R. K. Soost, (1977). Adventitious embryogenesis in *Citrus* and its relation to pollination and fertilization, *Ann. J. Bot.*, **64**(6) 607−614.

26. Fadl, M. S., and H. T. Hartmann, (1967). Relationship between seasonal changes in endogenous promotors and inhibitors in pear buds and cutting bases and the rooting of pear hardwood cuttings, *Proc. Amer. Soc. Hort. Sci.*, **91**, 96−112.

27. Fiorino, P. (1972). Nuove techniche per ottenere barbatelle di pesco. IV. Ricerche sulla "margotta di ceppaia," *Revista della Ortoflorofrutticoltura Italiana*, **56**(2), 100−104.

28. Frecon, J. (1978). Stark Bro's product development program for size-controlled trees on apples and pears, *Compact Fruit Tree*, **2**, 94−97.

29. Garner, R. J. (1979). *The Grafter's Handbook*, 4th ed. rev., Faber and Faber, 319 pp.

30. Guerriero, R., and F. Loreti (1975). Relationships between bud dormancy and rooting ability in peach hardwood cuttings, *Acta Hort.*, **54**, 51−59.

31. Harrison-Murray, R. S., and M. Fieldsend (1982). Air/water relations of cutting composts, *Rep. E. Malling Res. Stn. for 1981*, pp. 64−65.

32. Harrison-Murray, R. S., and B. H. Howard (1981). Mechanisms of cutting propagation: Etiolation, *Rep. E. Malling Res. Stn. for 1980*, p. 60.

33. Harrison-Murray, R. S., and B. H. Howard (1982). Preconditioning shoots for rooting: Etiolation and blanching of M.9 and other apples, *Rep. E. Malling Res. Stn. for 1981*, pp. 57–60.

34. Harrison-Murray, R. S., and J. C. McNeil (1984). Practical alternatives to peat-based media, *Rep. E. Malling Res. Stn. for 1983*, p. 81.

35. Hartman, H. T., and D. E. Kester (1983). *Plant Propagation: Principles and Practices*, 4th ed., Prentice-Hall, Englewood Cliffs, NJ, pp. 84–198.

36. Hartmann, H. T., and D. E. Kester (1983). *Plant Propagation: Principles and Practices*, 4th ed., Prentice-Hall, Englewood Cliffs, NJ, pp. 523–594.

37. Higazy, M. K., and M. F. Ashmawy (1974). Studies on vegetative propagation of pecan, *Egyptian Journal of Horticulture*, **1**(1), 23–29.

38. Hilborn, M. T., and J. H. Waring (1941). Terminal-shoot growth of apple varieties as apparently stimulated by Virginia Crab and Hibernal intermediate stocks, *Proc. Amer. Soc. Hort. Sci.*, **38**, 316–320.

39. Hooykaas, P. J. J., G. Ooms, and R. A. Shilperoot (1982). Tumours induced by different strains of *Agrobacterium Tumefaciens*, in *Molecular Biology of Plant Tumours*, G. Kahl and J. S. Schell, Eds., Academic, New York, pp. 373–390.

40. Howard, B. H. (1965). Increase during winter in capacity for root regeneration in detached shoots of fruit tree rootstocks, *Nature* (London), **208**, 912–913.

41. Howard, B. H. (1968). Effects of bud removal and wounding on rooting in hardwood cuttings, *Nature* (London), **220**, 262–264.

42. Howard, B. H. (1968). The influence of 4(indolyl-3) butyric acid and basal temperature on the rooting of apple rootstock hardwood cuttings, *J. Hort. Sci.*, **43**, 23–31.

43. Howard, B. H. (1971). Propagation techniques, *Sci. Hort.*, **23**, 116–126.

44. Howard, B. H. (1977). Effects of initial establishment practice on the subsequent productivity of apple stoolbeds, *J. Hort. Sci.*, **52**, 437–446.

45. Howard, B. H. (1977). Some approaches to propagation problems in the early multiplication of new clones of woody fruit plants, *Proc. Int. Plant Prop. Soc. for 1976*, **26**, 113–115.

46. Howard, B. H. (1978). Field establishment of apple rootstock hardwood cuttings as influenced by conditions during a prior stage in heated bins, *J Hort Sci.*, **53**, 31–37.

47. Howard, B. H. (1980). Moisture change as a component of disbudding responses in studies of supposed relationships between bud activity and rooting of leafless cuttings, *J. Hort. Sci.*, **55**, 171–180.

48. Howard, B. H. (1984). The effects of NAA-based Tipoff sprays on apple shoot production in MM.106 stoolbeds, *J. Hort. Sci.*, **59**, 303–311.

49. Howard, B. H. (1985). The contribution to rooting in leafless winter cuttings of IBA applied to the epidermis, *J. Hort. Sci.*, **60**, 153–159.

50. Howard, B. H. (1985). Factors affecting the response of leafless winter cuttings to IBA applied in powder formulation, *J. Hort Sci.*, **60**, 161–168.

51. Howard, B. H., and V. H. Oehl (1981). Improved establishment of *in vitro*–propagated plum micropropagules following treatment with GA_3 or prior chilling, *J. Hort. Sci.*, **56**, 1–7.

52. Howard, B. H., and H. R. Shepherd (1980). Possibilities for producing cheap trees, *Rep. E. Malling Res. Stn. for 1979*, pp. 81–82.

53. Howard, B. H., R. S. Harrison-Murray, and S. B. Arjyal (1985). Responses of apple summer cuttings to severity of stockplant pruning and to stem blanching, *J. Hort. Sci.*, **60**, 145–152.

54. Howard, B. H., R.S. Harrison-Murray, and C. A. Fenlon (1983). Effective auxin treatment of leafless winter cuttings, *British Plant Growth Regulator Group, Monograph 10*, pp. 73–85.

55. Howard, B. H., R. S. Harrison-Murray, and K. A. D. MacKenzie (1984). Rooting responses to wounding winter cuttings of M.26 apple rootstock, *J. Hort. Sci.*, **59**, 131–139.

56. Howard, B. H., D. S. Skene, and J. S. Coles (1974). The effects of different grafting methods upon the development of one-year-old nursery apple trees, *J. Hort. Sci.*, **49**, 287–295.

57. Jackson, J. E. (1980). Light interception and utilization by orchard systems, *Hort. Rev.*, **2**, 208–267.

58. James, D. J. (1983). Adventitious root formation "in vitro" in apple rootstocks (*Malus pumila*). 1. Factors affecting the length of the auxin-sensitive phase in M.9, *Phys. Plant.*, **57**(1), 149–153.

59. Jones, O. P., and M. E. Hopgood (1979). The successful propagation *in vitro* of two rootstocks of *Prunus*: The plum rootstock Pixy (*P. insititia*) and the cherry rootstock F12/1 (*P. avium*), *J. Hort. Sci.*, **54**, 63–66.

60. Jones, O. P., M. E. Hopgood, and D. O'Farrel (1977). Propagation *in vitro* of M.26 apple rootstocks, *J. Hort. Sci.*, **52**, 235–238.

61. Kawecki, Z., M. A. Kiszczak, J. Wazbinska, and S. Tarkowian (1978). Changes in protein fractions and in the activity of proteolytic enzymes of walnut seeds (*Juglans regia* L.) during their after-ripening, *Fruit Science Reports*, **5**(1), 1–6.

62. Kim, Yong-Koo, B. H. Howard, and J. D. Quinlan, (1984). Apparent ABA-induced inhibition of the lower laterals of one-year-old cherry trees, *J. Hort. Sci.*, **59**, 35–44.

63. Kim, Yong-Koo, B. H. Howard, and J. D. Quinlan (1984). Growth responses to different grafting and manipulating treatments in one-year-old fruit trees, *J. Hort. Sci.*, **59**, 23–33.

64. Kleinschmidt, G. D., and J. W. Gerdemann (1972). Stunting of citrus seedlings in fumigated nursery soils related to the absence of endomycorrhizae, *Phytopathology*, **62**(12), 1447–1453.

65. Knight, J. N. (1983). Chemical defoliation of nursery stock using chelated forms of copper and iron, *J. Hort Sci.*, **58**, 471–476.

66. MacKenzie, K. A. D. (1981). *Anatomy, Rep. E. Malling Res. Stn. for 1980*, pp. 63–64.

67. Medina Pena, J. P. (1977). A new method for pecan propagation (*Carya illinoensis* Kock.), *Plant Propagator*, **23**(2), 8.

68. Ministry of Agriculture, Fisheries and Food, Bull. 207, *Apples*, Her Majesty's Stationery Office, London, 205 pp.

69. Minotta, G. (1981). Ricerche sull' impiego di differenti carboidrati nei substrati di

micropropagazione del susino, *Revista della Ortoflorofrutticoltura Italiana*, **65**(5), 343–352.

70. Nahlawi, N., and B. H. Howard (1971). Effect of position of IBA application on the rooting of plum hardwood cuttings, *J. Hort. Sci.*, **46**, 535–543.

71. Nahlawi, N., and B. H. Howard (1972). Rooting responses of plum hardwood cuttings to IBA in relation to treatment duration and cutting moisture content, *J. Hort. Sci.*, **47**, 301–307.

72. Nicolin, P. (1953). Das "Nicolieren," eine neue Veredlungsmethode, *Deutsche Baumschule*, **5**, 186–187.

73. Novak, F. J., and Z. Juvona (1983). Clonal propagation of grapevine through *in vitro* axillary bud culture, *Scientia Hort.*, **18**(3), 231–240.

74. Parry, M. S. (1974). Depth of planting and anchorage of apple trees, *J. Hort Sci.*, **49**, 349–354.

75. Parry, M. S. (1980). Effect of height of budding on different rootstocks, *Rep. E. Malling Res. Stn. for 1979*, p. 50.

76. Pastyrik, L., and J. Ivanicka (1977). Influence of winter dormancy on the rooting of hardwood cuttings of the wild apricots *Armoniaca vulgaris* and *Armoniaca manshurica, Phys. Plant.*, **14**, 87–95.

77. Pontikis, C. A., K. A. D. MacKenzie, and B. H. Howard (1979). Establishment of initially unrooted stool shoots of M.27 apple rootstock, *J. Hort. Sci.*, **54**, 79–85.

78. Quinlan, J. D. (1980). Recent developments in the chemical control of tree growth, *Acta Hort.*, **114**, 144–151.

79. Robinson, J. C., and W. W. Schwabe (1977). Studies on the regeneration of apple cultivars from root cuttings. II. Carbohydrate and auxin relations, *J. Hort. Sci.*, **52**, 221–233.

80. Schmidt, H., and J. Krüger (1983). Fruit breeding at the Federal Research Centre for Horticultural Plant Breeding, Ahrensburg/Holstein, *Acta Hort.*, **140**, 15–33.

81. Sewell, G. W. F. (1981). Effects of *Pythium* species on the growth of apple and their possible causal role in apple replant disease, *Annals of Applied Biology*, **97**, 31–42.

82. Sewell, G. W. F., and G. C. White (1979). The effects of formalin and other soil treatments on the replant disease of apple, *J. Hort. Sci.*, **54**, 333–335.

83. Sewell, G. W. F., and J. F. Wilson (1975). The role of *Thielaviopsis basicola* in the specific replant disorders of cherry and plum, *Annals of Applied Biology*, **79**, 149–169.

84. Sewell, G. W. F., B. H. Howard, and J. F. Wilson (1974). Increased productivity of F12/1 layer-beds with benomyl treatments, *Rep. E. Malling Res. Stn. for 1973*, 193.

85. Shanmuganathan, N., and J. E. Crosse (1963). Experiments to test the resistance of plum rootstocks to bacterial canker, *Rep. E. Malling Res. Stn. for 1962*, pp. 101–104.

86. Shepherd, U. M. (1979). Effect of tree quality at planting on orchard performance, *Rep. E. Malling Res. Stn. for 1978*, p. 40.

87. Shepherd, H. R., and B. H. Howard (1984). Degradable rubber budding ties, *Rep. E. Malling Res. Stn. for 1983*, p. 86.

88. Shirzad, B. M., and N. W. Miles (1977). Seasonal and IBA induced changes in endogenous growth substances related to rooting of K-14 apple (*Malus sylvestris* (L.) Mill) rootstock, *Compact Fruit Tree* June 10, 100.

89. Shreve, L., and N. W. Miles (1972). Propagating black walnut clones from rooted cuttings, *63rd Annual Report Northern Nut Growers Association Ames, Iowa State Univ.*, pp. 45–49.

90. Skene, D. S., H. R. Shepherd, and B. H. Howard (1983). Characteristic anatomy of union formation in T- and chip-budded fruit and ornamental trees, *J. Hort. Sci.*. **58**, 295–299.

91. Smir, I., and A. Erez (1980). In-vitro propagation of Malling Merton apple rootstock, *HortScience*, **15**(5), 597–598.

92. Smith, N. G., and W. S. Rogers (1961). Higher budding of apple rootstocks, *Rep. E. Malling Res. Stn. for 1960*, pp. 49–51.

93. Smith, I. E., B. N. Wolstenholme, and P. Allan (1975). Carbohydrate relations of pecan (*Carya illinoensis* (Wang.) K. Koch) stem cuttings, *Agroplantae*, **7**(3), 61–66.

94. Sriskandarajah, S. (1984). Induction of adventitious roots in some scion cultivars of apple (*Malus pumila* Mill), Ph.D. thesis, Dept. of Agronomy and Horticultural Science, University of Sydney.

95. Sriskandarajah, S., and M. G. Mullins (1982). Micropropagation of apple scion cultivars, *Proc. Int. Plant Prop. Soc. for 1981*, **31**, 209–213.

96. Sriskandarajah, S., M. G. Mullins, and Y. Nair (1982). Induction of adventitious rooting *in vitro* in difficult-to-propagate cultivars of apple, *Plant Science Letters*, **24**(1), 1–9.

97. Terent'ev, S. N. (1968). Wet disinfection of grafting and planting material against coccids and mites (Russian), *Himija Sel'. Hoz.*, **61**(11), 36–38.

98. Tucker, D. P. H., and C. O. Youtsey (undated). *Citrus Nursery Practices*, Florida Cooperative Extension Service, Institute of Food and Agricultural Sciences, Circular 430, University of Florida, Gainesville, FL.

99. Van Der Meer, F. A. (1976). Observations on apple stem grooving virus, *Acta Hort.*, **67**, 293–304.

100. Vasek, J., and B. H. Howard (1984). Effects of selective and biennial harvesting on the production of apple stoolbeds, *J. Hort. Sci.*, **59**, 477–485.

101. Webster, A. D., and J. D. Quinlan (1984). Research update with the growth regulator PP333, *Compact Fruit Tree*, **17**, 133–141.

3

COMPATIBILITY AND STOCK—SCION INTERACTIONS AS RELATED TO DWARFING

Roy K. Simons
University of Illinois
Urbana, Illinois

INTRODUCTION

The impact of dwarfing rootstocks upon the changing character of the fruit industry has been significant (13, 17, 18). New developments are being made as a result of economic and societal demands for increased fruit production with decreased use of land and labor.

Dwarfing of fruit trees is a primary concern for contemporary fruit growers. Much has been accomplished in the manipulation of growth by management practices and by the use of many different plant materials (approximately 40 apple rootstocks and interstems are available to the industry) for dwarfing, precocity, winter hardiness, and disease resistance (16, 18, 19, 32, 33, 36, 102, 103).

Not all stock—scion combinations have proven to be compatible; however, some cultivars exhibiting incompatibility may still prove to be a successful production unit for the grower. For example, M.7 exhibits incompatibility traits with some cultivars, but remains one of the premier rootstocks being used throughout the industry. The Winesap cultivar on M.7 shows such extreme incompatibility symptoms between the stock and scion that it would not be wise to select this combination.

Another example of compatibility problems between the stock and scion is Granny Smith/M.26. Growth characteristics would indicate that this combination might be incompatible. However, with careful tree training and staking, production of 1200 to 2000 32lb boxes per acre and a tree life of 25 years have been reported.

1. TERMINOLOGY CONCERNING COMPATIBILITY

The symptoms of incompatibility become involved when asexual propagation has been made on a large scale (14, 15, 37, 38, 109, 111). Incompatibility in grafts between distantly related plants is indicated by a complete lack, or a very low percentage, of successful unions.

Incompatibility, uncongeniality, and dwarfing have been defined (104). The word congenial means "suited or adapted in nature or character; tolerant; kindred." One author states that it is possible to have degrees of congeniality, whereas it is not possible to have degrees of incompatibility and compatibility (104). Many factors are involved when the stock and scion are united. Trial and error may be the ultimate in testing the performance of the two components.

Stages of incompatibility have been outlined (68) into three groups, distinguished by responses to the introduction of a third, mutually compatible variety. These groups are those where incompatibility is cured by the interposition of an intermediate between two incompatible components, those where incompatibility is not cured by an intermediate, and those where incompatibility is induced between two normally compatible graft components by top working with another variety.

Further definition of incompatibility describes two categories (68). The first is characterized by starch blockage, that is, the accumulation of starch above the union and its almost complete absence below the union; phloem degeneration; different behavior of reciprocal grafts; normal vascular continuity at the union (although sometimes there may be marked overgrowth of the scion, with a consequent development of a crease containing compressed bark tissues); and early effects on growth. The second category (localized incompatibility) is associated with characteristic breaks in cambial and vascular continuity (although normal union structure may occur in a few instances); similar behavior of reciprocal combinations; and gradual starvation of the root system, with slow development of external symptoms proportional in severity to the degree of vascular discontinuity at the union.

In addition to poor growth characteristics, incompatibility is most evident in unsatisfactory unions, often resulting in breakage. Anatomical investigations (68) indicate that the most characteristic breakage is due to a defective union structure summarized as follows: normal processes of bud union and the earliest indications of incompatibility, the initiation of vascular discontinuity, the possibility of forecasting incompatibility from union structure, and anatomical abnormalities developing in incompatible trees but not directly connected with the structure of the graft union.

Another study (39) suggests four categories of incompatibility: graft combinations where the bud failed to grow out; graft combinations where incompatibility was due to virus infections; graft combinations with mechanically weak unions in which the cause of death was usually breakage (ill health, if shown, was due to mechanical obstruction at the union); and graft combinations where ill health was not directly due to abnormal union structure, but usually associated with abnormal starch distribution.

2. DWARFING MECHANISMS

Dwarfing may be achieved by varying horticultural practices such as selection of spur types and strains of scion cultivars; interstems and varying interstem lengths in relation to the scion cultivar; pruning (low heading of the tree); summer pruning; root pruning; use of growth regulators; fruiting and its relation to dwarfing; control of nutrient elements; girdling, scoring, bark inversion; twisting, spreading, bending; and the use of dwarfing or size-control rootstocks (104). By far, the most common method is the grafting of a size-controlling rootstock with a scion chosen for its fruiting potential.

There are different viewpoints concerning the mechanism of dwarfing. These are anatomical, physiological, or chemical methods (58). In reviews pertaining to dwarfing, the subject may be exploited according to the author's expertise. However, the objective of researchers and nurserymen should be to develop rootstocks with specific plant adaptibility characteristics, compatible both for the industry and for nurserymen, including the ability to produce economically over a wide range of environmental disease conditions (20−23, 32).

Research concerning dwarfing mechanisms in apples (50−52, 55−57, 108, 112, 113, 115) includes top−root relationships, endogenous factors affecting top−root relationships and growth, auxins, cytokinins, gibberellins, abscisic acid, nitrogen compounds, and phenols. It was concluded that the previous concepts of the dwarfing mechanism in apples have not explained many of the known facts about dwarfing, especially the effect of the interstock. The strongest evidence indicates that the bark is the key to the dwarfing mechanisms. This was found by the insertion of M.26 bark grafts in the scion of Gravenstein/MM.111, producing dwarfed trees. Other studies concerning phenolics have also been reported (44).

Certain rootstocks are related to hardiness of deciduous fruit trees (49, 107). The mechanisms inducing dormancy and rest in plants are important because actively growing tissues are damaged more than dormant ones at a given temperature. While these mechanisms are partially understood for simple plants, their interactions in compound genetic systems are poorly defined (107).

It has been postulated that dwarfing results when auxin produced by the shoot tip is translocated down the phloem. The amount arriving at the root influences root metabolism and affects the amount and kind of cytokinins synthesized and translocated to the shoot through the xylem (54).

Other physiological aspects of scion−rootstock combinations have been emphasized in research reports such as behavior of the root system, behavior of the scion, response of the cultivar from the dwarfing interstem, and the role of phenolics (2, 11, 57, 78, 79, 84, 85).

Studies of rootstock–scion interaction in apples, with special reference to root anatomy (3, 4, 6), show that the root structure of unworked Malling rootstocks is constant and may be correlated with potential vigor of the selections. When stocks and scions representing extremes of vigor are combined, modifications of the inherent root structure of the clonal rootstocks are apparently attributable to scion influence (45–47).

Bark thickness (53, 94, 98–100, 103) of scions of plants on M.9 rootstock was greater than that on MM.111, which had the thinnest (103). Among the interstock types, the bark percentages on new growth (scions and interstocks) were all higher on plants with M.9 interstocks than those with Red Delicious or MM.106 rootstocks. Other studies (88) found an interaction between dying stock and scion tissue of M.9 and Golden Delicious that changed the orientation of the phloem-ray cells and sieve tubes of the scion from their normal conducting patterns. The sieve tubes were necrotic in the immediate vicinity of the union. Similar malformations were found with Golden Delicious on M.7 (85), although it should be emphasized that the material was not virus free. Cross sections of the bark showed more active phloem in trees growing on a seedling rootstock than on M.7. The latter had thicker bark and almost twice as many phloem-fiber cells, but these were compressed into a smaller tree circumference. Other studies (53, 86, 87, 90–100, 103, 114) have found similar growth abnormalities.

Anatomical studies of stems and roots of hardy fruit trees (3, 4) indicate that vigorous rootstocks produced a larger number of vessels and twice as many xylem fibers as the dwarfing rootstocks. The dwarfing influence possessed by apple rootstocks was correlated with the amount of living tissue in the roots, the most dwarfing type of rootstock having the greatest proportion of living tissue in the root.

The structure of the wood of the rootstock roots (in particular, the amount of fiber elements and ray cells) was shown to be related to the vigor and fruitfulness of the scion variety (3, 4, 60, 73–75). Also, the anatomical structure of Golden Delicious and Golden Delicious hybrids was found to be different (43). Other studies recorded the rupture strength of Golden Delicious on different dwarfing rootstocks (70, 77).

Nutrient-element uptake must be considered when dwarfing rootstocks are involved (83, 101), significant differences were found between stock–scion combinations as to vascular tissue development associated with malfunctions (94–96). More Ca appeared throughout the conducting phloem than in the nonconducting phloem, cambium, or xylem. Excessive nonconducting phloem was found in the extreme dwarfing rootstocks, and Ca accumulation was associated with necrotic tissue. Rootstocks with a thick bark (such as M.9) have a greater proportion of nonfunctioning phloem as compared with M.7 or MM.106. Also, interstem combinations of Granny Smith/Ottowa 3/MM.111 and Granny Smith/M.27/MM.111 had the most Ca-containing crystals in the bark where the different rootstock combinations were evaluated (98).

In some T-bud combinations phloem development varied as much as 50% in thickness. Less abnormal phloem growth was found in chip-budded trees and confirmed previous studies (40).

Nutrient-element studies have shown that McIntosh on 16 different rootstocks produced more leaf Ca and Mg as contrasted to those on M.4 with the lowest leaf Mg, Ca, Na, and Al (72).

Starch distribution between different parts of grafted trees suggests that the study of its metabolism may be important in connection with incompatibility between the rootstock and scion (61). However, in a study of *Prunus avium* L./Mahaleb (*P. mahaleb* L.) exhibiting incompatibility symptoms (12), no differences were found in the amount of starch above and below the graft union or above and below the mechanical treatments.

Etiolation effects on stem anatomy and starch reserves as related to root initiation of layered *Malus* clones showed that nodal rooting was closely associated with the parenchymatous bud gaps of the stem (27). The highest starch concentration was in these tissues and the outer ring of pith cells. This was confirmed by scanning electron microscopic studies (94, 98, 99). Etiolation during layering increased stem starch and decreased sclerification of the cortex. Rooting success was negatively correlated with the degree of sclerification.

Vigor and dwarfism of apple trees in an *in vitro* tissue culture system showed that cultures of M.13 displayed a higher growth rate than those of M.9 at the optimum concentration of growth regulators. The cells of M.9 contained a greater amount of starch and protein nitrogen than those of M.13. These materials varied in quantity between the two types during the growth period (62).

Paper chromatography of tissue extracts showed the presence of a greater amount of phenolic substances in the tissues of M.9 (62).

3. CHARACTERIZATION OF THE GRAFT UNION

The use of grafting to unite two different components of plant tissue is a long-recognized practice. This is usually accomplished by budding techniques where the scion is reduced in size to one bud. Theoretical aspects of grafting and budding have been described, as well as the wound healing associated with establishment of the graft union (38, 106–110). Matching of the cambia, when the scion is attached to the stock, facilitates the establishment of cambial continuity and the successful junction of the subsequently formed vascular tissues (28–31). Callus arises from various living cells in the vascular region, among which phloem-ray cells and immature xylem-ray cells may be particularly active.

Graft union development involves many difficulties (28–31, 81). The causes of success or failure of specific grafts are not completely known. The effects of using interstocks in dwarfing apple trees can be advantageous or lead to incompatibility problems. When an incompatible rootstock variety is used as an interstock (81), the incompatibility is determined by the interaction of the scion variety and the root system rather than by a poor graft union. The dwarfing effect of certain dwarfing rootstocks, and of inverted rings of bark, may be due to checking the flow of auxin to the roots.

Clean breaks at the union attributed to viruses have been observed in some Wayne/M.9 and Wayne/M.26 plantings (22).

The difficulties of successful stock–scion interactions may not be due to the specific combinations, but rather to the compound genetic systems being exposed to a hostile environment for which they are not adapted (107). Environmental factors such as reflected light on sand or bare soil in which the newly formed bud is "blasted" or desiccated throughout the delicate tissues may limit the success of a graft union. Other hostile environmental factors include herbicide management and the alternate freezing and thawing of bare soil, resulting in temperature fluctuations that might cause abnormal development of the scion or winter injury of these tender tissues. In some cases, this may be only a marginal injury but will decrease the normal growth pattern of the scion. The stress of abnormal temperature or moisture should be avoided in order to establish a normal translocation system between the rootstock and scion.

Roots determine rootstock influence in apples (3–5, 44, 47, 60). The presence of a piece of rootstock stem tends to enhance the rootstock effect, but the stem effect was much smaller than that of the roots.

4. SURVEYS ON INCOMPATIBILITY

A survey was made to summarize rootstocks currently being propagated and to coordinate future research concerning stock–scion relationships where incompatibilities may arise or where adverse environmental conditions may be a limiting factor. This report included the East Malling series, hardy rootstocks, seedling rootstocks, and interstocks. In addition, peaches, red-tart cherries, sweet cherries, and pears were included. The trend toward using dwarfing rootstocks has been expanded since this report was made (89).

Another survey (69) was prompted by the failure of scion varieties to make satisfactory growth when budded or grafted to certain rootstocks. This difficulty has long been confronting growers and nurserymen.

The problem of incompatibility concerning pome and stone fruits was documented (1). The factors governing the success or failure of grafts or buds to form a union with the stock are listed as incompatibility, varietal peculiarities and the method of propagation, climatic conditions, size and age of stock, position of the top bud of the graft, and faulty manipulation in grafting. Other studies on graft incompatibility in fruit trees include reference 68 and the review of the Malling apple rootstocks in America (116).

5. NEW DEVELOPMENTS

The development of rootstocks for fruit trees has been progressing at a rapid rate. There may be as many as 40 different apple rootstocks thus far developed for the industry, and new ones are currently under research for introduction. The constant introduction of these new rootstocks has been brought about by the industry's demand for more efficient production, more trees per production unit, improved disease and cold resistance, and increased adaptability to certain adverse soil and climatic conditions (107).Thirteen pear rootstocks have been introduced to the industry. The stone fruits have a smaller number of rootstocks, but much interest has been initiated in this area (180).

6. RESEARCH CONCERNING COMPATIBILITY

An estimated 500,000 trees died in peach areas of the southeastern United States (82). Serious losses have prompted greater interest in rootstocks as a possible solution. At the present time, there is not a completely satisfactory rootstock for peach (80).

The nature and development of structural defects in the unions between pear and quince were reported as follows (64–66): all pear–quince unions (irrespective of the horticultural compatibility of the intermediate used) showed a well-marked, dark brown separation layer between the outer bark of the pear and quince; all pear–quince unions showed some microscopic evidence of uncongeniality in the form of abnormal proliferation of the phloem rays, and of necrotic cells occurring singly (or in groups) in the bark at (or very close to) the union. The largest amount of necrotic tissue occurred in the two-year-old and older phloem. The damage appeared to originate in older parts of the one-year-old phloem and to spread in the rays. The extent of microscopic abnormalities (and the spread of the separation layer into the inner phloem and up to the cambium) appeared to be related to the degree of horticultural compatibility of the pear varieties used as an intermediate.

These observations concerning compatibility between pear and quince (65, 66) advance the hypothesis that breaks in continuity of the woody tissues arise when lethal substances, which accumulate in the older phloem near the union, are able to reach the cambium. It was also shown that application of IAA at 100 mg/L plus GA_3 at 500 mg/L to the base of the scion before grafting improved the pear–quince union (10).

Sieve-tube necrosis has been associated with pear decline (7). Since the reviews conducted in the past were based upon research involving the early introduction of dwarfing rootstocks which contained viruses, future work will include those rootstocks plus the newly introduced scion materials which are virus free and which will produce different types of growth response. An example of this was confirmed in studies of Cox's Orange Pippin on EMLA.9 rootstock free from known viruses. The resulting trees were 50% larger than those on M.9a, which contained a latent virus (41, 71). Other studies show that apple union necrosis and decline was found to be associated with recovery of tobacco ringspot virus from infected rootstocks (48). Also, adventitious rooting in dormant hardwood cuttings was depressed by a latent infection of five different viruses (40).

The relationship of cyanogenic glycosides to graft incompatibility between the peach and plum has been investigated (8, 9). The combination Fay Elberta/ Marianna 2624 showed foliar symptoms of graft incompatibility. Prunasin, the only cyanogenic glycoside detected in both species, accumulated in young scion bark of Marianna 2624 and peach to nearly equal levels. Amounts of prunasin in leaves and bark of Fay Elberta on peach were usually greater than those on plum. The prunasin level in the plum rootstock bark (immediately below or 18 cm from the union) was unaffected by the scion species or by signs of ill health in the peach top. Although the rootstock was shown capable of affecting the accumulation of prunasin in scion tissues, the stability of the level of this glycoside in the peach–plum combination suggests that cyanogensis is not closely linked with their incompatibility.

The compatibility of peach cultivar–almond seedling rootstocks was compared to their cyanogenic content (8, 9). Incompatible cultivars contained more glycoside than compatible ones. In young plants, the cyanogenic glycoside content of the almond rootstock and the peach scion were directly correlated, the glycoside apparently moving from the scion into the rootstock. Almond seedling types obtained from trees which are relatively rich in cyanogenic glycoside were found less suitable as rootstocks for peaches than a poorer type.

The use of almond rootstocks greatly increases the resistance of peach trees to lime-induced chlorosis, and the growing of peaches on these rootstocks is of considerable importance with calcerous soils (24). Most peach cultivars show some degree of graft incompatibility with almond rootstocks, causing an early decline accompanied by an anatomical disturbance at the union between rootstock and bark.

A study with 10 apricot cultivars, each grafted onto six rootstocks (63), showed that the peach line GF 305 proved to have the best overall compatibility, with an average bud take of 76%. But even the least successful rootstock (Regina Claudia GF 1380) gave an average take of 58%. The stocks considered worthy of further study were GF 305, Canino apricot seedling, and Marianna GF 8/1.

Some varieties of apricot (*Prunus armeniaca*) present a localized graft incompatibility with lignification occurring imperfectly near the graft union. Peroxidases (especially syringaldazide) can be implicated in lignification and, therefore, be involved in graft incompatibility (76). Other studies with *Prunus* involve peroxidase activity and lignification (25, 26).

An electrophoresis method for water-soluble protein of *Prunus* has been determined (42) in an effort to make a rapid assay for graft compatibility.

The stage corresponding to the establishment of vascular connections was used to differentiate between compatible and incompatible shield-budded combinations of apricot on myrobalan rootstocks in a histophysiological study of the graft union of compatible and incompatible *Prunus* combinations (67). In compatible combinations (such as apricot cv. Polonais 1331 on myrobalan) a functional vascular connection was established 8 to 10 months after budding.

At the same period of development, incompatible combinations (with apricot cv. Canino 1343) showed an imperfect vascular connection, apparently the result of the abnormal functioning of the newly formed cambium in the union zone. Both compatible and incompatible combinations showed peroxidase activity in persistent necrosed cells at the interface of the newly formed tissues, differentiated by both stock and scion. This activity could be linked, in part, with the lignification of necrosed cells, but the persistence of necrosed cells did not seem to be responsible for the observed incompatibility. Myrobalan roots produced larger trees and heavier bloom but lower yield efficiency in prunes than did peach roots (110).

In vitro micrografting was utilized for a grafting incompatibility study with different species of *Prunus* (59). It allowed the early detection of two types of incompatibilities often manifested only years after grafting had been done: localized (apricot−myrobalan) and translocated (peach−apricot and peach−myrobalan).

Incompatibility symptoms were observed on every apricot−tomentosa union, but none on apricot−myrobalan unions (34, 35). They were expressed as a severe overgrowth of the apricot with weakness and breakage of the graft union. Every apricot−tomentosa tree was lost due to breakage of the graft union. In contrast, all trees on myrobalan rootstocks had strong graft unions and grew normally. Also, no *P. tomentosa* seed sources produced seedlings compatible with apricots; therefore, no compatible seedling could be selected for further trial. Thus it appears that *P. tomentosa* is completely unsatisfactory as a rootstock for apricots (34, 35).

7. MORPHOLOGICAL AND ANATOMICAL ASPECTS OF GRAFT UNION DEVELOPMENT

Trunk diameters of the stock and scion may grow at different rates and produce unequal growth. For example, a one-year-old T-budded Granny Smith/M.26 shows the results of removing the rootstock top (arrows, Fig. 1) and allowing the scion to grow. The differences in trunk diameter may be noted by the decreased growth of Granny Smith, which has already been affected by the rootstock. Also, the upright growth is not straight, which often occurs at this stage of development and persists throughout the tree life.

In Figure 2, a six-year-old Granny Smith/M.26 shows an overgrowth of the graft union. These trees were staked to or supported by a slender spindle type of training and were producing good commercial crops in the sixth producing year. However, they were irrigated and were not grown under stress conditions. The graft union has constricted abnormal vascular tissue development (88). This is an example where growth of the scion variety has been altered from an uncongenial situation to permit good economic production with a minimum of effort.

It has been found that chip budding is desirable for better bud take and the

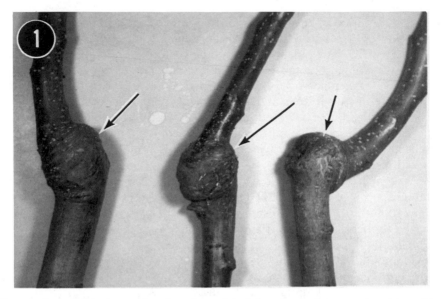

Figure 1. Irregularities in growth at the union and the failure to grow in a normal-upright position as a result of an unsatisfactory T-bud graft union of a one-year-old Granny Smith/M.26 apple tree.

Figure 2. Overgrowth of the graft union of Granny Smith/M.26 in commercial production. Decrease in scion and rootstock growth was evident on both sides of the union.

production of a straight-upright tree (41). Redchief/MM.111 was T-budded (left, Fig. 3) and was compared with a chip-budded tree (right, Fig. 3). These trees gave a straight-upright growth, with the chip-budded tree being of a higher quality. These may be compared to Figure 1, where the rootstock–cultivar differences are noted.

Breakage between the unions occurs when some compound genetic systems (the use of interstems between stock and scion) are employed. Figure 4 illustrates a union between M.9 interstem and MM.111 rootstock. This type of

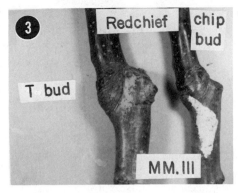

Figure 3. Comparison of budding techniques (Redchief/MM.111) showing uniformity of scion and rootstock growth of the chip-budded tree. Left: T bud; right: chip bud.

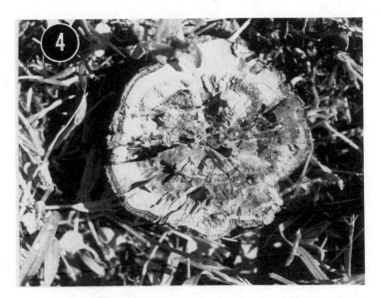

Figure 4. Cross section of broken graft (M.9/MM.111) that was incompatible. The parenchymatous xylem-ray cells developed abnormally, thus preventing a union of vascular tissues between the stock and scion.

development occurs in the formative stages, soon after the scion has been combined with the rootstock. Any abnormal growth occurring during this period will perpetuate new wound-regenerative tissue with subsequent necrosis. A continuation of this process will accentuate breakage as the tree comes into bearing. This stage of development is a likely time for stress problems to appear.

The breakage in Figure 4 is the result of development of xylem- and phloem-ray parenchyma in abnormal alignment between the rootstock and scion.

Failure of bud take and abnormal development after budding may be accentuated when a bud is placed in contact with swirling tissue such as shown in Figure 5. This is not necessarily a compatibility problem, but may be due to bud quality, stage of development, or specific tissue development at the time the rootstock was budded. Figure 5 is typical unspecialized wound-regenerative tissue likely to develop burrknots, lateral shoots, or root initiation. The tissue illustrated would be expected to produce an abnormal union.

Another example of abnormal growth at the union is the Golden Delicious /M.7 shown in Figure 6. This tree was from a planting using a short-bud stick, resulting in a shallow root system that required staking. Although these trees were precocious and heavy bearing, trunk twisting was evident.

The Smoothee/M.26 combination has exhibited breakage problems in Midwest orchards. The poor stock/scion union of which a 13-year-old tree died is illustrated in Figure 7.

An example of a successful union (Red Delicious/M.9) shows the continuity of the tissues uniting the stock and scion (Fig. 8) (90). Callus development was uniform and was without any necrotic tissue as a result of the graft wound. However, in Figure 9 (Golden Delicious/M.7a) there was some necrosis developing at the union area (arrows). The newly formed xylem and phloem were disorganized on a horizontal axis rather than in a normal longitudinal pattern originating from the rootstock to the scion. Arrows indicate sites of calcium accumulation, which is associated with necrotic tissue.

There is a redistribution of immobilized calcium in apple trees where at least a portion of calcium in the trees at dormancy is in an exchangeable form (101).

A longitudinal section of a one-year-old graft union of Granny Smith/M.26 shows a distinct demarcation between the stock and scion (arrows, Fig. 10). When staining with safranin and fast green, the safranin stained the Granny Smith phloem red, indicating there was a senescing effect occurring at this early date. It has been found that approximately 60% of the phloem tissues in EMLA.27 (the most dwarfing rootstock) may be classified as nonconducting (105). However, calcium accumulation has been found in the outer, nonfunctioning area of the phloem (98). Necrotic phloem will accentuate this accumulation. Another example of phloem discontinuity (Granny Smith/EMLA.27) is shown in Figure 11. Development of the vascular tissues between the stock and scion has a continuous swirling effect from the original growth, and these tissues become necrotic as subsequent growth forms in the future. Arrows indicate potential sites of Ca accumulation.

A longitudinal section of a dormant Granny Smith/M.26 (Fig. 12) shows tissue development on the Granny Smith side of the union in which the epidermis, nonfunctioning phloem, functioning phloem, cambium, and xylem are designated. The calcium crystals are located in the functioning phloem as indicated by arrows. Other studies concerning tissue development in the graft union as related to dwarfing in apple have verified the location and that these crystals contain calcium (98).

Radial-ray development contiguous to the graft union shows that many rays develop within this newly formed wound parenchyma (Fig. 13A, B, C). This

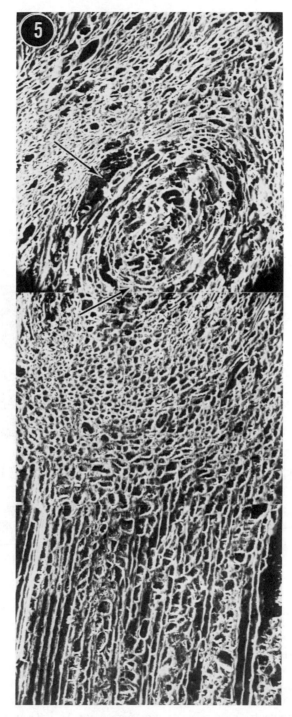

Figure 5. Swirling within the graft union (Granny Smith/M.26) prevents successful unions. This type of growth is a precursor of burrknot or bud initial (× 102).

Figure 6. Growth cracks in the scion and burrknot development on the rootstock of an 11-year-old Golden Delicious/M.7 that was propagated by T-budding with a short rootstock shank.

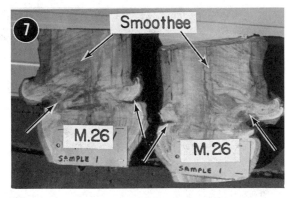

Figure 7. A longitudinal section through a dying 13-year-old graft union (Smoothee/M.26) showing overgrowth of the scion (arrows) and the constriction of vascular tissue at the union.

example is from a Granspur/M.7a graft union. The arrows (Fig. 13A, and enlarged in Fig. 13C) show radial-ray parenchyma forming adjacent to xylem vessels that have been initiated in the regenerative-wound parenchyma of the graft union. Radial-ray parenchyma may persist and form an abnormal union where breakage occurs (Fig. 4).

Another type of cellular development may be found where there is suspected incompatibility within the graft union. Figure 14A, B, C, shows excessive cell wall thickening of the xylem rays. Development was extensive (Fig. 14A), limiting the number of xylem vessels throughout this area. Excessive cell wall thickening (arrow, Fig. 14A, enlarged in Fig. 14B) with cells containing starch (S, Fig. 14C) contiguous to the thickened xylem parenchyma.

It has been found (99) that starch accumulation is significant in the outer layer of pith contiguous to the secondary xylem. However, in developing tissues of the graft union of a dormant Granspur/M.7a (Fig. 15A, B, C), there

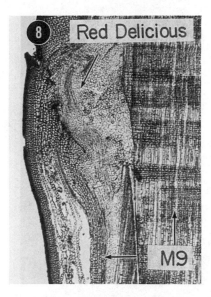

Figure 8. Continuity of the vascular tissues developing between the stock and scion are shown in a longitudinal section through a one-year-old Red Delicious/M.9 graft union (× 7).

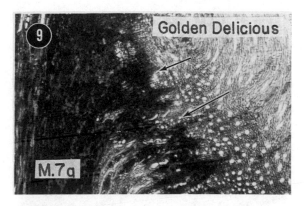

Figure 9. Juncture of developing vascular tissue at the union and disruption of the normal apical orientation of these tissues (arrows) are shown in a longitudinal section of a Golden Delicious/ M.7a graft union (× 8).

was significant starch storage in relation to the xylem elements that have been recently developed in response to the graft union activity. The growing point development of Granspur is shown by arrows in Figure 15A. Apical orientation has not been established, but should be accomplished with future growth. Starch accumulation (S) was from pith cells (Fig. 15C) contiguous to the xylem elements in Figure 15B.

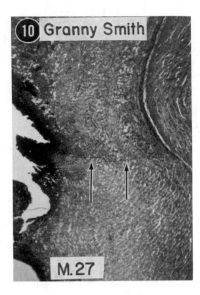

Figure 10. Senescence is apparent in scion tissue (stained dark red with safranin). The stock tissue is meristematic (stained green by a fast green) in a longitudinal section through the graft union of a one-year-old Granny Smith/M.27 (arrows indicate the juncture of stock and scion tissue) ($\times$ 12).

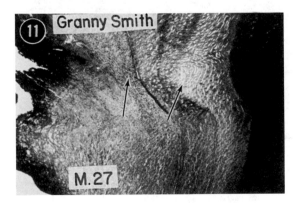

Figure 11. Orientation of vascular tissues are shown in a longitudinal section through a two-year-old Granny Smith/M.27 in the union area. Arrows indicate sites of Ca accumulation contiguous to necrotic tissues. A large portion of the rootstock phloem was nonfunctioning ($\times$ 8).

Growth effects of the graft union upon xylem-ray development is illustrated by a normal Granny Smith/MM.106 graft union in Figure 16A, B, C. Arrows in Figure 16A indicate the effect distally to the union; Figure 16B shows the horizontal orientation of the rays to the axis. In Figure 16C, xylem rays were

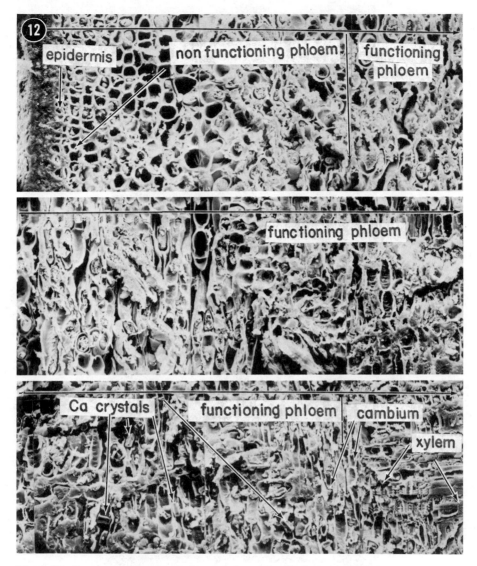

Figure 12. Composite of rootstock tissue sectioned longitudinally through a chip-budded graft union (Granny Smith/M.26)(× 283).

also affected by the union (arrows), which would indicate the persisting effect of the immediate tissues between the stock and scion.

If the types of growth illustrated in Figures 1, 2, 5, 7, 10, 11, 13, 14, and 16 persist, abnormal unions will develop. They may break or at least cause undue stress between the rootstock and scion. Incompatibilites may result in abnormal crops of low quality, and eventual death of the trees.

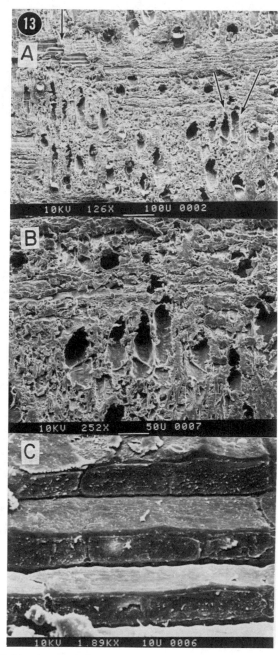

Figure 13. *A,B,C.* Vasular tissue development in a one-year dormant graft union (Granspur/M.7a). Both the number of xylem vessels varied (arrows, upper right, 13*A*) and the distance between vessels (13*B*). The arrows (upper left in 13*A* and enlarged in 13*C)* show excessive xylem-ray parenchyma that will lead to breakage as illustrated in Figure 4 (*A* , × 83; *B*, × 166; *C*, × 1250).

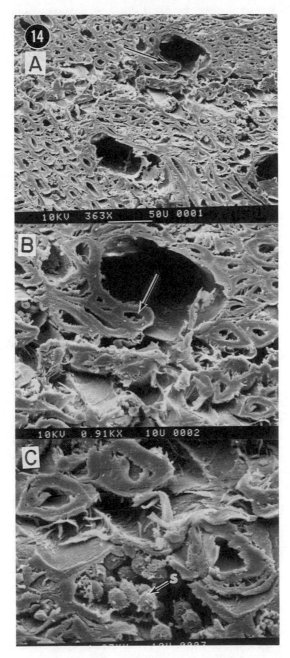

Figure 14. *A,B,C.* Thickening of xylem-ray parenchyma is shown in a longitudinal cross section through the graft union (Granspur/M.7a). The entire area between the xylem vessels consisted of this type of tissue (arrow in 14*A* and enlarged in 14*B*). Starch accumulation (S) was contiguous to the xylem vessels in 14*A,B* (*A*, × 240; *B*, × 600; *C*, × 1207).

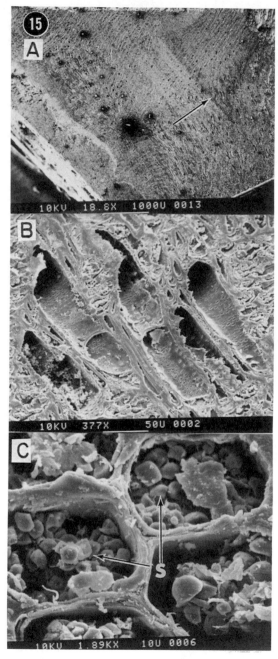

Figure 15. *A,B,C.* A longitudinal section of the scion growth point at juncture of a dormant graft union (Granspur/M.7a; arrow, 15*A*). Xylem vessels from the base of this area on the rootstock side of the union (15*B*) with starch accumulation (S, 15*C*) (*A*, × 13; *B*, × 248; *C*, × 1250).

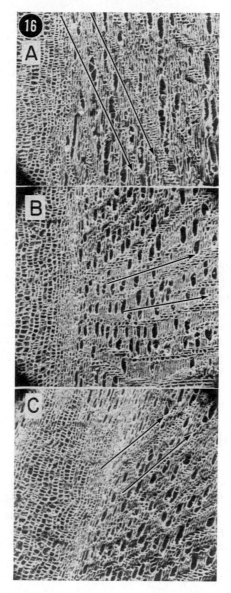

Figure 16. *A,B,C.* Effects upon orientation of xylem tissue as indicated by arrows in a longitudinal section through the graft union of a one-year-old Granny Smith/MM.106 tree. *A*, distal; *B*, union; and *C*, proximal to the union (*A*, × 54; *B*, × 53; *C*, × 51).

ACKNOWLEDGEMENTS

The author acknowledges support of the Illinois Agricultural Experiment Station and the Center for Electron Microscopy of the University of Illinois, Urbana, IL.

REFERENCES

1. Argles, G. K. (1937). A review of the literature on stock–scion incompatibility in fruit trees, with particular reference to pome and stone fruits, *Imp. Bur. Fruit Prod. Tech. Com.*, **9**, 1–115.
2. Barlow, H.W.B. (1979). Variation within the clonal apple rootstock crab C, *J. Hort. Sci.* **54**, 57–61.
3. Beakbane, A. B., and E. C. Thompson (1939). Anatomical studies of stems and roots of hardy fruit trees. II. The internal structure of the roots of some vigorous and some dwarfing apple rootstocks and the correlation of structure with vigor, *J. Pomol. Hort. Sci.*, **17**, 141–149.
4. Beakbane, A. B., and E. C. Thompson (1947). Anatomical studies of stems and roots of hardy fruit trees. IV. The root structure of some new clonal apple rootstocks budded with Cox's Orange Pippin, *J. Pomol.*, **23**, 206–211.
5. Beakbane, A. B. (1952). Anatomical structure in relation to rootstock behavior, *13th Intl. Hort. Cong.*, **1**, 152–158.
6. Beakbane, A. B., and W. S. Rogers (1956). The relative importance of stem and root in determining rootstock influence in apples, *J. Hort Sci*, **31**, 99–110.
7. Blodget, E. C., H. Schneider, and M. D. Aichele (1962). Behavior of pear decline disease on different stock–scion combinations, *Phytopath.*, **52**, 679–684.
8. Breen, P. J. (1974). Cyanogenic glycosides and graft incompatibility between peach and plum, *J. Amer. Soc. Hort. Sci.*, **99**, 412–415.
9. Breen, P. J., and T. Muraoka (1975). Seasonal nutrient levels and peach plum graft incompatibility, *J. Am. Soc. Hort. Sci.*, **100**, 339–342.
10. Brian, C., and M. Duron (1972). A contribution to the study of pear–quince graft incompatibility. II. Modifications of the union process caused by the application of growth substances to herbaceous material, *Annales de l'Amelioration des Plantes.*, **22**, 81–93.
11. Carlson, R. F. (1967). The incidence of scion-rooting of apple cultivars planted at different soil depths, *Hort. Res.*, **7**, 113–115.
12. Carlson, R. F., and K. Yu (1969). Starch content in cherry stems near loci of graft, banding and scoring, *HortScience*, **4**, 246–247.
13. Carlson, R. F. (1973). New systems of apple tree culture, *HortScience*, **8**, 358–361.
14. Carlson, R. F. (1974). Some physiological aspects of scion/rootstocks, *XIXth Intl. Hort. Cong. Warszawa*, pp. 293–302.
15. Carlson, R. F. (1975). Current fruit rootstock, *Compact Fruit Tree* **8**, 26–30.
16. Carlson, R. F., and S. D. Oh (1975). Influence of interstem lengths of M.8 clone *Malus sylvestris* Mill. Growth, precocity and spacing of 2 apple cultivars, *J. Amer. Soc. Hort. Sci.*, **100**, 450–452.

17. Carlson, R. F., and J. Hull, Jr. (1975). *Rootstocks for Fruit Trees.*, Michigan State University Extension Bul. E-851.

18. Carlson, R. F. (1977). Dwarf rootstocks—propagation and usage, *Intl. Plant. Prop.*, **27**, 371–374.

19. Carlson, R. F. (1978). Apple scion behaviour when dwarfed with interstems, *Acta Horti.*, **65**, 167–171.

20. Cummins, J. N., and H. S. Aldwinckle (1974). Breeding apple rootstocks, *Hort-Science*, **9**, 367–372.

21. Cummins, J. N., and H. S. Aldwinckle (1983). "Breeding apple rootstocks," *Plant Breeding Reviews*, Vol. 1., Jules Janick, Ed., AVI Publishing Co., Westport, CT, Chapter 10, pp. 294–394.

22. Cummins, J. N., and R. L. Norton (1974). Apple rootstock problems and potentials, N.Y. Food and Life Science Bul. No. 41, pp. 1-15.

23. Cummins, J. N., and H. S. Aldwinckle (1982). New and forthcoming apple rootstocks, *Fruit Var. J.*, **36**, 66–73.

24. Day, L. H. (1953). *Rootstocks for Stone Fruits*, Bull. No. 736 Ca. Agr. Exp. Sta.

25. Deloire, A., and C. Hebant (1982). Peroxidase activity and lignification at the interface between stock and scion of compatible and incompatible grafts of *Capsicum* on *Lycopersicum*, *Ann. Bot.*, **49**, 887–891.

26. Deloire, A., and C. Hebant (1983). Histophysiological study of the graft union of compatible and incompatible *Prunus* combinations, *Agronomie*, **33**, 207–212.

27. Doud, S. L., and R. F. Carlson (1977). Effects of etiolation stem anatomy and starch reserves on root initiation of layered *Malus* clones, *J. Am. Soc. Hort. Sci.*, **102**, 487–491.

28. Esau, K. (1964). "Structure and development of the bark in dicotyledons," *The formation of Wood in Forest Trees*, M. H. Zimmerman, Ed., Academic, New York, pp. 37–50.

29. Esau, K. (1965) *Plant Anatomy*, Wiley, New York, 767 pp.

30. Esau, K. (1969). *The Phloem* , Gebruder Borntraiger, Berlin, Stuttgart, 505 pp.

31. Esau, K. (1977). *Anatomy of Seed Plants*, 2nd ed., Wiley, New York, pp. 295–394.

32. Ferree, D. C. (1982). Multi-state cooperative apple interstem planting established in 1976, *Fruit Var. J.*, **36**, 2–7.

33. Fideghelli, C., G. D. Strada, and R. Quarta (1984). Breeding program by ISF of Rome to develop genetic dwarf trees, *Acta Horti.*, **146**, 47–57.

34. Fridlund, P. R. (1979). Incompatibility between apricot and *Prunus tomentosa* seedlings, *Fruit Var. J.*, **33**, 90–91.

35. Fridlund, P. R. (1981). Introgeneric graft compatibility in relation to graft transmission of prunus viruses, *Luck. Stiint Inst. Agron. "Nicolea Balcesu" Agron*, **23**, 5–10.

36. Gur, A., and A. Blum (1975). The water conductivity of defecting graft unions in pome and stone fruits, *J. Amer. Soc. Hort. Sci.*, **100**, 325–328.

37. Hartman, W. (1973). The problem of walnut grafting. Studies from the technical, histological and physiological viewpoint, Dissertation, Universitat Hohenhein, German Federal Republic, 136 pp.

38. Hartman, H. T., and D. E. Kester (1983). *Plant Propagation, Principles and*

Practices, Chapter II, "Theoretical aspects of grafting and budding." 4th ed., Prentice-Hall, Englewood Cliffs,NJ.

39. Herrero, J. (1951). Studies of compatible and incompatible graft combinations with special reference to hardy fruit trees, *J. Hort. Sci.*, **26**, 186–237.

40. Howard, B. H. (1972). Depressing effects of virus infection on adventitious root production in apple hardwood cuttings, *J. Hort. Sci.*, **47**, 255–258.

41. Howard, B. H., D. S. Skene, and J. S. Coles (1974). The effects of different grafting methods upon the development of one-year-old nursery apples trees, *J. Hort. Sci.*, **49**, 287–295.

42. Huang, F. H., S. Tsai, and R. C. Rom (1984). An electrophoresis method for water-soluble protein of *Prunus*, *HortScience*, **19**, 242–243.

43. Jaumien, F., and M. Faust (1984). Stem anatomical structure of Delicious and Golden Delicious apple hybrids with various growth dynamics, *Acta Horti.*, **146**, 69–74.

44. Jones, O. P., and S. G. S. Hatfield (1976). Root initiation in apple shoots cultured *in vitro* with auxins and phenolic compounds, *J. Hort. Sci.*, **51**, 495–499.

45. Jones, O. P., and J. S. Pate (1976). Effect of M.9 dwarfing interstocks on the amino compounds of apple xylem sap, *Ann. Bot.*, **40**(170), 1237.

46. Jones, O. P., and J. S. Pate (1976). Effect of dwarfing interstocks on xylem sap composition in apple trees: Effect on nitrogen, potassium, phosphorus, calcium and magnesium content, *Ann. Bot.*, **40**(170), 1231–1235.

47. Jones, O. P. (1984). Mode-of-action of rootstock/scion interactions in apple and cherry trees, *Acta Horti.*, **146**, 175–182.

48. Lana, A. F., J. F. Peterson, G. L. Rouselle, and T. C. Vrain (1983). Association of tobacco ringspot virus with a union incompatibility of apple, *Phytopath*, **106**, 141–148.

49. Lapins, K. 1963. Cold hardiness of rootstocks and framebuilders for fruit trees, Bul. B.6. Canada Dept. of Agr. Res. Sta., Summerland, B.C., pp. 1–36.

50. Lasheen, A. M., and R. G. Lockard (1972). Effects of dwarfing rootstocks and interstems on the free amino acid and protein levels in apple roots, *J. Amer. Soc. Hort. Sci.*, **97**, 443–445.

51. Lockard, R. G., and A. M. Lasheen (1971). Effects of rootstock and length of interstem on growth of 1-year-old apple plants in sand culture, *J. Amer. Soc. Hort. Sci.*, **96**, 17–21.

52. Lockard, R. G. (1974). Effects of rootstocks and length and type of interstock on growth of apple trees in sand cultures, *J. Amer. Soc. Hort.*, **99**, 321–325.

53. Lockard, R. G. (1976). The effect of apple dwarfing rootstocks and interstocks on the proportion of bark on the tree, *Horti. Res.*, **15**, 83–94.

54. Lockard, R. F., and G. W. Schneider (1978). Advances in the concept of the dwarfing mechanism in apple trees, *HortScience*, **13**, 349 (Abstr.).

55. Lockard, R. G., and G. W. Schneider (1981). Stock and scion growth relationships and dwarfing mechanism in apple, *Hort. Rev.*, **3**, 315–375.

56. Lockard, R. G., G. W. Schneider, and T. R. Kemp (1982). Phenolic compounds in 2 size controlling apple rootstocks, *J. Amer. Soc. Hort. Sci.*, **107**, 183–186.

57. Lockard, R. G. and G. W. Schneider (1982). Phenols and the dwarfing mechanism

in apple rootstocks, Proc. of the 4th Intl. Sym. on Growth Regulators in Fruit Prod, *Acta Horti.*, 107–112.

58. Makhmet, B. M., A. A. Bulakh, P. A. Kolesnichenko and P. A. Ukhainskaya (1980). Anatomical indices for determining incompatibility between components of interspecific grafts of trees and shrubs. *Fiziologiyai Biokhimiya Kul' turnykh Rastenii*, **12**, 179–185.

59. Martinez, J., J. L. Poessel, J. Hugard, and R. Jonared (1981). In-vitro micro grafting utilization for grafting incompatibility study, *C. R. Seances Acad. Sci. Ser. III Sci. Vie.* **292**(16), 961–964.

60. McKenzie, D. W. (1961). Rootstock–scion interaction in apples with special reference to root anatomy, *J. Hort. Sci.*, **36**, 40–47.

61. Mendel, K., and A. Cohen (1967). Starch level in the trunk as a measure of compatibility between stock and scion in citrus, *J. Hort. Sci.*, **42**, 231–234.

62. Messer, G. S., and S. Lavee (1969). Studies on vigor and dwarfism of apple trees in an *in vitro* tissue culture system, *J. Hort. Sci.* **44**, 219–233.

63. Monastra, F., D. G. Strada, and C. Damiano (1973). A study on the behavior in the nursery and the graft compatibility of some apricot cultivars on various rootstocks. *Annali dell' Istitute Sperimentale per la Fratticoltura*, **4**, 195–212.

64. Mosse, B. and J. Herrero (1951). Studies on incompatibility between some pear and quince grafts, *J. Hort. Sci.*, **26**, 238–244.

65. Mosse, B., and F. Scaramuzzi (1956). Observations on the nature and development of structural defects in the unions between pear and quince, *J. Hort. Sci.*, **31**, 47–54.

66. Mosse, B. (1958). Further observations on growth and union structure of double-grafted pear on quince, *J. Hort. Sci.*, **33**, 186–193.

67. Mosse, B. (1961). Graft incompatibility in plums: Interactions between the varieties Victoria, President and Myrobalan B., *Ann. Rept. East Malling Res. Sta.*, pp. 57–60.

68. Mosse, B. (1962). *Graft-incompatibility in Fruit Trees*, Tech. Comn. No. 28, Comm. Bur. of Hort. and Plantation Crops, East Malling, England.

69. Nelson, S. H. (1968). Incompatibility survey among horticultural plants, *Intl. Plant Propagators Soc.*, **18**, 343–407.

70. Palandzhyan, V. A., L. A. Apoyan, and I. Sosyan (1979). Inter influence on different components in grafts with interstocks on component anatomy, *Biol. Zhurnal Armenii*, **32**, 704–708.

71. Parry, M. S. (1980). Evidence of clonal variation and latent virus effect on the vigor of Cox's Orange Pippin apple trees on M.9 rootstocks, *J. Hort. Sci.*, **55**, 439–440.

72. Poling, E. B., and G. H. Oberly (1979). Effect of rootstock on mineral composition of apple leaves, *J. Amer. Soc. Hort. Sci.*, **104**, 799–801.

73. Proebsting, E. L. (1926). Structural weaknesses in interspecific grafts of *Pyrus*, *Bot. Gaz.*, **82**, 336–338.

74. Proebsting, E. L., and E. H. Barger (1927). The precipitin reaction as a means of determining the congeniality of grafts, *Sci.*, **65**, 573–574.

75. Proebsting, E. L. (1928). Further observations on structural defects of the graft union, *Bot. Gaz.*, **86**, 82–92.

76. Quessada, M.-P., and J.-J. Macheix (1984). Caracterisation d'une peroxydase impliquee specificequement dans la lignification, en relation avec l'incompatibilite au grettage chez l'abricotier, *Physiol. Veg.*, **22**(5), 533−540.

77. Rehkugler, G. E., J. N. Cummins, and E. D. Markwardt (1979). Rupture strength of unions of "Golden Delicious" apple with Malling 8, Malling 9, and vigorous rootstocks, *J. Amer. Soc. Hort. Sci.*, **104**, 226−229.

78. Robitaille, H. A., and R. F. Carlson (1970). Graft union behavior of certain species of *Malus* and *Prunus*, *J. Amer. Soc. Hort. Sci.*, **95**, 131−134.

79. Robitaille, H. and R. F. Carlson (1971). Response of dwarfed apple trees to stem injections of Gibberellic and abscisic acids, *HortScience*, **6**, 539−540.

80. Rom, R. C. (1983). The peach rootstock situation: An international perspective, *Fruit Var. J.*, **37**(1),3−4.

81. Sax, K. (1953). Interstock effects in dwarfing fruit trees, *Proc. Amer. Soc. Hort. Sci.*, **62**, 201−204.

82. Sharpe, R. H. (1974). Breeding peach rootstocks for the southern USA, *HortScience*, **9**, 362−363.

83. Shear, C. B. and M. Faust (1970). Calcium transport in apple trees, *Plant Physiol.*, **45**, 670−674.

84. Siegler, E. A. (1943). Anatomical and other studies on Mazzard cherry seedlings having excessive roots at the collar region, *J. Agr. Res.*, **67**, 1−16.

85. Simons, R. K. (1966). Abnormal phloem development and sieve-tube necrosis associated with Golden Delicious on EM VII rootstock, *Proc. Amer. Soc. Hort. Sci.*, **89**, 14−22.

86. Simons, R. K. (1967). Stereoscan Electron microscope—A useful tool for studying plant tissues, *HortScience*, **2**, 65.

87. Simons, R. K. (1967). Controlling the size of apple trees—Some problems and limitations of the East Malling VII rootstock, *Ill. Res.* (Spring 1967), 10−11.

88. Simons, R. K. (1968). Phloem tissue development in the stock/scion union of East Malling IX apple rootstock, *Hort. Res.*, **8**, 97−103.

89. Simons, R. K., and R. F. Carlson (1968). Characteristics and propagation of rootstocks for deciduous fruits in the North Central Region, *HortScience*, **3**, 221−225.

90. Simons, R. K., and R. F. Carlson (1971). Scion/rootstock tissue development of the apple as a result of asexual propagation and mechanical injuries, *Fruit Var. and Hort. Dig.*, **25**, 33−37.

91. Simons, R. K. (1972). Development of stock/scion tissues of "Starking"/MM104 when grown under adverse conditions, *Fruit Var. and Hort. Dig.*, **26**, 7−10.

92. Simons, R. K., and M. C. Chu (1979). Graft union characteristics: II. Scarlet Red Stayman/M.7a, Double Red Staymared/M.9 and Winesap/M.26, *Compact Fruit Tree*, **12**, 141−144.

93. Simons, R. K. (1980). Preliminary observations of apple tree growth in relation to interstems, *Compact Fruit Tree* **13**, 33−38.

94. Simons, R. K. (1981). Growth characteristics of graft unions—implications upon subsequent.tree performance. *Compact Fruit Tree* **14**, 164−166.

95. Simons, R. K. (1982). Scion/Rootstock incompatibilities in young trees. *Compact Fruit Tree,* **15**, 30–32.

96. Simons, R. K., and M. C. Chu (1983). Graft union development: Granny Smith/Em.26—specific growth characteristics between stock and scion, *Compact Fruit Tree*, **16**, 73–82.

97. Simons, R. K. (1983). Growth comparisons of Starkspur Supreme Red Delicious (Paganelli cv.) on nine different rootstocks in the pre-fruiting stage of development, *Compact Fruit Tree*, **16**, 89–95.

98. Simons, R. K., and M. C. Chu (1983). Growth characteristics of apple graft unions of stock/scion combinations in relation to dwarfing, *Acta Horti*, **140**, 79–86.

99. Simons, R. K., and M. C. Chu (1984). Tissue development within the graft union as related to dwarfing in apple, *Acta Horti.*, **146**, 203–210.

100. Simons, R. K., and M. C. Chu (1985). Graft union characteristics of M.26 apple rootstock combined with Red Delicious strains—Morphological and anatomical development, *Scientia Horti.*, **25**, 49–59.

101. Terblanche, J. H., L. G. Wooldridge, I. Hesebeck, and M. Joubert (1979). The redistribution and immobilization of calcium in apple trees with special reference to bitter pit, *Commun. Soil Sci. Plant Anal.*, **10**, 195–215.

102. Tubbs, F. R. (1967). Tree size control through dwarfing rootstocks, *XVVII Intl. Hort. Cong.*, **2**, 43–56.

103. Tubbs, F. R. (1980). Growth relations of rootstock and scion in apples, *J. Hort. Sci.*, **55**, 181–189.

104. Tukey, H. B. (1964). *Dwarfed Fruit Trees*, Macmillan New York, 562 pp.

105. Ussahatanonta, S. (1983). The morphological and anatomical comparison of the Golden Delicious and Granny Smith Apple on various rootstocks with reference to compatibility and dwarfing, M. S. thesis, University of Illinois, 180 pp.

106. Westwood, M. N. (1963). Some differences in growth, chemical composition and maturity between a spur mutant and standard-growing Delicious apples, *Wash. State Hort. Assoc. Proc.*, **59**, 119–120.

107. Westwood, M. N. (1970). Rootstock–scion relationships in hardiness of deciduous fruit trees, *HortScience*, **5**, 418–421.

108. Westwood, M. N., and A. N. Roberts (1970). The relationship between trunk cross sectional area and weight of apple trees, *J. Amer. Soc. Hort. Sci.*, **95**, 28–30.

109. Westwood, M. N., H. R. Cameron, P. B. Lombard, and C. B. Cordy (1971). Effects of trunk and rootstock on decline, growth, and performance of pear, *J. Amer. Soc. Hort. Sci.*, **96**, 147–150.

110. Westwood, M. N., M. H. Chaplin, and A. N. Roberts (1973). Effects of rootstock on growth, bloom, yield, maturity and fruit quality of Prune *(Prunus domestica* L.), *J. Amer. Soc. Hort. Sci.*, **98**, 352-357.

111. Westwood, M. N. (1978). *Temperate Zone Pomology*, Freeman, San Francisco, Chapter 4, "Rootstocks: Their propagation, function and performance."

112. Yadava, U. L., and D. F. Dayton (1972). The relation of endogenous abscisic acid to the dwarfing capability of East Malling apple rootstocks, *J. Am. Soc. Hort. Sci.*, **97**, 701–705.

113. Yadava, U. L., and R. G. Lockhard (1977). Abscisic acid and gibberellin in 3 ungrafted apple Malus sylvestris rootstock clones, *Plant physiol.*, **40**, 225–229.

114. Yadava, U. L., and S. L. Doud (1978). Effect of rootstock on the bark thickness of peach scion, *HortScience*, **13**, 538–539.

115. Yu, K., and R. F. Carlson (1975). Paper chromatographic determination of phenolic compounds occurring in the leaf, bark and root of *Prunus avium* and *P. mahaleb*, *J. Amer. Soc. Hort. Sci.*, **100**, 536–541.

116. Zeiger, D., and H. B. Tukey (1960). *An Historical Review of the Malling Apple Rootstocks in America*, Mich. State Circ. Bul. 226, 74 pp.

4

APPLE ROOTSTOCKS

David C. Ferree
Ohio Agricultural Research and Development Center
Ohio State University
Wooster, Ohio

Robert F. Carlson
Michigan State University
East Lansing, Michigan

1. ORIGIN

The history of mankind and the history of the apple appear to be anciently linked. The center of genetic diversity of *Malus* (apple) is in Asia Minor, likewise the cradle of human civilization.

Malus is a diverse genus and species identification within the genus is difficult. Man has collected, transported, hybridized, and selected over the millennia so that our common apple is not a naturally evolved species, but contains germplasm from many wild progenitors. The impetus for this interest by man was certainly initially as a food source, but then, over time, also for ornamental horticulture purposes, as unique tree growth forms were noted.

The first written record of dwarf apple trees can be traced back to three centuries before Christ, to two students of Aristotle in Greece (50). Alexander the Great, on one of his conquests into Asia Minor, sent back a low-growing type of apple to the Lyceum, a center of learning created by his teacher Aristotle. Theophrastus, who followed his teacher Aristotle in directing the Lyceum, recorded the acquisition and mentioned that the dwarf apple had probably long been grown in Asia Minor.

The Chinese during the Tun Dynasty (eleventh to thirteenth century A. D.) describe dwarf apple, pomegranate, apricot, and other plants. These dwarf plants are a recognized feature of Chinese art and culture. The culture of dwarf plants in Japan, as bonsai, is a more recent development, with earliest records about 1700.

By the fifteenth century, budding and grafting techniques were developed and the use of rootstocks became possible. European gardeners used dwarf apple selections as rootstocks and trained the trees to many elaborate shapes, such as pyramids, trellises, and espaliers. The recording of history also improved by this time and first mention was made of the very dwarfing 'Paradise' (French Paradise) and the less dwarfing 'Doucin' (English Paradise) apple rootstocks (50).

The classification of dwarfing stocks either as 'Paradise' or 'Doucin' persisted into the mid-1800s in Europe. By then, however, much variation existed in the plant material called by these common names. Many new stocks had been introduced inaccurately under these names and undoubtedly viruses and genetic mutations had also occurred in the plant materials. In 1870, Thomas Rivers mentioned 14 "kinds" of 'Paradise' in his widely circulated book, *The Miniature Fruit Garden* (50).

Research workers at the East Malling Research Station in Kent, England, began in 1912 to gather rootstocks named 'Paradise' and 'Doucin' from around the world and determine their trueness to name and identity. R. Wellington initated the work and was followed by R. Hatton. Hatton gathered 71 collections from 35 sources and confirmed the mixed nature of the available stocks. He reported that 36% of the original collections contained more than one phenotype, and that 66% were improperly named. The confusion in nomenclature led Hatton to abandon the common names of these rootstocks and to assign Roman numerals I through XXIV to the selections. Selections were numbered in no particular order with respect to tree size.

In 1917, the first apple rootstock breeding program (36) was initiated at the East Malling Research Station in England. Many orchards in parts of the British Empire were infested with woolly apple aphids, particularly in Australia and New Zealand. To develop a rootstock resistant to this pest, the John Innes Institute and East Malling Research Station started a breeding program. The "Merton Immune" (MI) series, M.I. 778−M.I. 793 was introduced in the 1930s, followed by the "Malling Merton" (M.M.) series, M.M. 101−M.M. 115 in 1952. Rootstocks from these two series derived their resistance to woolly apple aphid (*Eriosoma langerum* Hausman) from 'Northern Spy'.

Historically, the early apple production in the United States was based on seeds imported with the European settlers. Favorite cultivars were grafted onto seedling rootstocks. Little use was made of the native North American *Malus* species, which produced much less desirable fruits. Dwarfing rootstocks were apparently imported to the United States from Europe in the early 1800s (56), but little interest was shown by American horticulturists except for ornamental and home garden purposes.

San Jose scale threatened the U.S. fruit industry in the early 1900s and growers first used dwarfing rootstocks commercially to facilitate covering the tree for cyanide gas fumigation. Lime sulfur sprays more effectively solved the San Jose scale problem, and the interest in using size-controlling rootstocks declined.

In 1920, R. D. Anthony imported EM.12 to use as clonal material in a nutrition study at the Pennsylvania State College. Several years later, a range of the Malling stocks were imported for orchard trials in Pennsylvania and New York. In Canada in 1927, the Ontario Station set out an orchard of five cultivars on four of the Malling stocks.

In western Europe, the use of seedling rootstocks gave way to M.2 and MM.106, depending on planting density. In southwestern Europe, M.4 is widely used. The very intensive cultural systems, particularly the espalier developed in France and variations of the spindle bush developed in Holland, were dependent on M.9 rootstock. Many of the modern plantings in other parts of Europe adopted the Dutch systems.

2. ROOTSTOCK USAGE

The following comments are general in nature because in all regions, other clones and new rootstocks are being tried as well as the major ones mentioned. There is an almost universal dissatisfaction with the currently available rootstocks in some portion of each apple-growing region. Each stock seems to have an intolerance for a climatic or soil extreme or to be susceptible to a potentially serious insect or disease problem.

Worldwide, the most common rootstock used for apple are seedlings. In America, seeds of commercial cultivars, mostly 'Delicious', are secured from juice or other processing plants, or imported seeds of French Crab, Antonovka, or other cultivars are utilized. These have provided uniform stands of viorous rootstocks with good compatibility characteristics and a high percentage of salable trees. In Eastern Europe, seedlings of common Antonovka are widely used because of their cold-hardiness. In Nova Scotia, seedlings of hardy 'Beautiful Arcade' have resulted in hardy, well-adapted trees. The variability in tree size and performance in trees on seedling rootstocks in Western Europe precipitated the search, about 1900, for a rootstock with greater uniformity. Today many modern orchards in America and other apple-growing regions are planted on one or more of the clonal rootstocks

In North America, the most widely used clonal rootstocks are M.26, M.7, MM.106, and MM.111, with smaller numbers of M.9, M.9 interstems, and several others. In South America, most apples are on M.I. 793, with smaller numbers of MM.106 and MM.111. The intensive orchards in Western Europe are mostly on M.9, with some of the less intense plantings on MM.106. In northern Europe, M.2 and Alnarp 2 are widely used. Eastern Europe is still planting trees on Antonovka, but many new plantings are being planted on some of the Budagovsky and Polish rootstocks and M.26. In the far east, new plantings are using MM.106 and M.26. Australia and New Zealand depend on M.I. 793, with considerable plantings on MM.106.

3. ROOTSTOCK ADAPTATIONS

3.1. Soils

Soil Depth. Atkinson (2) in a recent review pointed out that rootstocks influence the size of the root system, its distribution in the soil, and its periodicity of growth and activity. All of these factors influence the adaptability of rootstocks to various soil types. It is generally true that large trees have a larger total root system than dwarf trees. However, there is a common misconception that dwarfing rootstocks such as M.9 are shallow rooted and more vigorous types are deep rooted. Coker (8) showed that some of the vigorous types are relatively shallow rooted and the dwarfing M.9 produced more dense roots below 120 cm than the much larger M.1. Hoekstra (26) observed no difference in the distribution of fine roots at various soil depths between semidwarfing M.4 and dwarfing M.9. Thus it appears that depth of rooting depends heavily on soil type and the genetic characteristics of the rootstock which cannot be generally classified by ultimate tree size. Depth of rooting is readily checked by imperfect drainage, soil compaction, or the presence of a high water table on most apple rootstocks.

Soil Compaction. It is known that soil compaction significantly influences apple tree growth, but very few studies have evaluated the response of rootstocks to soil compaction. Slowik (48) found that trees on M.26 were more adversely affected by controlled compaction in two soil types than trees on M.4. Antonovka seedlings withstood heavy compacted soils better than either M.26 or M.4.

Soil Temperature. Although the apple root system does not go dormant, only minimal growth occurs when soil temperatures are 1–4°C. It is generally accepted that active root growth begins when the soil temperature reaches about 7°C. A soil temperature above 28°C may begin to limit growth of trees on most rootstocks. Under controlled conditions, several investigators (23, 37) have shown that M.7, M.16, M.25, MM.109, and seedling are relatively resistant to high soil temperatures and M.1, M.2, M.9, and MM.104 perform poorly at temperatures of 25°C or above. These results are in accordance with the performance of apple orchards in warm climates where M.1, M.2, and M.9 have been found to be short-lived. Rootstocks such as M.7 and seedlings are known for their adaptation to a wide range of fruit-growing areas.

Soil Moisture. Although many poorly drained soils are high in organic matter and nutrients, trees on most rootstocks do very poorly on these soils unless they are systematically tile drained. Field experience indicates that trees on M.26 and MM.106 seem to be particularly sensitive to "wet feet" conditions, and serious tree losses have occurred when trees on these stocks are planted on problem soils. In controlled flooding studies comparing seedling with M.7,

M.26, MM.106, and MM.111, Rom (46) found M.7 and M.26 most tolerant of flooding during the growing season and MM.111, showing the greatest number of dead trees. It is of interest to note that MM.111, the least tolerant of flooding conditions, is reported as being the most drought resistant of the clonal rootstocks, while M.9 and M.26 are characterized as intolerant of drought. The MM.111 rootstock is characterized by having many fine roots which would have the ability to thoroughly extract moisture from a soil. Cultivars on MM.111 have survived summer drought conditions better than the same cultivar on M.2. However, oxygen depletion in a flooded environment would likely occur quicker with a fine root system, resulting in reduced respiration. Flooding has been shown to cause adventitious root formation on MM.106 (1), which may partially explain its greater tolerance under these conditions.

Makariev (34) ranked rootstocks resistant to asphyxia in the following descending order: MM.106, M.7, MM.111, Golden Pearmain seedling, M.2, M.4, MM.104, A.2, M.26, MM.109, and M.9. Several other studies have also shown MM.106 more tolerant of flooding than many other rootstocks. The vigorous M.1, M.13, and M.16 have shown field tolerance to wet soil conditions. However, neither M.1 or M.13 tolerate dry soil conditions.

Certain rootstock clones are more efficient in water utilization than others. For example, MM.104 and MM.111 consumed 18 and 22% less water, respectively, than seedlings, while making equivalent growth (6). Alnarp 2 and M.2 were less efficient, using more water than seedlings and making less growth. Six days of flooding caused a decrease in transpiration of M.2, M.7A, M.9, and MM.106, while 12 days were required for a decrease in MM.111. Transpiration rate in young 'Delicious' trees did not differ under conditions of no water stress on M.9, M.26, M.7, M.2, MM.106, and M.111. Work by Marro and Cereghini (35) suggests that guard cells were more responsive on trees on M.9 than on seedling, thus resulting in better transpiration control.

Soil Texture. In whole-tree excavation studies, Rogers and Vyvyan (44) found that depth of rooting increased from sand to loam to clay. In loam and sand, the root system had the same general conformation—a shallow scaffold with vertically descending roots. While in clay, most roots sloped downward and grew in the subsoil. The total amount of root/tree was greatest in loam soils, followed by clay, with the least amount of root in sandy soils. On all soil types, roots spread farther than the branches, and in sand, root spread was 2 to 3 times branch spread, but in loam and clay, it was only 1 to 1.5 times. In sand, roots were long, thin, straight, and spreading, while in clay the roots were fairly short, tapering, rapidly branching, and twisting in all directions. In the loam, the roots were intermediate in appearance between those in sand or clay. The rootstocks grown in these various soil types maintained their unique character with only slight changes on various soil types. For example, on the sandy soil, depth of rooting was greatest for the experimental rootstock (DF5) and least for M.9 and M.1, while on the loam, M.9 had the deepest roots. See Figure 1.

Figure 1. Compaction collar developed in clay soils caused by winds when soil is moist. If not corrected by application of gravel or sand at the tree base, growth can be stunted and anchorage impaired.

Soil Management System. Atkinson (2) provides many examples showing that different soil management systems modify and influence root distribution of fruit trees. Cultivation generally eliminates roots in the surface soil, while grass middles with herbicide strips increase the percentage of the root system close to the soil surface. Incorporation of organic material in the planting hole increases the number of fine roots (20). Generally, these responses to soil management occur similarly in all rootstock types. Studies (9, 22) have compared annual applications of various herbicides over several years with other soil management practices on a number of rootstocks, with the general conclusion that interaction between soil management practice and rootstock did not occur.

Soil Fertility and Mineral Uptake. Jones (28) has shown that xylem sap collected from different apple rootstocks contained different concentrations of minerals. Generally, exudate from the dwarfing rootstocks or interstocks contained depleted levels of nitrogen, phosphorus, and potassium. Kennedy et al. (30) reported that nitrogen and phosphorus were more influenced by environmental factors than by rootstock genotype. However, potassium, magnesium, and calcium levels were significantly influenced by rootstock, but relative levels of each were not presented. In subsequent work (31), it was reported that 'Northern Spy' and rootstocks with 'Northern Spy' as a parent had low calcium levels, with the exception of M.25, which had very high calcium levels. Consistently, the lowest calcium levels were found in MM.111, which has 'Northern Spy' as both a parent and grandparent. The rootstock showing the greatest inconsistency over the six years of the study was M.26. The authors (31) classified 49 apple rootstocks for their consistency of leaf calcium over the years and also by their genetic background and suggest good potential for breeding stocks to improve nutrient uptake. Work by Eaton and Robinson (15), using four scion cultivars as interstems, found minimal effect of interstem on leaf or fruit nutrient content, with the major influence being the scion cultivars.

3.2. Rootstock Adaptation to Climate

Hardiness. There are genetic differences among apple rootstocks in cold hardiness. Generally, however, in the principle apple-producing regions of North America, the commonly used Malling and Malling Merton rootstocks have proved sufficiently winter hardy for commercial use. In an occasional test winter, especially in areas of more extreme temperature fluctuations like the Great Plains, mortality of the rootstocks has occurred.

In New York State (27), field nursery observations following air temperatures of -31 to -34°C revealed a relatively more hardy group of rootstocks (M.1, M.2, M.3, M.4, M.5, M.7, M.12, and M.16), a relatively less hardy group (M.9 and M.13), and a group not hardy at these temperatures (M.6, M.8, M.10, and M.15). In western United States and Canada, Delicious on M.7 survived -44°C, while trees on M.2 showed severe injury (32). In other cold-hardiness trials (27), Ottawa 4, Ottawa 3, MM.106, and C52 have been tolerant of -35°C but intolerant of -40°C.

Predisposition of rootstocks may affect their response to cold temperatures. for example, M.26 has appeared to be very cold hardy in some tests and relatively sensitive in others. Roots discernible at the soil surface and arising from burrknots or nodes on M.26, M.27, MM.106, and CG.10 have been killed at approximately -15°C. Unfortunately, this type of injury may serve as infection sites for root-rotting fungi. Factors such as previous crop load and temperatures leading up to the extreme may be equally important in determining ultimate injury, and these factors no doubt account for much of the variability among rootstocks.

Fruit-producing areas of the world subject to extremely cold winter temperatures have served as rootstock cold-hardiness test sites. In Poland, in a test year without snow cover, and with a soil temperature of -11°C at a depth of 5 cm, nearly 75% of trees on M.9 were killed, while only 15−20% of trees on M.26 and Bud.9 were killed (14). The Polish rootstocks P.2 and P.22 were as hardy as Antonovka seedlings and retained considerable hardiness during fluctuating warm and cold periods (14). In Minnesota field studies that purposely removed snow from test sites, the rootstocks M.7, M.9, M.26, MM.104, and MM.106 were hardy at -18°C. Both root-freezing tests and field observations indicated that M.26 was more hardy and M.7 less hardy than the other clones.

Winter-hardy cultivars have been used as rootstocks and as trunk and scaffold systems for less hardy cultivars. During a test winter in Washington State with temperatures of -44°C, Robusta 5, Antonovka, 'Hibernal', and 'Red Astrachan' survived and demonstrated more cold tolerance than scaffold systems of 'McIntosh', 'Delicious', 'Haralson', and 'Canada Baldwin' (45). Apparently in addition to the hardiness provided by the trunk stock itself, some hardiness can be imparted to the scion cultivar by 'Hibernal' (45). In Oregon during a November freeze of -18°C, trunk stocks of 'Transparent', 'Hibernal', 'Haralson', 'Canada Baldwin', Antonovka, 'Charlamoff', and 'Hyslop Crab' prevented tree crotch and trunk injury. Lapins (32) classified the hardiness of various frame builders as follows: *Very Hardy*: 'Anis', 'Charlamoff', 'Heyer 12', 'Hibernal'; *Hardy*: Antonovka, 'Beautiful Arcade', 'Borovinka', 'Duchess', 'Haralson', *Malus robusta* No. 5, 'Robin', 'Yellow Transparent'; *Moderately Hardy*: 'Anoka', 'Canada Baldwin', 'McIntosh', 'Melba', 'Red Astrachan', 'Winter St. Lawrence'.

One of the most important ways that rootstocks can influence hardiness of the scion cultivar is by speeding maturity in the fall and delaying budbreak in the spring. Rootstocks that induce early maturity of the scion, such as M.9 and M.7, tend to protect trees from fall freezes, but this may not relate to protection against midwinter cold. M.1, M.4, M.5, and M.7 have been shown to attain cold hardiness nearly two months earlier than M.2, which is most cold sensitive early in the autumn (32). See Figure 2.

Several rootstocks, such as M.13 or M.16, when left to develop into fruiting trees, bloom as much as two weeks later than the bloom of most commercial apple cultivars. However, the time of bloom characteristic apparently isn't conveyed to the scion, as scions on these stocks generally bloom at their normal time. Another factor to be considered is precocity, or the earliness in the life of the tree that flowering begins. 'Northern Spy' is notoriously tardy, as a cultivar, to begin flowering and fruiting. However, as a rootstock, it tends to induce precocity similar to seedling rootstocks. M.9 tends to be precocious both as a scion and rootstock, and as a rootstock induces precocity in the scions. Thus it appears that the flowering characteristics of a given clone cannot be predicted to have the same influence as a scion as when the clone is used as a rootstock.

3.3. Rootstock Adaptation to Pests

Diseases. The most important diseases affecting apple rootstocks have been fire blight, caused by the bacterium *Erwinia amylovora* (Burr), and crown (collar) rot, caused by *Phytophthora cactorum*. Fire blight can cause rapid tree death to large numbers of trees on M.9 or M.26 rootstocks or where M.9 or C6 is used as an interstem. Limited data indicated less loss of M.9 interstem trees when MM.111 was the rootstock than when M.7, MM.106, or apple seedling served as the rootstock (21). A summary of the relative susceptibility to root diseases and insects of a number of rootstocks is shown in Table 1. Rootstock

Figure 2. Northern Spy on MM.106 with graft union planted too high, thus causing winter injury and collar rot because this combination grows late in the summer and fall.

infection can occur on root suckers or burrknots; however, in many cases it may be translocated or enter through bark cracks in the rootstock.

Rootstocks also have a role in influencing the sensitivity of the scion cultivar to fire blight infection (13, 51). For example, 'York Imperial' or 'Golden Delicious' appear much more susceptible to fire blight when they are on M.26 than on MM.111 or apple seedling. Some cultivars, such as 'Delicious', have resulted in minimal tree loss through fire blight infection in the rootstock or interstem, while more blight-susceptible cultivars, such as 'Rome Beauty' and 'Jonathan', have experienced severe tree loss on the same stock combinations (21). The rootstock influence on the scion may be due to the acceleration of the onset of flowering, so that more infection sites are available when the tree is

TABLE 1. Relative Susceptibility of Various Apple Rootstocks to Selected Diseases and Insects[a,b]

Rootstock	Crown Rot	Fire Blight	Apple Scab	Powdery Mildew	Latent Viruses	Woolly Aphid
Alnarp 2	MS	VS	MR	MS	T	S
Budagovsky 9	VR	S	M	MS	T	S
Budagovsky 490	MR	M	M	S	T	MS
Budagovsky 491	MS	S	M	MS	NT	S
Malling 2	MR	MR	M	MR	T	S
Malling 4	R	MR	M	M	T	S
Malling 7	MR	R	M	MR	T	S
Malling 9	R	S	M	MR	T	S
Malling 13	R	M	M	MR	T	S
Malling 26	MS	S	M	MR	MS	S
Malling 27	R	MS	M	MR	MS	S
Malling-Merton 104	S					R
Malling-Merton 106	MS	M	M	M	T	R
Malling-Merton 111	M	M	M	MS	T	R
Novole	R	R	R	MR	S	MR
Ottawa 3	R	MS	M	MR	S	VS
Poland 1	MR	MS	NT	MR	NT	MS
Poland 2	R	MS	NT	MR	NT	MS
Poland 18	R	MR	NT	MR	NT	S
Poland 22	R	MS	NT	MR	NT	MA
Robusta 5	MR	R	R	MR	M	VR

[a]From J.N. Cummins (1982).
[b]Rating system: VS = very susceptible; S = susceptible; MS = moderately susceptible; M = Intermediate; MR = moderately resistant; R = resistant; VR = very resistant; NT = not tested; T = tolerant.

small. Bloom period can also be prolonged when rootstocks (M.9, M.26) cause lateral flowering on one-year wood which bloom after the normal flowering period. The longer bloom period provides a greater opportunity for optimum weather conditions to combine with the fire blight bacterium. Some rootstocks (e.g., MM.106) cause scion cultivars to continue active vegetative growth for a longer period and thus increase the potential for these trees to develop shoot fire blight infections. The rootstock may also have altered the physiological processes in the tree in a presently unknown way and thus rendered the scion more susceptible. Although not proven experimentally, field experience demonstrates that the most susceptible period for tree loss due to fire blight infection of the rootstock occurs when the trees are young (three-five years) and bearing their first crop (21). Significant losses did not occur in older trees on M.9 or M.9 interstems during the epidemic fire blight years of 1979 and 1980 in Ohio.

Crown rot has been a serious disease of apple rootstocks when they are grown in soils with imperfect drainage or high moisture-holding capacity. Generally, infected trees weaken slowly and die over a period of several years. Tree loss often attributed to crown rot may be due to fire blight or girdling by wires, mice, or rabbits. The interaction of winter injury and crown rot is also a problem, particularly when cultivars that tend to grow late in the season are combined with especially susceptible rootstocks such as MM.106 (13). The crown rot organism has been difficult to isolate, and study of this disease is further complicated by different responses in susceptibility in laboratory versus field studies.

In Europe, many of the cultivars are extremely susceptible to collar rot, and rootstocks affect the scion resistance. With the cultivar 'Cox's Orange Pippin', rootstocks inducing high resistance were M.9, M.7, M.26, and MM.109, and those inducing low resistance were M.25, MM.104, and MM.109, with M.2 and MM.111 intermediate between the two groups (47). The influence of the rootstock on scion resistance was apparently unrelated to inherent rootstock resistance.

Rootstocks differ in susceptibility to foliar disease (Table 1), which is important to nurserymen. However, these difference generally are not manifest when they are used as rootstocks in commercial plantings.

Viruses. Tolerance to latent viruses varies with rootstock, and Cummins (Table 1) has classified relative tolerance of some of the most common.

Although some viruses are latent and seem to cause no adverse effects on apple trees, others have serious effects on tree growth or fruit quality and yield (Fig. 3). There has been a long-standing attempt to remove viruses from rootstocks to avoid the potential problems. An effort to remove rubbery wood, apple mosaic, star crack, and chat fruit resulted in an "A" designation of clones free of these three, but not necessarily the latent viruses (e.g., M.7A, M.9A). Generally rootstocks with the "A" designation produced trees nearly the same size as the original clone. Due to the heavy demand for heat-treated and virus-tested budwood of apple cultivars and rootstocks, the East Malling Research Station and Long Ashton Research Station provided tested material to be distributed through the Nuclear Stocks Association. The best clone from either station was given the original M. or MM. number with the EMLA suffix. The first main issues of the EMLA virus-free rootstocks went to the Nuclear Stock Asssociation in 1969–1970. Comparison of trees on EMLA status rootstocks with the original M. and MM. stocks indicates a slight increase in tree size with most of the stocks, with the largest change in M.9 EMLA compared to M. 9, which may be as much as 50% larger than the original clone. In part, this increased size is due to removal of viruses, but in addition to M.9 EMLA it appears to be a "subclone" of M.9 with greater growth potential.

A new virus problem, apple union necrosis and decline, has developed into a

Figure 3. Jonathan on M.8 showing stem pitting virus on the rootstock.

significant commercial problem in the eastern United States (12). Trees with union necrosis break over at the union with a sharp, clean break at least on one side. New growth on trees in the dormant state has a slight pinkish cast and during the growing season develops smaller, yellowish leaves. The disease is caused by the tomato ringspot virus and can be spread in the orchard through the soil by the dagger nematodes, *Xiphinema americanum*. Most of the problem has occurred on MM.106. M.26 also appears susceptible, while M.4, M.7, Ottawa 3, 'Novole', and Robusta 5 appear to be resistant. Of fruiting cultivars, 'Delicious' appears most sensitive; 'Empire', 'Golden Delicious', and 'York Imperial' appear tolerant, with many other cultivars in between. This new problem is particularly ominous because most previous virus problems were associated only with infected nursery stock, while this one has the potential to spread in the field and is a common virus that infects many other plants.

Insects. The wooly apple aphid (*Eriosoma lanigerum* Hausm.) is a problem on apple rootstocks grown in warmer climates. 'Northern Spy' is resistant to woolly aphid; the Malling-Merton (MM) rootstocks series resulted from a breeding program using 'Northern Spy' as the parent to provide resistance to woolly aphid. The resistance of various rootstock clones are shown in Table 1. A serious infection of wooly aphid can stop tree growth.

Various borers have found in declining trees that have been weakened by other causes. Ambrosia beetle (*Xylosandrus germanus*) was found in trees declining from fire blight on M.9 and also in interstem trees of C6 and M.9. The limited data indicated a greater infestation when M.7 was the rootstock with M.9 interstem trees than when MM.106, M.111, or apple seedling was the rootstock. Infestations of dogwood borer (*Synanthedon scitula* Harris) and others have been found in burrknots. Stocks such as M.9, M.26, M.7, and MM.111, which were prone to produce burrknots, have been most often infected.

Rodents. The pine vole (*Microtus pinetorum* Lecone) and the meadow vole (*M. pennsylvanicus* Ord.) cause serious economic losses to apple orchards in the eastern United States. Several studies (e.g.,54) have compared extensive collections of apple clones for their sensitivity to vole feeding. Most of the commonly used rootstocks are very susceptible, and dwarfing clones with thick rootstock bark such as M.9 seem particularly attractive to voles. Recently 'Novole' (PC286613) has been introduced as a rootstock with resistance to vole feeding and also to apple scab, crown rot, and fire blight.

3.4. Rootstock Anchorage and Planting Depth

Malling 8, M.9, and most of their dwarfing derivatives, including Bud. 9 and M.26, have short fibers and a greater proportion of root bark and, therefore, are quite brittle. Under severe windy conditions even large structural roots will break cleanly at the trunk base. This phenomenon can be readily observed when older trees on M.9 are grubbed. The trunks often come out, leaving the majority of large roots in the soil. Because of the brittle nature of their roots, trees on M.9 and, in some soils, M.26 should be permanently supported either by a stake or some form of trellis. Tree leaning and inadequate anchorage have also been associated with M.7, M.4, M.2, and MM.109, but generally only young trees that begin to lean are supported, rather than supporting the entire planting.

On particularly windy exposed sites, the use of a temporary support stake has often been recommended. However, Preston (43) found that with M.2 and MM.104 on an exposed site there was no evidence to suggest that trees from which stakes were removed after 10 years were more stable than those that had never been staked. He suggested planting MM.106 and M.25, which have superior root systems and are better anchored. Other rootstocks that have a long history of excellent anchorage are apple seedling and MM.111.

In an attempt to improve anchorage and decrease suckering on problem rootstocks, some nurseries have budded higher than standard on the rootstock shank and recommend planting the trees deeper than they grew in the nursery (Fig. 4). When this practice is successful, a new root system forms on the shank above the original root system, anchorage is improved, and suckering lessened.

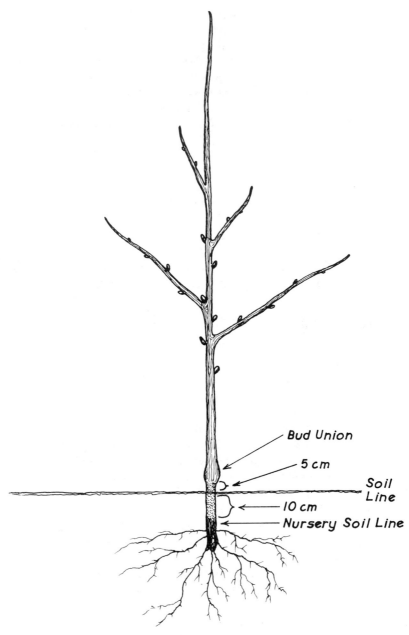

Figure 4. In planting apple trees on clonal size-controlling rootstocks it is critical (1) that the graft union be above the soil line to prevent scion rooting and (2) that on clay and silt loam soils they not be planted more than 7–10 cm deeper than they were in the nursery.

However, in heavier clay loam and silt loam soils and particularly if the season is wet after planting, the original root system may die due to inadequate oxygen before a new one can form and serious loss and stunting of the trees may occur. For this reason, it is suggested that trees not be planted more than 7−10 cm deeper than they were in the nursery on problem soils (Fig. 5).

It should be pointed out that for the desired influence of size-controlling rootstocks, the graft union must be maintained above the soil. If the trees are carelessly planted and the graft union buried, scion rooting can occur and the rootstock influence lost. See Figure 6. Sometimes trees in Western Europe are budded high and planted with 30−60 cm of the rootstock exposed as a protection against *Phytophthora* splash on the trunks of susceptible scions and to gain

Figure 5. Burrknot development on M.7 rootstock planted with 25 cm of the rootstock exposed above the soil line.

Figure 6. Variability in root formation on M.9 interstems (outlined by white tape) on MM.106 rootstock when half of the interstem is buried below the soil line (black). Some form no roots (left), many form fine roots (center), and a few form structural roots (right).

additional rootstock influence. High budding with a significant portion of the rootstock exposed above the soil line has been shown to increase dwarfing, decrease branching on the leader, and tend to reduce fruit size (52). Some rootstocks, such as M.7, M.26, and MM.111, form excessive burrknots (aerial root intitials) if the rootstock is exposed. These burrknots can cause fluting of the trunk and, in severe cases, partial girdling and stunting of the trees. With these problem stocks, extra care should be taken to expose a minimal amount of the rootstock above the soil surface.

Training System. Anchorage, tree size, and precocity are factors that determine the adaptability of a particular variety/rootstock combintion to a training system (13). However, a combination that would be unsatisfactory for one grower may work quite well for another grower more skilled in pruning and training techniques. Thus it is difficult to generalize as to which rootstocks are best suited for a particular training system. Generally, as tree density increases, tree size decreases and the ranges of tree densities in Table 2 have been successful in research and commercial plantings.

Generally, trees managed as supported trees on one of the versions of the spindle systems (slender spindle, north Holland spindle, axe, etc.) depend on M.9 as the rootstock. With vigorous cultivars on fertile soils M.27 has been used. The most common rootstocks used in orchards of freestanding central leader trees in hedgerows are: Alnarp 2, M.2, M.4, M.7, MM.106, MM.111, M.I.793, or domestic or Antonovka seedling. The choice depends on climate, soil type, and cultural history of the fruit-growing region. It is important to

TABLE 2. Ranges of Tree Density That Have Been Successful for Various Rootstocks in Research and Commercial Plantings

Rootstock	Density Range (trees/acre)	Support
M.27	1500+	Supported
M.9	350–1500	Supported
M.26 & M.9/MM.111	150–350	Occasional
M.7	180–250	Occasional
MM.106 & MM.111	90–180	None
Seedling	60–120	None

match rootstock vigor and cultivar precocity with the management abilities of the grower if the training system is to achieve its potential.

In medium-density orchards of freestanding trees, it is important to quickly produce a productive canopy volume and a tree structure with strong primary scaffolds. To achieve this goal, relatively heavy pruning is used in the formative years on trees on well-anchored rootstocks, which delays cropping. Optimum limb structure is achieved through using toothpicks or wooden spring-type clothespins the first year to achieve wide crotch angles and spreaders or tie downs in subsequent years to position limbs (Fig. 7). These training techniques tend to reduce growth and encourage flower initiation. Rootstocks such as M.7 or MM.106 greatly assist in moderating tree growth and enhance the efficiency of these training systems.

The intensive training systems in high-density orchards are dependent on the growth control of dwarfing rootstocks such as M.9. In systems with more than 400 trees/acre, pruning alone is not an economical method for the grower to avoid overcrowding. Therefore, it is best to use a fully dwarfing rootstock. A pruning severity that would cause a tree on a more vigorous rootstock to be excessively vegetative will not have the same invigorating effect on trees on M.9 or M.27. Trees on the fully dwarfing rootstocks appear to be more tolerant of pruning with mechanical hedgers (53) than trees on more vigorous rootstocks (17, 18).

In defining the role of the rootstock in the overall efficiency of an orchard management system, it is often difficult to separate the direct genetic benefit and the change in canopy arrangement to improve light exposure. A number of studies (19, 25) have shown that as tree size is decreased, the percentage of leaves and proportion of the canopy receiving adequate light increases. This effect is confounded with a direct rootstock effect if a series of rootstocks were used to reduce tree size. Ferree (19) compared two systems on M.9 and found the trellis system to have greater efficiency then the slender spindle, as judged by fruit/100cm^2 spur or shoot leaves and yield per unit trunk cross-sectional area. However, if cumulative yield was used to measure efficiency of the

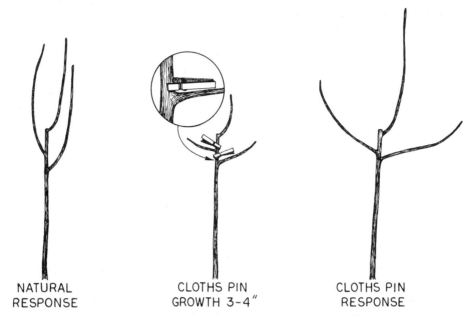

NATURAL   CLOTHS PIN   CLOTHS PIN
RESPONSE   GROWTH 3-4″   RESPONSE

Figure 7. Improving branch angle and ultimate canopy structure of semidwarf or semistandard freestanding trees by inserting clothespins when growth is 3−4 in. long.

overall system, the slender spindle produced 24% more fruit per hectare in the first six years in the orchard. Thus it is important to define how efficiency is measured and be aware that both rootstock and management system have an important role in determining efficiency.

4. ROOTSTOCK RELATIONSHIP TO TREE SIZE

Apple rootstocks produce tree sizes ranging from trees larger than on seedling rootstocks (e.g., M.25) to 15 to 20% the size of trees on seedling (e.g., M.27). Relative tree sizes of many currently available and potential rootstocks for the future are presented in Table 3. The wide diversity in tree size has allowed apple growers to tailor the tree to the training and/or orchard management system they want to use.

The relative tree size relationships are usually consistent within a cultivar. However, differences between scion cultivars on the same rootstock can vary markedly. For example, spur 'Delicious' on MM.111 may be slightly smaller than on MM.106, while with other cultivars the reverse is true. Cultivars that are strong, vigorous growers will normally be larger on all rootstocks than weaker-growing cultivars. Long-term studies at the East Malling Research Station comparing numerous rootstocks have shown that relative tree size among apple rootstocks of M.26 size or smaller doesn't change after the trees

TABLE 3. Influence of Selected Apple Rootstocks on Scion Vigor

Rootstocks	Vigorous	Semi-vigorous	Semi-dwarf	Dwarf	Sub-dwarf	Ref
			Level of Vigor Control of Scion			
Malling series						
M.1	X					39
M.2		X				39
M.4		X				39
M.7			X			39
M.8				X		39
M.9				X		39
M.13	X					39
M.20					X	41
M.25	X					40
M.26				X		41
M.27					X	41
Malling Merton series						
MM.106		X				40
MM.111		X				40
Polish series						
P-1			X			38,33
P-2				X		38,33
P-16				X		38,33
P-18		X				33
P-22				X		38,33
Russian series						
Bud.9				X		33,22
Bud.118						22
Bud.146						22
Bud.490						22
Bud.491						22
Michigan apple clone series						
MAC-1			X			7
MAC-4	X					7
MAC-9 (MARK)				X		7
MAC-24	X					7
MAC-39					X	7
MAC-46				X		7
Ottawa series						
0.1		X				36
0.2		X				36
0.3				X		36,22
0.8						22
0.11	X					22
Kentville stock clone series						
KSC 3	X					35

(continued)

TABLE 3. (*Continued*)

Rootstocks	Level of Vigor Control of Scion					
	Vigorous	Semi-vigorous	Semi-dwarf	Dwarf	Sub-dwarf	Ref
KSC 6	X					35
KSC 7		X				35
KSC 11		X				35
KSC 25		X				35
KSC 26				X		35
KSC 28		X				35
Others						
Domestic seedling	X					22
Alnarp 2	X					22
Anis		X				35
Antonovka seedling	X					22
Beautiful Arcade		X				35
Robusta 5	X	X				22
Novole	X					22
J9				X		22
OAR 1				X		22
Northern Spy			X			40
Merton 793						

are 7 years old, but for rootstocks producing larger trees the age is 15 years before they no longer change their relative positions.

The rootstock effect on tree size generally becomes evident as the trees begin to crop. Very dwarfing rootstock types like M.9, M.26, and M.27 cause most scions to begin cropping at a very young age, which slows vegetative growth. In several studies, Avery (3) at East Malling has shown that trees on dwarfing M.9 cause more photosynthate to go to fruit than do vigorous rootstock types such as M.16. These studies, conducted under equivalent crop loads, imply an internal mechanism controlled by the rootstock that directs the distribution of the products of photosynthesis.

One of the theories proposed by Jones (29) to explain these effects suggests that graft unions of dwarfing rootstocks or interstocks with the scions will deplete the solutes of the xylem sap, and this effect increases with the dwarfing effect of the rootstock or interstock. The depletion appears nonselective, affecting all the nutrients and probably also the growth regulators of the sap. Lockard and Schneider (33), in their review on the dwarfing mechanism in apple, postulated that auxin produced by the shoot tip is translocated down the phloem and the amount arriving at the root influences root metabolism and affects the amounts and kinds of cytokinins synthesized and translocated to the shoot through the xylem. Auxin in the phloem would be oxidized or otherwise degraded by compounds in the bark, thus varying the amounts reaching the roots in different rootstocks. Both theories appear to explain many of the observed phenomena and may be interrelated when additonal research reveals the true mechanism.

Dwarfing rootstocks have also been used as interstems to control tree size. Many of the dwarfing rootstocks (M.8, M.9, M.27, Bud.9) are not freestanding, but a freestanding dwarf tree may be obtained by placing a piece of stem called an "interstock" or "interstem" between the rootstock and scion cultivar. This has been a very old practice used to dwarf fruit trees and came into commercial use with the introduction of Clark Dwarf trees by Stark Brothers Nursery, Louisiana, MO. In recent years, stem pieces of M.9 and C6 have been used in commercial plantings. Numerous studies have shown that the longer the stem piece, the greater the amount of dwarfing. Studies also indicate that the degree of dwarfing is increased if the entire stempiece is above the soil line, as opposed to the lower (rootstock−interstem) union being buried under the soil. Generally, a 15−20-cm length of interstem has been used to produce trees slightly larger than the same scion propagated directly on M.9. Rootstock also makes a difference, and with equivalent interstem lengths trees on MM.106 are smaller than those on MM.111. The interstem combination of M.9 on MM.106 appears to be particularly efficient and productive compared to other combinations. One of the problems with interstem trees is the profuse production of root suckers produced even on normally nonsuckering rootstocks such as MM.106 and MM.111. This problem is particularly severe when rootstocks prone to suckering, such as apple seedling and M.7, are used, and for this reason these rootstocks are not recommended for interstem trees.

5. ROOTSTOCK CHARACTERISTICS

5.1. The Malling Apple Rootstock Series

This rootstock series, described and classified at the East Malling Research Station, Kent, England (1912−1918), was the first successful effort to test and evaluate rootstocks in an orderly fashion. This first series was made up of 9 different stocks and each was given a Roman numeral with the prefix M. Since that series 15 more were added, giving a total of 24, and these latter ones were given an Arabic number. From the woolly apple aphid resistance breeding program, M.25 was introduced in 1956 as a productive, semivigorous stock. M.26 and M.27 evolved from the M.IX crosses and were subsequently added.

The abbreviated letter/numbers were changed from Roman to Arabic numbers about 10 years ago. The early nomenclature was, for example, type IX, then M.IX, and currently M.9.

5.2. Malling Rootstocks in the United States

Sample testing of the East Malling rootstocks began in the early 1920s, primarily in the eastern states such as Massachusetts, New York, and Pennsylvania.

As more of the Malling series were tested in England and they became available in greater numbers, wider ranges of testing were conducted at research stations in major apple-producing areas. Growers also began to evaluate rootstocks in the 1930–1940s. There was an obvious note of skepticism among growers for some time because they were conditioned to produce fruit on large trees.

The word *dwarf* meant "very small in proportion to the standard large trees" and was thought of as a tree for the backyard. However, as the words *semidwarf* and *semistandard* came into use, the idea of apple rootstocks gradually became accepted among progressive growers. In other words, the Malling rootstocks were not all dwarf but represented a wide range of vigor from truly dwarf to standard size. Although a few growers contended that the "dwarf trees could not carry a payload," the enthusiasm for smaller trees gradually gained in the next three decades.

Out of the 24 East Malling rootstocks, some have stood the test of time in the United States and served the industry well as smaller, more manageable apple trees. These are, in order of increasing vigor, M.9, M.7, M.2, M.1, and M.13. M.3, M.5, and M.6 had some faults, such as nonprecocity, weak-rootedness, and overvigorousness. M.8 has many of the M.9 characteristics by having brittle roots, being very dwarfing, and exhibiting durability. It has been reported that M.8 and the 'Clark' interstem are identical (4). M.12 and M.16 were used in limited ways but, due to their vigor, are not propagated now. The rest—M.10, M.11, M.14, and M.15—were introduced to research stations, but did not, due to vigor, receive extensive grower trials.

M.2.   M.2 produces a tree slightly larger than M.7 (75% size compared to seedling) and has been precocious and productive. In a long-term Ohio trial with 'Delicious', M.2 significantly out-produced trees on M.7, M.12, M.16, and apple seedling. Unfortunately, it was "dropped" from the recommended list just prior to the introduction of the varied strains of spur-type 'Delicious.' It does not propagate as well as MM.106 and is tolerant of most soil types. In the nursery its distinguishing characteristics are obtuse leaf serrations, dark brown bark, and small lenticels.

M.4.   M.4 is similar in size to M.2, but induces earlier and heavier production. One major fault of M.4 is that roots tend to develop mainly on one side, thus it is not well anchored. It is especially resistant to collar rot. For this and other reasons, it is being brought back in the regional testing programs. In the nursery it has sharp sawtooth leaf serrations, light brown bark, and is prone to spur with small leaves at the base of the leaf petioles. M.4 may return as a useful rootstock for some cultivars.

M.7.   This rootstock produces a tree 55 to 65% the size of apple seedling and has been the backbone of the U.S. fruit industry during the first 50-yr introductory period of rootstocks in the United States. It has been, and still is, the most commonly used stock, the most disease tolerant, and the most adaptable to a range of soil types and climates. In the nursery, it is easily identified due to its slender, spindly shoots with pale brown bark, sharply deep-serrated and wavy leaf margins with small, almost minute stipules and whitish lenticels. It propagates readily and so is an ideal stock for both nurserymen and growers. The most significant disadvantage of M.7 is its tendency to sucker from the roots.

M.8. M.8 was disseminated to the United States in at least two ways. At East Malling it was collected as 'French Paradise' and listed as type VIII in the Malling series. Another source, perhaps as an "escape" from the French Paradise, came via South America to Iowa and was named the 'Clark Dwarf.' This "stray" was the only stock used in the four-piece dwarf interstem trees in the 1930s and 1940s. Karl Brase at the New York State Agricultural Experiment Station (4) later keyed this 'Clark Dwarf' to be identical to M.8. This rootstock is interesting because it has many of the M.9 qualities. In addition, it is slightly more dwarfing and more winter hardy than M.9. For this reason it is very dependable when used as an interstem for dwarfing.

M.8 has dark brown wood with sparsely spaced lenticels; its leaves are smooth, flat, and dark green with inconspicuous stipules at the base of the leaf petiole. Like M.9, it does not root easily in stoolbeds or by cuttings.

M.9. This is the most commonly used dwarf rootstock throughout the world both for one-union trees, producing a tree $25-35\%$ the size of seedling, and for interstem trees. It was selected as a chance seedling in 1879 in France and called 'Jaune de Metz.' Under normal culture the shoots of M.9 are thick and stout, slightly curving to one side at the base in stoolbeds. The bark is reddish to silvery, slightly pubescent, and also forming small nodules at each side of the bud nodes. The leaves are ovate to oblong, leathery thick and dark, with a shiny upper surface. Under favorable stoolbed cultures M.9 roots fairly well, but is most difficult to root via hardwood cuttings. The roots are brittle and break easily, so trees on M.9 need to be supported. M.9 tends to sucker from the roots, which is a special problem due to the susceptibility of this stock to fire blight. Detailed studies have shown that M.9 trees will set fruit with fewer seeds than the same cultivar on other stocks.

M.9 is used in high-density plantings, most extensively on the European continent because of its precocity and productivity. Currently several extensive European tests are comparing subclones of M.9 that differ in tree size, fruiting, and other characteristics, but to date the most desirable has not been identified. In North America it is used for smaller trees in pick-your-own plantings and for producing dwarf interstem trees.

M.13. This stock produces a full-size tree that is no more precocious than apple seedling, but is more tolerant of wet soil conditions than other rootstocks. In the nursery it is easy to distinguish when in full leaf because of its large, wavy, deeply serrated and sawtooth leaf margins. The rather large stipules at the leaf petiole base make foliar shoots appear full and leafy. The dark brown bark of the shoots has sparsely spaced whitish lenticels. M.13 roots easily and tends to produce a spreading, shallow root system with cultivars. It may in the future have a place in shallow soils with a significant clay content, or those soils underlain with hardpan. There is a trend to bring this rootstock into use under certain soil types where other rootstocks may fail to perform. It is especially useful under a dwarfing interstem.

M.25. This rootstock is the result of a cross of Northern Spy and M.2 made at East Malling in the late 1920s. It makes nearly a full-size tree, but is more precocious and productive than seedling. In stoolbed it grows erect with a slight curvature at each node, giving a zigzag appearance. The bark is dark green and the leaf blades are large with irregular margins. Due to its vigor, it has not come into use in this country.

M.26. This is from a cross of M.16 and M.9. After its introduction in 1959, it became popular very rapidly because of its precocity, dwarfing (40−50% compared to apple seedling), and productivity. M.26 is more vigorous and better anchored than M.9 with most cultivars. It was extensiveley propagated and planted so that by mid-1970 it equaled M.7. It is nonsuckering, but has demonstrated partial incompatibility with some cultivars such as Blaxtayman, Holiday, etc. M.26 is intolerant of extended wet soil conditions.

In the nursery M.26 has large leaves that are leathery in appearance with wavy sawtooth margins. The leaves have rather slender stipules and subtending smaller leaflets often forming lateral rosettes. The stems of the shoots are stocky, erect, and brown with sparse, conspicuous lenticels.

In a well-drained, deep soil, trees on M.26 will not need support. However, cultivars that tend to be vigorous need support such as post or trellis. It is prone to collar rot and fire blight. M.26 is the most winter hardy of the Malling rootstocks now used commercially. The attributes of midwinter hardiness and late spring bud break were derived from its M.16 parent.

M.27. This rootstock is the most recent introduction. It is a cross of M.13 and M.9 made in 1929. It is very dwarfing, producing a tree smaller than M.9, very precocious, and needs support. At this time it is not widely used commercially as a single-grafted tree. However, it is used as interstem material for grafting on vigorous clones.

In the nursery the shoots of M.27 are spindly, erect, with a brown bark. The leaves are oval with minute, serrate margins and cupping. It is productive in the layer beds. It may find a place in future high-density planting systems, especially on good, deep, well-drained soils.

5.3. Malling-Merton (MM.) Rootstock Series

During the decade of 1920, two research stations, the East Malling and the John Innes Horticultural Institute at Merton, England, collaborated in hybridizing the 'Northern Spy' and some of the Malling rootstocks. The major objective was to develop woolly-aphid–resistant rootstocks, much needed in some fruit-growing areas of the world. Other rootstock qualities, such as precocity, dwarfing, propagation, performance, and productivity, were also to be included in the final selections.

From some 3300 resulting seedlings, 15 clones were chosen, all resistant to woolly aphid. So as not to confuse these with the Malling series, the numbering began with 101 and the prefix was MM from the Malling and Merton research stations.

Out of these 15 only 4 were chosen in 1952 for testing in other countries; however, the entire series was introduced to several research stations in the United States for testing under different climates and soil conditions. The series MM.101 to MM.115 varied in performance and ranged in vigor from semidwarf to standard. The ones extensively tested with several cultivars were MM.104, MM.106, MM.109, and MM.111, whereas the other 11 were not adaptable to U.S. conditions. For example, MM.103 had the most dwarfing characteristics, resembling M.7, but tended to be poorly anchored.

MM.104. This was a selection from a cross of M.2 × 'Northern Spy'. Budded with cultivars, it produced trees 80–90% the size of seedling on strong soils. In early tests it was considered to be fairly resistant to collar rot; however, in test plantings in Michigan it proved to be very susceptible. On clay loams, 50% of the trees died in years four to eight in the orchard. For this reason the use of MM.104 was discontinued about 1965.

MM.106. Due to its inherent precocity, this rootstock rapidly gained favor after its introduction. It is from a cross of 'Northern Spy' × M.1. In the nursery stoolbed, shoots are stocky, upright, with pubescence and conspicuous nodes. The leaves are large, flat, with a glossy upper surface. Stipules are large, almost leaflike, and for these reasons easy to identify.

Trees on MM.106 are well anchored, do not sucker, are semidwarfing (60–75% the size of trees on apple seedling), and very productive. It is more vigorous than M.7 but, due to its fruitfulness, vigor can be controlled. European studies with 'Cox Orange Pippin' (normally a small-fruited cultivar) have shown a reduction in fruit size in trees on MM.106 compared to other rootstocks. It has shown a tendency for collar rot on poorly aerated soil. Cultivars on MM.106 tend to grow late in the season, defoliate late, and develop dormancy slowly, which can result in increased trunk injury due to fall temperature changes. The tendency to grow late in the season has also caused an increase of fire blight in susceptible scions.

MM.109. This rootstock was tested in research and commercial plantings but, due to its vigor (equal in size to apple seedling) and weak root anchorage, it was discontinued and is no longer propagated. Early studies indicated greater drought tolerance compared to other, similar-sized rootstocks.

MM.111. Even though this rootstock is semistandard (90% of a seedling tree), it is still in use on lighter soils and for interstem trees. This stock was selected from a 'Northern Spy' × MI.793 cross. It is easy to propagate. In the nursery the leaves are coarsely serrate and the stems are rather slender and erect. In field trials under different soil moisture levels it proved to be most tolerant of droughty soil conditions. It is not precocious, but is much more productive in long-term trials than apple seedling, and seems adapted to many soil and climatic types and has survived well compared to others.

5.4. Polish Series

The Research Institute of Pomology and Floriculture at Skierniewice, Poland, initiated an apple rootstock breeding program in 1954 because the M. and M.M. rootstocks were not sufficiently hardy (39, 55). Antonovka was crossed with M.9 and, after preliminary screening, 28 clones were selected. Of these, 4 show particular promise as dwarfing rootstocks and have survived at least two test winters and are reported to have considerable resistance to crown rot. However, tests in New York (22) indicate that these rootstocks are susceptible to fire blight and woolly apple aphid. P.2 and P.22 show particular promise as dwarfing interstems.

P.1. Studies indicate that trees on this rootstock are approximately M.9 in size, with similar production efficiency, but superior to M.9 in propagation ability and winter hardiness.

P.2. Trees are larger than trees on M.9 and smaller than trees on MM.106, with slightly lower production efficiency than M.9. It is equal to M.9 in propagation ability, but surpasses all the M. and MM. stocks in hardiness.

P.16. Trees on this stock are approximately M.9 in size, similar in winter hardiness and production efficiency, but superior in propagation abilities.

P.18. A cross of M.4 × Common Antonovka exhibited a hardiness similar to Antonovka, with less susceptibility to fire blight than the other rootstocks of the Polish series. Trees on P.18 are similar in size to those on M.4.

P.22. Trees on this stock are consistently smaller than M.9 and equal or superior in production efficiency. P.22 surpasses all the M. and M.M. rootstocks in hardiness and has propagation ability similar to M.9. It remains dormant longer than any of the dwarfing stocks except M.26.

5.5. Budagovsky Series

The Michurin College of Horticulture introduced the Budagovsky series to provide adaptability to the severe climate of central Russia (38).

Bud.9. Red-Leafed Paradise resulted from a cross of M.8 × 'Red Standard' and has similar dwarfing potential as M.9 and tests in New York (13) indicate similar susceptibility to fire blight, woolly aphids, and brittleness. Bud.9 is hardier than M.9, has greater resistance to crown rot, and is somewhat difficult to propagate. It has been widely used in Poland as a dwarfing interstem with good success.

Bud.490. Trees on this stock are in the MM.106 vigor class, very winter hardy, induce early production, and are easy to propagate, even by hardwood cuttings, with resistance to crown rot.

Bud.491. This rootstock produces trees smaller than on M.9, with brittle roots; it is very winter hardy and propagates easily in the stoolbed.

5.6. Ottawa Series

There are two Ottawa Series: the Ottawa Hybrid Seedlings (OH) and the Ottawa Clonal Series (0).

Ottawa Hybrid Seedlings. Development of this series began in 1961, and the following six rootstocks were introduced in 1971: OH-1 (Heyer 12 × Robusta 5), OH-2 (Osman × Heyer 12), OH-3 (Robusta 5 × Antonovka), OH-4 (Osman × Robusta 5), OH-5 (Osman × Antonovka), and OH-6 (Heyer 12 × Antonovka). Most of the Ottawa Hybrids are tolerant of latent viruses commonly found in commercial apple varieties and they have performed well in the nursery row. Long-term field trials have not yet been completed for these rootstocks, but all appear to be very vigorous.

Ottawa Clonal Rootstocks. This was a series (0.1 to 0.14) of hardy apple rootstocks selected at the Ottawa Research Station in Canada, with test plantings established in several locations in 1967. Varieties on 0.1, 0.2, 0.4, 0.7, 0.11, 0.12, and 0.14 appeared to be slightly more vigorous than when grown on MM.106. In a seven-year trial, 0.4 had the highest productive efficiency; 0.1, 0.2, and Robusta 5 had greater efficiency than 0.3 or MM.106 (49). Because of poor performance or incompatibility with at least one of the scion cultivars tested, 0.5, 0.6, 0.9, 0.10, and 0.13 were discarded.

0.3. From a progeny of 'Robin' × M.9 ('Robin' is a hardy crab) come trees approximately M.26 in size and of equal productivity (49). They are better anchored than M.9, not as brittle nor as resistant to crown rot, and just as susceptible to fire blight and woolly apple aphids. It is very difficult to propagate in a stoolbed, and suckering in the orchard is rare. It is sensitive to the virus causing brownline decline.

0.8. Trees on this rootstock are equivalent in vigor and productivity to trees on

MM.106, but it is much hardier. It originates from a cross of *M. baccata* Gracilis × M.7.

5.7. Michigan Apple Clone Series

Open-pollinated seed from a planting of the Malling rootstocks (M.1 through M.16) plus Alnarp 2 and Robusta 5 were collected and planted in 1959 at Michigan State University. The goals of the program were to screen for rootstocks that were more tolerant of U.S. conditions. The seedlings were screened in the field for freedom of woolly aphids and desirable nursery characteristics. Cultivars were propagated on 56 of these selections and evaluated in the field (7).

MAC 1. (open-pollinated seed from M.1). Trees on this rootstock are approximately M.7 in size, but do not sucker and are well anchored. In the nursery, trees are nonspurred and propagate well.

MAC 9. (open-pollinated seed from M.9). Trees on this rootstock will be similar in size to M.9, or slightly larger, and equivalent in productivity. MAC 9 is better anchored and does not sucker. It propagates easily in nursery stoolbeds. Trees on this stock tend to have an open, spreading character. In 1979, MAC 9 was named 'Mark', which is an abbreviation of Michigan Apple Rootstock Clone.

MAC 24. (open-pollinated seed from Robusta 5). Trees on this stock are vigorous and in the MM.111 size class. This stock has good nursery characteristics and roots well by hardwood cuttings as well as in the stoolbed. It has a shallow spreading root system and is well anchored, but may produce root suckers.

MAC 39. (open-pollinated seed from M.11). MAC 39 produces trees that are smaller than M.9, very precocious, but not well anchored.

MAC 46. (open-pollinated seed from M.9). MAC 46 produces trees that are slightly larger than M.9, precocious, but not well anchored.

5.8. Kentville Stock Clone Series

This series originated from 30 seedlings of 'Beautiful Arcade' in a population of 9000 that survived wide midwinter temperature changes in a nursery in 1970. These survivors were planted at the Agricultural Canada Kentville (Nova Scotia) Station and were propagated by root cuttings. After preliminary evaluation, the following appear productive, efficient, and worthy of additional testing (% shown is % of the largest clone with a 'McIntosh' scion): KSC 3 (99%), KSC 6 (100%), KSC 7 (81%), KSC 11 (88%), KSC 25 (79%), KSC 28 (57%).

5.9. Cornell-Geneva Series

The Cornell-Geneva (CG) series was developed at Geneva, New York, in 1953 from 158 open-pollinated seedlings of the very dwarfing M.8 with M.1− M.16 of the Malling series, 'McIntosh', and 'Northern Spy' as likely parents (11). Suckering has been heavy on all clones. Most clones proved unacceptably susceptible to fire blight.

CG60. Produces a tree smaller than M.9, very precocious, but excessive root suckering.

CG80 and CG10. These stocks produce trees equal in size to M.9, but were less productive.

CG44. This stock is very productive and approximately the size of M.26.

5.10. Miscellaneous Rootstocks

Bemali. This is a cross of 'Mank's Codlin' × M.4 and the first introduction from the apple rootstock breeding program at Balsgard, Sweden. Bemali produces trees about the size of M.26 and was selected for ease of propagation, freedom from woolly apple aphid, and high productivity. Bemali is the only dwarfing rootstock yet tested at Geneva that has significant resistance to fire blight.

Jork 9 (J9). This is an open-pollinated M.9 seedling released by the Jork Research Station in West Germany that has the following advantages over M.9: hardier, more easily propagated, with equal productivity.

'Novole' (Pl286613). This rootstock was introduced by the Geneva, New York, station in 1982 as a rootstock to produce a full-size, vigorous tree that is relatively unpalatable to voles. It is resistant to crown rot, fire blight, and tomato ringsport virus. It is easily propagated by cuttings. 'Novole' is suggested for use under a dwarfing interstem.

Alnarp 2. This is an introduction from the Alnarp Fruit Tree Station in Sweden that is winter hardy, produces full-sized trees, and induces early bearing and productivity in the scion cultivar.

MI.793. One of the Merton Immune Series (MI) selected from 80 seedlings of the cross 'Northern Spy' × M.2. MI.793 produces a tree slightly smaller than apple seedling and larger than 'Northern Spy', but is resistant to woolly apple aphid and collar rot. It has been adapted to a wide range of soils and has induced earlier and heavier cropping than 'Northern Spy'.

K-14. This stock was a French Crab seedling selected at the Kansas Agricultural Experiment Station in 1938. It was selected from a population that survived two intense summer droughts and a severe freeze. It is used by Stark Brothers Nurseries as a hardy trunk stock.

Antonovka. Seedlings of this group of varieties have been widely used where winter hardiness is required. Currently, some clonal selections are being tested with specific resistances to problem diseases.

C.6. This rootstock was originated to be used as a dwarfing interstem by Stark Brothers Nursery from open-pollinated M.8. It reportedly produces trees the size of M.26 and is very susceptible to fire blight.

OAR 1. An Oregon planting of 'Gravenstein' on seedling rootstocks made in 1943 had many trees blown over by a severe storm in 1962. Among the survivors was a conspicuously dwarfed tree that was very productive. Suckers of this stock were propagated and are now under test.

Robusta 5. This was originated by the Canadian Department of Agriculture in 1928 from seed obtained from Siberia. It produces trees equal in size to apple seedling and has proven winter hardy, resistant to fire blight, and easy to propagate.

6. NEW AND FUTURE ROOTSTOCKS

6.1. Rootstock Breeding Programs

The earliest breeding programs to develop new apple rootstocks were initiated in England at John Innes and at East Malling in 1922 and 1924, respectively. These early breeding efforts resulted in the introduction of the Malling–Merton stocks and M.25 in 1952, M.26 in 1965, and M.27 in 1976. The English programs have focused on productivity and tree size control with emphasis on adaptation to soil and pest problems. Currently, substantial effort is directed at developing fruiting cultivars that will root readily and will be suitable for use in the ultra-high-density systems.

Recent summaries (11, 13, 36) of rootstock breeding programs indicate active research in Germany, Sweden, Canada, the Soviet Union, Poland, Czechoslovakia, Romania, the United States, and, most recently, Japan. The projects in Canada, Sweden, Germany, Poland, Czechoslovakia, Romania, and the Soviet Union have emphasized winter hardiness, ease of propagation, and dwarfing. The program in the United States was initiated in 1968 in Geneva, New York, and is directed toward developing rootstocks with desirable horticultural characteristics as well as tolerance to important diseases such as fire blight and collar rot and insect and rodent problems such as woolly aphids and voles.

A number of these breeding programs have a series of rootstocks that have gone through local tests for horticultural traits such as propagatability, precocity induction, and tree size control. However, they have not been widely tested in orchards in other fruit-growing areas and objective assessments of responses to limiting factors of biotic and physical environments are sadly lacking (13).

6.2. Rootstock Testing

In the past, rootstock testing has been accomplished with locally important cultivars and cultural management chosen by the investigator. Most early trials were based on the long-term trials conducted under the rather mild maritime climate at southwestern England. Rootstock trials in areas with more demanding climates soon demonstrated problems with some stocks. Although susceptibility to some insect and disease problems can be partially evaluated by laboratory and greenhouse screening techniques, these tests generally do not fully reflect field response. Thus there is no substitute for rather long-term field tests to evaluate rootstock performance.

To improve the efficiency of rootstock testing, a group of researchers from across the United States and Canada interested in testing new rootstocks organized under the auspices of the North Central Region 140 Committee. The goals of this group are quickly to expose new rootstocks to widely varying soil and climatic conditions. The cooperative plantings were established under uniform cultural conditions of spacing, training, and data collection. Plant material for the trials has been developed for all sites by Oregon Rootstocks Inc. for uniformity with some of the costs supported by the International Dwarf Fruit Tree Association.

The first cooperative plantings were established in 10 states in 1976, and subsequent plantings have been made in 20–30 U.S. states, provinces of Canada, and Mexico which contain the most important fruit-growing areas. Other regions with particular extreme soil and climatic conditions are also included. Initially, the scope of this group was limited to apple rootstocks, but now pear and stone fruit rootstocks are being similarly tested. These trials should greatly shorten the time needed to test the rootstocks adequately and lessen the cost for each researcher because each location has only one planting and the information from all plantings is shared. Results from these plantings will provide growers, extension specialists, and researchers an opportunity for comparing results from other locations as well as their own and facilitate an understanding of the type of spacing and cultural recommendations appropriate for each area.

Since the use of size-controlling rootstocks and more intensive management systems changes many aspects of orcharding, a group of fruit growers and research and extension personnel gathered in 1958 to share information. From these humble beginnings the International Dwarf Fruit Tree Association developed into an organization that collects and disseminates funds to support rootstock research, as well as disseminate current grower experiences and research findings. This association has been active in sponsoring tours to all parts of the world where tree fruit is grown and annually invites international speakers to the conference.

6.3. Own-Rooted Trees

Generally, apple does not root easily from cuttings, and in the past, rootstocks that rooted easily through layering were used with little hope of inducing roots on important scion cultivars. However, recent scientific advances in propagation techniques and tissue cultures now provide the opportunity to compare cultivars on their own roots with those on accept rootstocks. Most of these trials are too young to evaluate fully the long-term effects on tree size, precocity, and other horticultural characteristics. Reports from England indicate that own-rooted trees in the first year or two grow more slowly than maiden trees of the same cultivar an either M.9 EMLA or MM.106 EMLA. There also appears to be a shift in growth pattern, with own-rooted trees producing fewer but longer shoots than trees on M.9 EMLA. Own-rooted trees bore their first substantial crop a year later than trees on MM.106 and M.9 EMLA. Yields of own-rooted 'Greensleeves' and 'James Grieve' were similar to those of trees on M.9 EMLA, but smaller than those on MM.106. 'Cox Orange Pippin' on its own roots produced very few fruits in comparison with trees on M.9 EMLA and MM.106. Early evidence from Beltsville indicates that trees from tissue culture may be as vigorous as trees on seedling rootstock, if special handling techniques are used in handling the trees out of tissue culture.

Thus these early reports indicate the expected difference in performance of cultivars on their own roots. Trials seem to indicate that trees seem to grow quite rapidly after the slow start in the first couple of years. This may indicate that most cultivars on their own roots will be relatively large. The East Malling Research Station has placed significant emphasis on breeding apple cultivars that are easy to root by hardwood cuttings. Currently, several groups in the United States and Europe are active in this exciting area of research, and this effort may provide good insight into the potential for own-rooted apple trees in the future.

INTERNATIONAL CATALOG OF ROOTSTOCK BREEDING PROGRAMS

Canada
> Kentville Research Station
> Agriculture Canada
> Kentville, Nova Scotia, Canada
> A.D. Crowe, C. Embree

Czechoslovakia
> Institute of Fruit Growing and Breeding
> 50751 Holovousy
> Breeding Station Techobuzico
> Antonin Dvorak

German Democratic Republic
Institute for Obstforschung
Dresden–Pillnitz
8057 Dresden
Manfred Fischer

Japan
Ibaraki Prefectual Horticultural Experimental Station
Yatabe, Ibaraki, Japan 305
Masao Yoshida

Peoples Republic of China
Center for Agri. Technics
Jin-xian County, Liasoning Province
PRC
Deng Jai Qi

Fruit Research Institute
Jinling Academy of Agriculture
Sci-Gongzhuling Jilling
PRC
Gu-Mo

Poland
Research Institute of Pomology and Floriculture
96-100 Skierniewice, Poland
S. Zagaja, T. Jakubowski

Sweden
The Swedish University of Agricultural Science
S 230 Alnarp, Sweden
Victor Trajkovski

United States
New York State Agricultural Experimental Station
Geneva, NY 14456
James N. Cummins

USSR
Armenian Research Institute of Horticulture and Viticulture
L. A. Apoyan

Crimean Horticultural Experimental Station, USSR
J.G. Borisenko

Moldavian Research Institute of Horticulture
Kishiner, USSR

D. P. Andryuscenko

North-Caucasian Research Institute of Horticulture and Viticulture
Krasnodar, USSR

G.V. Trusevich

I.V. Michurin Experimental Station
Michurinsk, USSR

D.S.N. Stepanov

REFERENCES

1. Anderson, P. C., P. B. Lombard, and M. N. Westwood (1984). Leaf conductance, growth, and survival of willow and deciduous fruit tree species under flooded soil conditions, *J. Amer. Soc. Hort. Sci.*, **109**, 132–138.
2. Atkinson, D. (1980). The distribution and effectiveness of the roots of tree crops, *Horticultural Reviews V2*, 424–490.
3. Avery, D. J. (1970). Effects of fruiting on the growth of apple trees on four rootstock varieties, *New Phytol.*, **69**, 19–30.
4. Brase, K. D. (1954). Similarity of the Clark dwarf and East Malling rootstock VIII, *Proc. Amer. Soc. Hort. Sci.*, **61**, 95–98.
5. Campbell, A. I. (1980). The effects of viruses on the growth, yield and quality of three apple cultivars on healthy and infected clones of four rootstocks, *Acta Hort.*, **114**, 185–191.
6. Carlson, R. F. (1967). Growth response of several rootstocks to soil water, *HortSci.*, **2**, 109–110.
7. Carlson, R. F. (1980). The Michigan apple clones—An update, *Acta Hort.*, **114**, 159–169.
8. Coker, E. G. (1958). Root Studies XII. Root systems of apple on Malling rootstocks in five soil series, *J. Hort. Sci.*, **33**, 71–79.
9. Crabtree, G. D., and M. N. Westwood (1976). Effects of weed control method and rootstock on flowering, growth and yield of apples, *J. Amer. Soc. Hort. Sci.*, **101**, 454–456.
10. Cummins, J. N., and H. S. Aldwinckle (1982). New and forthcoming apple rootstocks, *Fruit Var. J.*, **36**, 1–10.
11. Cummins, J. N., and H. S. Aldwinckle (1983). "Breeding apple rootstocks," in J. Janick, Ed., *Plant Breeding Reviews*, AVI Publishing Westport, CT, pp. 294–394.
12. Cummins, J. N., J. K. Uyemoto, and R. E. Stouffer (1978). "*Union Necrosis and Decline,*" a Newly Recognized Disease of Apple Trees, Associated with Tomato ringspot virus, *N.Y. St. Hort. Soc. 123.*
13. Cummins, J. N. and R. L. Norton (1974). Apple rootstock problems and potentials, *NY Food and Life Sci. Bul.*, **41**, 1–15.

14. Czynczyk, A., and T. Holubowicz (1984). Hardy, productive apple tree rootstocks in Poland, *Compact Fruit Tree*, **17**, 19–31.

15. Eaton, G. W., and M. A. Robinson (1977). Interstock effects upon apple leaf and fruit mineral content, *Can. J. Plant Sci.*, **57**, 227–234.

16. Embree, C. G. (1985). The new KSC (Kentville Stock Clones) apple rootstock series—their current and future use, *Compact Fruit Tree*, **18**, 34–38.

17. Ferree, D. C. (1976). Influence of slotting saw mechanical pruning and Alar on apple fruit size and quality, *OARDC Res. Circ.*, **220**, 3–8.

18. Ferree, D. C. (1980). Canopy development and yield efficiency of 'Golden Delicious' apple trees in four orchard management systems, *J. Amer. Soc. Hort. Sci.*, **105**, 376–380.

19. Ferree, D. C., and A. N. Lakso (1979). Effect of selected dormant pruning techniques in a hedgegrow apple orchard, *J. Amer. Soc. Hort. Sci.*, **104**, 736–739.

20. Ferree, D. C., and M. A. Ellis (1984). Influence of rootstock and collar rot treatment on growth, yield and root development of Golden Delicious apple trees, *Fruit Crops 1984: A Sum. of Res., OARDC Res. Circ.*, **283**, 1–4.

21. Ferree, D. C., M. A. Ellis, and F. R. Hall (1983). Tree loss due to fire blight infection of rootstocks and interstems in Ohio apple orchards, *Compact Fruit Tree*, **16**, 116–120.

22. Ferree, D. C., and R. G. Hill, Jr. (1982). Influence of six rootstocks and herbicides on growth, cropping and fruit quality of Blaxtayman apple trees, *Fruit Crops 1982: A Sum. of Res., OARDC. Res. Circ.*, **272**, 3–6.

23. Gur, A. Y. Mizrahi, and R. M. Samish (1976). The influence of root temperature on apple trees. II. Clonal differences in susceptibility to damage caused by suproaoptimal root temperature, *U. Hort. Sci.*, **51**, 195–202.

24. Hatton, R. G. (1928). The influence of different rootstocks upon the vigor and productivity of the variety budded or grafted thereon, *J. Pomology*, **VI**, 1–23.

25. Heincke, D. R. (1964). The micro-climate of fruit trees. III. The effects of tree size on light penetration and leaf area in 'Red Delicious' apple trees, *Proc. Amer. Soc. Hort. Sci.*, **85**, 33–41.

26. Hoekstra, C. (1968). Fruit growing on holdings with a marked variation in soil suitability [in Dutch], *Fruitteelt*, **58**, 482-484.

27. Holubowicz, T., J. N. Cummins, and P. L. Forsline (1982). Response of *Malus* clones to programmed low-temperature stresses in late winter, *J. Amer. Soc. Hort. Sci.*, **107**, 492–496.

28. Jones, O. P. (1971). Effects of rootstock and interstocks on the xylem cap composition in apple trees: Effects on nitrogen, phosphorus, and potassium content, *Ann. Bot.*, **35**, 825-836.

29. Jones, O. P. (1983). Mode-of-action of rootstock/scion interactions in apple and cherry trees, *Acta Hort.*, **146**, 175–182.

30. Kennedy, A. T. , P. W. Rowe, and T. J. Samuelson (1980). The effects of apple rootstock genotypes on mineral content of scion leaves, *Euphytica*, **29**, 477–482.

31. Kennedy, A. T., R. Watkins, and J. M. Werts (1986). Variation in leaf calcium in a range of apple rootstocks, *HortSci.* (in press).

32. Lapins, K. (1963). *Cold Hardiness of Rootstocks and Frame Builders for Tree Fruits* SP 32, Can. Dept. of Ag., Summerland, BC.

33. Lockard, R. G., and G. W. Schneider (1981). Stock and scion growth relationships and the dwarfing mechanism in apples, *Hort. Rev.*, **3**, 315–375.

34. Makariev, Z. (1977). Response of apple to excessive soil moisture and resistance of rootstocks to asphyxia. Ovoshcharstro: **56**(6),30–31.

35. Marro, M. and F. Cereghini (1977). Observations on various morphological and functional apsects of spurs on Richard apples on seedling and M9 rootstocks, *Rivista della Ortoflorofruticoltura Italian.*, **60**, 1–14 (*Hort. Abst.*, **47**, 1169 [1977]).

36. Moore, J. N., and J. Janick (1983). *Methods in Fruit Breeding*, Purdue University Press, West Lafayette, IN.

37. Nelson, S. H., and H. B. Tukey (1955). Root temperature affects the performance of East Malling apple rootstocks, *Mich. Agr. Exp. Sta. Quart. Bul.*, **38**, 46–51.

38. Pieniazek, S. A., A. Czyncyzk, and S. W. Zagaja (1976). Apple rootstocks from other socialist countries evaluated in Poland, *Compact Fruit Tree*, **V9**, 52–57.

39. Pieniazek, S. A., S. W. Zagaja, and A. Czynczyk (1976). Apple rootstock breeding programs in Poland, *Compact Fruit Tree*, V9, 15–19.

40. Preston, A. P. (1958). Apple rootstock studies: Thirty-five years results with Cox's Orange Pippin on clonal rootstocks, *J. Hort. Sci.*, **33**, 194–201.

41. Preston, A. P. (1966). Apple rootstock studies: Fifteen years results with Malling-Merton clones, *J. Hort. Sci.*, **41**, 349–360.

42. Preston, A. P. (1967). Apple rootstock studies: Fifteen years results with some M.IX crosses, *J. Hort. Sci.*, **42**, 41–50.

43. Preston, A. P. (1974). Apple rootstock studies: Anchorage of trees on two clones staked and not staked at planting, *Exp. Hort.*, **26**, 40–43.

44. Rogers, W. S., and M. C. Vyvyan (1934). Root studies. V. Rootstock and soil effect on apple root systems, *J. Pom. Hort. Sci.*, **12**, 110–150.

45. Rollins, H. A., Jr., F. S. Howlett, and F. H. Emmert (1962). *Factors Affecting Apple Hardiness and Methods of Measuring Resistance of Tissue to Low Temperature Injury*, Ohio Agr. Exp. Sta. Res. Bul. 901.

46. Rom, R. C., and S. A. Brown (1979). Water tolerance of apples on clonal rootstocks and peaches on seedling rootstocks, *Compact Fruit Tree*, **12**, 30–33.

47. Sewell, G. W. F., and J. F. Wilson (1973). Phytophora collar rot of apple: Influence of the rootstock on scion variety resistance, *Ann. Appl. Biol.*, **74**, 159–169.

48. Slowik, K. (1970). Influence of machinery compaction on soil physical properties and apple tree growth, *Prase. Inst. Sadownictwa W. Skierniewicach Tom.*, **XIV**, 140.

49. Spangelo, P. S., S. O. Fejer, S. J. Leuty, and R. L. Granger (1974). Ottawa 3 clonal apple rootstock, *Can. J. Plant Sci.*, 601–603.

50. Tukey, H. B. (1964). *Dwarfed Fruit Trees*. Macmillan, New York.

51. Van der Zwet, T., and H. L. Keil (1979). *Fire Blight, a Bacterial Disease of Rosaceous Plants*, U.S. Dept. of Agr. Handbook 510.

52. Van Oosten, H. T. (1978). High working for control of vigor of apple trees, *Acta Hort.*, **65**, 157–166.

53. Westwood, M. N., A. N. Roberts, and H. O. Bjornstad (1976). Influence of in-row spacing on yield of 'Golden Delicious' and 'Starking Delicious' apple on M9 rootstock in hedgerows, *J. Amer. Soc. Hort. Sci.*, **101**, 309–311.

54. Wysolmeiski, J. C., R. E. Byers, and J. N. Cummins (1980). Laboratory evaluation of some *Malus* clones for susceptibility to girdling by pine voles, *J. Amer. Soc. Hort. Sci.*, **105**, 675–677.

55. Zagaja, S. W. (1980). Performance of two apple cultivars on P. series dwarf rootstocks, *Acta. Hort.*, **114**, 162–169.

56. Zeiger, D. and H. B. Tukey (1960). *An Historical Review of the Malling Apple Rootstocks in America*, Mich. State Univ. Circ. Bul. 226.

5

PEAR ROOTSTOCKS[1]

Porter B. Lombard and Melvin N. Westwood
Oregon State University
Corvallis, Oregon

1. ORIGIN OF PEARS

There are about 22 primary *Pyrus* species, all of which originated in regions of either Asia or Europe (Table 1). The pear has been cultivated in China for at least 3000 years (19). No doubt the first species to be domesticated was *Pyrus pyrifolia*, because the wild type is edible without selection. The hardy northern *P. ussuriensis* probably became cultivated later, as the wild type is small, filled with stone cells, and is so astringent that it is inedible by man. Natural hybridization between these two wild species likely occurred in central China, and the best of them may have been carried even farther north. Most modern "ussuri" cultivars are complex hybrids of *P. ussuriensis* and *P. pyrifolia*. They are much hardier than *P. pyrifolia* cultivars but less hardy than pure wild *P. ussuriensis*. All Asian cultivars are crisp in texture (in contrast to the melting texture of European pears) and will ripen on the tree like apples.

In other parts of the world, cultivated pears have been derived from *P. communis*. Cultivation of pear in Western civilization was first mentioned by Homer nearly 3000 years ago (18). *P. communis* var. *pyraster* and/or *P. communis* var. *caucasica* were probably the ancestors of the common pear, but the French cultivars are probably complex hybrids with these two varieties and *P. nivalis*, which is known as the snow pear (9). The snow pear has been used in parts of Europe for making pear cider. Most of the present commercial cultivars of European pears—for example, 'Williams' 'Bon Chretien'—were chance seedlings.

World distribution of pears in surface area and production, major cultivars and rootstocks, and current production trends are given in Table 2.

[1]Oregon State University Agricultural Experiment Station Technical Paper Number 7936.

TABLE 1. World Distribution of Pyrus Species[a]

Species	Distribution
European	
P. communis. L.	West, southeast Europe, Turkey
P. caucasica Fed.	Southeast Europe
P. nivalis Jacq.	West, central, south Europe
P. cordata Desv.	France, Spain
Circum-Mediterranean	
P. amygdaliformis Vill.	Turkey, Greece, Yugoslavaia, Sardinia
P. elaeagrifolia Pall.	Turkey, Russia, southeast Europe
P. syriaca Boiss.	Tunisia, Libya, Israel, Syria, Lebanon
P. longipes Coss. & Dur.	Algeria
P. gharbiana Trab.	Morocco
P. mamorensis Trab.	Morocco
Mid-Asian	
P. salicifolia Pall.	North Iran, south Russia
P. regelii Rehd.	Afghanistan
P. pashia D. Don.	Pakistan, India, Nepal
East Asian	
P. pyrifolia (Burm.) Nak.	China, Japan, Korea, Taiwan
P. pseudopashia Yü	Northwest China
P. ussuriensis Max.	Siberia, Manchuria, north China, Korea
P. hondoensis Kik. & Nak.	Japan
P. calleryana Dcne.	Central, south China
P. betulifolia Bunge	Central, north China, south Manchuria
P. fauriei Schneid.	Korea
P. dimorphophylla Makino	Japan
P. kawakamii Hayata	Taiwan, south China

[a]From M. N. Westwood (35).

2. ROOTSTOCK USAGE

Pear production and rootstock practices for various countries are briefly described in Table 2. Generally, seedling rootstocks are the most widely used. In Eastern Europe, Taiwan, and India, local wild seedlings are used, while seedlings of commercial cultivars, particularly 'Williams' and 'Winter Nelis,' are used in Canada, the United States, New Zealand, South Africa, and Chile. In Japan, seedling stock of various cultivars of *P. pyrifolia* are used. In some cases, cultivars are grafted *in situ* on wild *Pyrus* trees in Turkey, Greece, and Syria. Other *Pyrus* species used for seedling stock are *P. pashia* in India; *P. calleryana* in China, the United States, and Australia; *P. betulifolia* in China, the United States, and Israel; *P. elaeagrifolia* in Turkey; *P. amygdaliformis* in Yugoslavia; and *P. ussuriensis* in China.

TABLE 2. World Pear Production, Trends, Major Cultivars, and Rootstocks, 1976–1983

Region and Country	Production $\left(\dfrac{MT}{1,000}\right)$	(ha)	Trend	Cultivars	Rootstocks
Africa					
Algeria	17[a]				
Egypt[b]	70	5,840	increase (400 ha/yr)	Le Conte	P. communis and P. calleryana seedlings (sdlgs.)
Lybia	1[a]				
Morocco	12[a]				
South Africa[c]	128	9,300		Williams, P. Triumph, Bosc., Hardy, Clapp Favorite	Williams sdlg.; QA, BP1, BP3
Tunisia	6[a]				
Total Region	206				
Asia					
China[d]	1500		Increase	Yali, Tzik, Chioubaili, Shiyue Huali, Pinguoli, Tsuli, Jingbaili	Sdlgs. of P. betulifolia (N), P. ussuriensis (NE), P. calleryana (S), P. pashia (SW), P. xerophila (NW)
India[e]	62[a]	2,500		Williams, W. Nelis, Kieffer, P. pyrifolia cvs.	sdlgs. of P. pashia, P. pashia, var. kumaoni, P. pyrifolia, & domestic

(continued)

TABLE 2. (*Continued*)

Region and Country	Production $\left(\dfrac{\text{MT}}{1,000}\right)$	Production (ha)	Trend	Cultivars	Rootstocks
Japan[f]	497	19,000	Slight Increase	Nijiseiki, Chojuro, Kosui, Hosui, Kikisui, Shinseiki, Shin-sui, Williams	*P. pyrifolia* and *P. betulifolia* sdlgs.
Korea, North	55[a]				
Korea, South[g]	102	10,000	Slight Increase	Chojuro, Niitaka, Okusankichi, Imamuroaki, Waseaka, Shinko, Jioseng-Paldal	*P. ussuriensis,* *P. pyrofolia* and *P. serotina* sdlgs.
Pakistan	37[a]				
Taiwan[h]	250	12,000	Slight Increase	Hengshan, Shinseiki, Nijiseiki, Chojuro	*P. pyrifolia* × *P. kawakamii* sdlg.
Total Region	2503				
Asia Minor					
Cyprus	2[a]				
Iran	42[a]				
Iraq	3[a]				
Israel[i]	9	1,000	Decrease	Spadona, Coscia	QA, self-rooted cvs., Q, Q C.I.7, *P. betulifolia* sdlg.

					P. elaeagrifolia sdlg. & cvs.
Lebanon	18[a]				
Syria	8[a]				
Turkey	260[a]			Akca, Ankara, Ayasu, Goksulu, Limon, Mustafabey, Williams	
Total Region	342				
Europe					
Austria[j]	129[a]				
Belgium	90	3,500	Slight Increase	Conference, Comice, Durondeau	QA, QC, QAdams (Belgium selection)
Bulgaria	112[a]				
Czechoslovakia	46[a]				
Denmark[k]	13[a]				
France[k]	380	22,600	Slight Increase	J. Guyot, Williams, P. Crassane	PQ, PQBA 29
Germany, East	50[a]				
Germany, West[l]	388[a]				
Greece[l]	130		Slight Increase	P. Crassane, Tsakoniki, A. Fetel	PQ, QA, Q BA 29
Hungary[m]	102	4,500	Slight Increase	Bosc, Hardy, Clapp, Williams, Diel	QC, wild pear sdlg.
Italy[n]	1494	47,000	Slight Increase	Williams, A. Fetel, P. Crassane, Kaiser, Conference	QA, sdlg., PQ, PQBA 29
Ireland[o]	10				
Netherlands[p]	100	5,500	Slight Increase	Conference, Comice, Cooking, Hardy	QA, QC

(*continued*)

TABLE 2. (*Continued*)

Region and Country	Production (MT/1,000)	(ha)	Trend	Cultivars	Rootstocks
Norway	11[a]				
Poland[q]	105[a]		Slight Increase	Clapp Favorite, B. Alexander Lucas, Williams, Conference, Coloree de Juliet, Hardy, Princesse Marianne, Comice, Comtesse de Paris	*P. caucasica* and *P. communis*, QA, QC, QS1 (Poland selection)
Portugal	45[a]				
Romania[r]	100	14,400	Slight Increase	Cure, Clapp Favorite, Williams, Bosc, Comtesse de Paris, P. Crassane	Harbuzesti, Cu Miezul rosu, Pepenii, and Alamii sdlgs., QA, Hardy, Cure interstock
Spain[s]	512	37,100	Slight Increase	J. Guyot, Blanquilla, Coscia, Williams, P. Crassane, Conference, Comice	Q & PQ-type sdlg.
Sweden	16[a]				
Switzerland[t]	50	869	Slight Decrease	Williams, Louis Bonne, Conference	QA, *P. communis* sdlg., PQBA 29, QC
United Kingdom[u]	50	4,000	Slight Decrease	Conference, Comice	QA, QC, PQBA 29
U.S.S.R.	590[a]				

Region / Country					
Yugoslavia[v]	123		Increase	Cure, Williams, Bosc, P. Crassane	QA, wild sdlgs. of P. communis, var. pyraster, & P. amygdaliformis
Total Region	4636				
Oceania					
Australia[w]	137	4,900	Slight Decrease	Williams, P. Triumph, Bosc, Josephine	P. communis, Kieffer, P. calleryana, and W. Nelis sdlgs., P. calleryana D6 clone, QA
New Zealand[x]	12		Slight Decrease	Williams, P. Triumph, W. Cole, W. Nelis, Bosc	Williams sdlg., QA, QC
Total Region	149				
North America					
Canada[y]	28	876	Slight Decrease	Williams, Anjou, Bosc	Williams sdlg., OH × F clones
Mexico[z]	46[a]	2,200	Slight Decrease	Williams, Anjou, Max Red Bartlett	Williams sdlg.
United States	750	37,700	None	Williams, Anjou, Bosc, Comice, Kieffer, Seckel, Winter Nelis, Forelle, Red Bartlett, Nijiseiki, Chojuro, Red Clapp	Williams, W. Nelis, P. calleryana, P. betulifolia sdlgs., OH × F clones, PQA, PQBA 29, QC
Total Region	824				

(continued)

TABLE 2. (*Continued*)

Region and Country	Production ($\frac{MT}{1,000}$)	(ha)	Trend	Cultivars	Rootstocks
South America					
Argentina[aa]	154		Increase (300ha/yr)	Williams, P. Triumph Anjou, Winter Bartlett, Red Bartlett, Guiffard, Clapp, Winter Nelis	*P. communis* sdlg.
Bolivia	5[a]				
Brazil[bb]	40	4,200	Increase	P. Triumph, Williams, D'Agua, Nijiseiki, Yali	*P. communis* sdlg., PQBA 29
Chile[cc]	43	3,700	Increase	P. Truimph, W. Nelis, Williams, Bosc, Anjou	W. Nelis sdlg., QA, *P. calleryana* & *P. betulifolia* sdlgs. betulifolia sdlgs.
Equador	4[a]				
Peru	8[a]				
Uruguay	6[a]				
Total Region	260				
TOTAL	8920				

[a]From FAO statistical report in 1979.

[b]A-Talaat-El Wakeel, Senior Research Seiemties, Horticulture Research, Institute, Cairo, 1985.

[c]H. J. van Zyl, Stellenbosch, 1984.

[d]Wang Yu-lin, Zhengzhow Fruit Institute, Honan, 1985.

[e]S. S. Randhawa, Amartara, Simla, 1984.

[f]K. Kumashiro, Shinshu University, 1984.

[g]Jung-Ho Kim, Suweon, 1984.

[h]Cheng Nee, Chung Hsing University, 1984.

[i]P. Spiegel-Roy, The Volcani Center, 1984, R. Assaf, Newe-Ya'ar, 1984.

[j]W. Porreye, Gorsem, 1984.

[k]I. Pochon, B. Thibault, Angers, 1984.

[l]M. Vasilakaksi, University of Thessaloniki, 1984.

[m]J. Nyeki, Kerteszeti Egyetem, Budapest, 1984.

[n]S. Sansavini, G. Costa, University of Bologna, 1984.

[o]N. D. O'Kennedy, Agriculture Institute, Dublin, 1984.

[p]H. J. van Oosten, Wilhelminadorp, 1984.

[q]A. Czynczyk, Institute, Skierniewice, 1985.

[r]Parnia, Bodi, Cociu, Pitesti-Maracineni, 1982.

[s]M. Carrera, SIA, Zaragosa, 1984.

[t]G. Perrardin, W. Pfammatter, Conthey, 1984.

[u]G. Browning, East Malling, 1984.

[v]D. M. Stankovic, Belgrad, 1984.

[w]L. A. G. van Heek, L. G. Issell, Tatura, 1984.

[x]S. Austin, DSIR, Havelock North, 1984.

[y]H. Quamme, Summerland, B.C., 1984.

[z]J. Siller-Cepeda, INIA, Chihuahua, 1984.

[aa]H. Castro, INTA, General Roca, Rio Negro, 1985.

[bb]A. J. Feliciano, EMBRAPA, Pelatos, 1984.

[cc]G. Gil, Catholic University of Chile, 1984

Clonal rootstocks for pears have been limited primarily to selections of quince *Cydonia oblonga* L. that were developed since 1930. Extensive use of Angers Q and Provence quince PQ for pears are found in Western Europe and on a limited basis in milder regions of Eastern Europe, the United States, South Africa, and Israel. More recently, clonal stocks of *P. communis* have been used in the United States and South Africa, and *P. calleryana* has limited use in Australia.

Rootstocks used in Asia vary by country and latitude. *P. pyrifolia* seedlings are used primarily in Japan for cultivars of the same species. In Taiwan, the evergreen pear *P. kawakamii* was said to be the main rootstock. However, inspection of numerous orchards there indicated that most of the seedling stocks are hybrids between *P. pyrifolia* and *P. kawakamii*. These seem to have arisen from open pollination between domestic *P. pyrifolia* cultivars and native evergreen pear. In south China, *P. calleryana* is used for the *P. pyrifolia* cultivars. In the more northern Chinese regions of Beijing and the Bohai Gulf area, *P. betulifolia* seedlings are used for both *P. pyrifolia* and *P. ussuriensis* hybrid cultivars. In some of the soils of these areas, trees show considerable lime-induced chlorosis. In far northeast China and Manchuria, the hardier, more chlorosis-resistant *P. ussuriensis* stocks are used. *P. pashia* in southwest China and *P. xerophila* in north China are used for rootstocks. In northern India, *P. pashia* and *P. pashia* var. *Kumaoni*, *P. pyrifolia*, and local cultivars are used as a seed source for pear rootstocks.

Pear rootstocks in Europe have been principally domestic or wild seedlings for the colder regions of Eastern Europe, while quince has been used in Italy, Greece, and the milder regions of the United Kingdom and the Netherlands. The development of the Angers Quince A(QA) at East Malling in the 1930s and, later, the selection of PQ in France and Italy were important for the planting of moderate-density orchards in Western Europe. However, most cultivars have poor compatibility with quince, with the exception of 'Passe Crassane,' 'Comice,' and 'Hardy.' Many trees were either propagated with a 'Hardy' interstock or supported by stakes, posts, or trellis wires to improve anchorage and prevent breakage at the union. The QA, QC, and QAdams have been widely planted in the more northern regions of Europe (the United Kingdom, the Netherlands, and Germany). The PQ Lapage C, PQBA29, and various local selections of PQ have been used as the principal rootstocks in Italy, France, Spain, and Greece.

Seedling stocks from domestic varieties are still used widely throughout Europe, but particularly in the eastern regions. Also, scion cultivars are propagated on seedlings from wild pears in Hungary, Romania, and Yugoslavia.

Pears in Asia Minor and the Mediterranean regions have been frequently grafted on wild trees of *P. elaeagrifolia* in Turkey, of *P. syriaca* in Syria and Lebanon, of *P. amygdaliformis* in Yugoslavia, Turkey, and Greece, of *P. salicifolia* in Iran and southern Russia, and of *P. longipes* in Algeria and Morocco. However, most commercial orchards use seedling stock from domestic cultivars or wild trees. Pear orchards in Israel have used QA, QC.1.7, and *P. betulifolia* seedlings, but generally quince rootstocks have been less than satisfactory in the hotter areas of the Mediterranean region.

Current commercial rootstocks in the United States derive mostly from *P. communis*. These rootstocks include 'Williams' ('Bartlett') seedling, 'Winter Nelis' seedling, and clonally propagated 'Old Home' × 'Farmingdale' (OH × F). *P. calleryana* seedling is used in areas with relatively mild winters, such as the southern states, California, and southern Oregon. It is not hardy enough to be used in the northern fruit areas. Also, it is sensitive to calcareous soils and may develop chlorosis. Seedlings of *P. betulifolia* are used to a limited extent in California and southern Oregon, where high vigor is needed in clay soils and where soil drainage is poor. Clonal quince stocks (QA, PQBA29, PQ Lepage C, QC, etc.) also are used to some extent in California and southern Oregon. Such stocks are not hardy and are susceptible to both fire blight and lime-induced chlorosis, but in suitable situations they are precocious and productive.

Canadian growers use almost entirely *P. communis* stocks, primarily 'Bartlett' seedlings. Some efficient growth-controlling OH × F clones are now being used to a limited extent. *P. calleryana*, *P. betulifolia*, and quince are not hardy to most pear districts there. Mexican pear production is grown primarily on 'Bartlett' seedling.

In South Africa, the principal pear stocks used have been 'Williams' seedling and QA. Several *P. communis* clonal stocks have been developed for pear at Stellenbosch: BP-1 (dwarfing) and BP-3 (standard). However, the BP-3 has been difficult to propagate. Australia's pear industry used French seedlings (*P. communis*) before 1930, followed by Kieffer seedlings until the mid-1950s, and then *P. calleryana* seedling and the D6 clone of *P. calleryana* on clay soils and 'Winter Nelis' seedling on lighter soils. Most pear trees in Australia are on the 'Kieffer' seedling rootstock. Trees in New Zealand are principally on 'Williams' seedling and to a lesser degree on QA and QC.

The rootstock practice in Argentina, Chile, and Brazil has been the use of seedling 'Williams' and 'Winter Nelis,' but some orchards have been recently planted on QA and PQBA29. However, considerable tree failure has occurred with the quince rootstocks even though compatible interstems 'Hardy' and OH were used. Some pear trees have been propagated on *P. calleryana* and *P. betulifolia* seedlings in Chile.

3. ROOTSTOCK ADAPTATIONS AND PERFORMANCE

Adaptation of pear rootstocks of the world to climates, soils, and pests are shown in Tables 3 and 4.

In most cases there is an inverse relationship between adaptation to a hot climate and to cold winters. There is a similar inverse relationship between tolerance to low and high pH. Some stocks, however, show a balanced toler-

TABLE 3. Pear Rootstock Adaptation to Temperature Extremes and Soil Conditions[a]

	Temperature Adaptation		Soil Adaptation and Tolerance					
	Warm Winter Hot Summer[b]	Winter Cold[c]	Low pH[d]	Chlorosis[e]	Wet[f]	Dry[g]	Sandy[h]	Clay[i]
European								
P. caucasica	2	4	4	4	4	4	4	4
P. communis								
seedling								
Kirschensaller	4	4	3	3	4	3	4	4
Williams (Bartlett)	4	4	3	3	4	3	4	4
Winter Nelis	4	3	3	3	4	3	4	4
clonal								
Anjou	4	4	3	3	4	3	4	4
Bartlett	4	4	3	3	3	3	4	2
Old Home (OH)	4	5	3	3	4	3	4	4
OH × Farmingdale (F)								
OH × F 51	3	3	3	3	3	3	4	4
OH × F 34	3	4	3	3	3	3	4	4
OH × F 69	3	4	3	3	3	3	4	4
OH × F 87	3	4	3	3	3	3	4	4
OH × F 230	3	4	3	3	3	3	4	4
OH × F 333	3	4	3	3	3	3	4	4
OH × F 217	3	4	3	3	3	3	4	4
OH × F 267	3	4	3	3	3	3	4	4
OH × F 361	3	4	3	3	3	3	4	4
OH × F 18	3	4	3	3	3	3	4	4
OH × F 97	3	4	3	3	3	3	4	4
OH × F 112	3	4	3	3	3	3	4	4
OH × F 198	3	4	3	3	3	3	4	4
Oregon 1	4	4	3	3	4	4	4	4

Species							
P. cordata	4	2	2	4	4	4	1
P. nivalis	2	4	2		2	4	
Circum-Mediterranean							
P. amygdaliformis	5	2	2	5	2	4	3
P. elaeagrifolia	2	4	3	5	3	4	4
P. gharbiana	3	2		4		4	
P. longipes	3	4		4		4	
P. mamorensis	3	2	4			3	
P. syriaca	3	2	2	4	4	3	
Mid-Asia							
P. pashia	5	1	5	1		4	4
P. regelii		3		3		3	
P. salicifolia		3	3	3	3	3	
East Asia							
P. betulifolia							
seedling							
Talent (Oregon)	4	4	4	2	5	4	5
Italian	4	4	4	2	5	4	5
clonal							
Oregon 260, 261, 264	4	3	4	2	5	4	5
P. calleryana							
seedling	5	2	4	2	5	5	5
clonal							
Oregon 211	5	2	4	2	4	5	5
Oregon 249	5	2	4	2	4	5	5
D6	5	2	4	2	4	3	4
P. dimorphophylla	2	3	4	2	4		
P. fauriei	4	4	4	2	2	3	5
P. hondoensis	1	3	3	3			
P. pseudopashia	2	4					
P. kawakamii	5	1	3	3	3	3	3
P. pyrifolia	1	3	3	2	2	4	2
P. ussuriensis	2	5	3	2	3	3	3

(continued)

TABLE 3. (*Continued*)

	Temperature Adaptation			Soil Adaptation and Tolerance				
	Warm Winter Hot Summer[b]	Winter Cold[c]	Low pH[d]	Chlorosis[e]	Wet[f]	Dry[g]	Sandy[h]	Clay[i]
Related Genera								
Amelanchier species		4	3	3			3	3
Crataegus species								
OH/Crataegus		4	3	3	3			4
Cydonia oblonga (Q)								
Provence Q (PQ) Lapage C	2	2	4	1	5	2	3	5
PQBA 29	2	2	4	1	5	2	3	5
QA	1	2	3	1	4	2	3	5
QC	1	2	3	1		2	3	5
Adams Q	1	2	3	1		2		
Pillnitz Q		2	3	1		2		
Palestine Q		2	3	1		2		
Caucasian Q		3	3	1		2		
Fontenay Q		3	3	1		2		
Hardy/Q	2	3	3	1		2		
OH/Q	3	3	3	1		2		
Malus species								
Winter Banana/M26	2	4	3	4	1	2	4	1
Sorbus species		4	2	3			3	4

[a] Rating from 1–5: 1 = susceptible, 2 = low tolerance, 3 = moderate tolerance, 4 = high tolerance, 5 = very tolerant. Based on data from field plots in the United States (2, 25, 29, 39–41).

[b] Warm winter, hot summer; sunburning or poor tree survival in hot temperatures, rating of 1.

[c] Winter cold: no survival at -20°C is rated 1, survival at -30°C is rated 4, survival at -40°C is rated 5.

[d] Low pH, acid soils: limited root and shoot growth at pH 4 is rated 1–2, good–excellent growth at pH 4 are rated 4–5.

[e] Chlorosis or high pH: chlorosis symptoms and poor growth in lime-induced chlorosis, pH 7.5–8.5 are rated 1–3.

[f] Wet soils–high water table: low survival and growth is rated 1–2, fair–excellent survival and growth are rated 3–5.

[g] Dry–shallow soils: poor growth is rated 1, excellent growth is rated 5.

[h] Sandy soils: moderate growth and production are rated 4–5, excessive growth is rated 3.

[i] Clay soils: adequate growth is rated 4–5, low growth is rated 1–2.

TABLE 4. Pear Rootstock Resistance to Diseases and Insect Pests[a]

	Diseases								To Pests	
	Pear Decline[b]	Fire Blight[c]	Bacteria Canker[d]	Leaf Spot[e]	Powdery Mildew[f]	Crown Gall[g]	Collar Rot[h]	Oak Root Fungus[i]	Wooly Aphid[j]	Root Lesion Nematode
European										
P. caucasica	3	2	3	3	3	2	3	5	3	1
P. communis										
seedling										
Kirschensaller	4	1	3	3	3	2	3	4	2	1
Williams (Bartlett)	4	1	3	3	3	2	3	4	2	1
Winter Nelis	4	1	3	3	3	2	3	4	2	1
clonal										
Anjou	5	3	4	3	3	2	3	4	2	1
Bartlett	5	1	3	3	3	2	3	4	2	1
Old Home (OH)	5	5	2	3	3	2	3	4	2	1
OH × Farmingdale (F)										
OH × F 51	5	5	4	3	3	3	3	4	2	1
OH × F 34	5	5	4	3	3	3	3	4	2	1
OH × F 69	5	5	4	3	3	3	3	4	2	1
OH × F 97	5	5	4	3	3	3	3	4	2	1
OH × F 230	5	5	4	3	3	3	3	4	2	1
OH × F 333	5	5	4	3	3	3	3	4	2	1
OH × F 217	5	5	4	3	3	3	3	4	2	1
OH × F 267	5	5	4	3	3	3	3	4	2	1
OH × F 361	5	5	4	3	3	3	3	4	2	1
OH × F 18	5	5	4	3	3	3	3	4	2	1
OH × F 97	5	5	4	3	3	3	3	4	2	1
OH × F 112	5	5	4	3	3	3	3	4	2	1
OH × F 198	5	5	4	3	3	3	3	4	2	1
Oregon 1	5	2	4	3	3	2	3	4	2	1
P. cordata	3	1		3	4				1	3
P. nivalis	4	2	3		4				4	

(continued)

159

TABLE 4. (*Continued*)

	Diseases								To Pests	
	Pear Decline[b]	Fire Blight[c]	Bacteria Canker[d]	Leaf Spot[e]	Powdery Mildew[f]	Crown Gall[g]	Collar Rot[h]	Oak Root Fungus[i]	Wooly Aphid[j]	Root Lesion Nematode
Circum-Mediterranean										
P. amygdaliformis	4	2	4	4	5				3	2
P. elaeagrifolia	4	2	4	4	4		4		4	2
P. gharbiana		1	4							1
P. longipes	3	1								
P. mamorensis		1							2	
P. syriaca	3	2	4	4	4		3		4	4
Mid-Asia										
P. pashia	4	2	4	1	3		3		3	2
P. regelii		2	2							
P. salicifolia		2	2							
East Asia										
P. betulifolia										
seedling										
Talent (Oregon)	5	3	4	4	4	4	4	4	4	5
Italian	5	1	4	4	4	4	4	4	4	5
clonal										
Oregon 260, 261, 264	5	3	4	4	4	4	4	4	4	5
P. calleryana										
seedling	4	5	3	5	4	4	4	4	5	5
clonal										
Oregon 211	4	5	3	5	4	4	4	4	5	5
Oregon 249	4	5	3	5	4	4	4	4	5	5
D-6	3	5	3							
P. dimorphophylla	4	3	3	4	5		4		1	3
P. fauriei	3	4	3		4	3	4		2	3
P. hondoensis										
P. pseudopashia			4							

Species/Rootstock	Pear decline[b]	Fire blight[c]	Bacterial canker[d]	Leaf spot[e]	Powdery mildew[f]	Crown gall[g]	Collar rot[h]	Oak root fungus[i]	Wooly pear aphid[j]
P. kawakamii	3	4	3		3		1	1	
P. pyrifolia	1	3	3	2	4		1	1	2
P. ussuriensis	1	5	3	2	4		1	4	4
Related Genera									
Amelanchier species	3	4	4	4	5	4	4		5
Crataegus species									
OH/Crataegus	5	4	4	4	4	4	4		5
Cydonia oblonga(Q)									
Provence Q Lapage C	5	1	3	1	5	5	4	3	4
PQBA 29	5	1	3	1	5	5	4	3	4
QA	5	1	3	1	5	5	4	3	4
QC	5	1	3	1	5	5			
Adams Q	5	1	3	1	5	5			
Pillnitz Q	5	1	3	1	5	5			
Palestine Q	5	1	3	1	5	5			
Caucasian Q	5	1	3	1	5	5			
Fontenay Q	5	1	3	1	5	5			
Hardy/Q									
OH/Q									
Malus species									
Winter Banana/M26		2	5	4	4	4	2		
Sorbus species	5	4	4	4	4	4	4		5

[a] Rating from 1–5; 1 = susceptible, 2 = low tolerance, 3 = moderate tolerance, 4 = high tolerance, 5 = very tolerant. Based on data from research in the United States since 1911 (4, 26, 27, 30, 32, 38, 39, 41).

[b] Pear decline, a mycoplasma-caused graft union failure.

[c] Fire blight, *Erwinia amylovora* (Bur.) Winslow et al.

[d] Bacterial canker, *Pseudomonas syringae* van Hall.

[e] Leaf spot, *Fabraea maculata* Atk.

[f] Powdery mildew, *Podosphaera leucotricha* (Ell. & Ev.) Salm.

[g] Crown gall, *Agrobacterium tumefaciens* (E. F. Sm. & Towns) Conn.

[h] Collar rot, *Phytophthora cactorum* (Leb. & Cohn) Schroet.

[i] Oak root fungus, *Armillaria mellea* (Jahl) Quel.

[j] Wooly pear aphid, *Eriosoma pyricola* Baker & Davidson.

ance to extreme conditions. For each climate and soil, certain critical genetic traits are required for success. The choice of a rootstock suitable to a given set of conditions is crucial, yet obviously no single "best" rootstock is found that is suitable to all conditions where pears are grown. Thus a number of superior stocks are sought, each suited to a particular set of conditions. In cases where a lethal disease, such as pear decline, is related to rootstock, there is an absolute requirement for resistance. Tables 3 and 4 should be studied to determine the desired combination of traits for a give location.

The size of trees on a given rootstock determines in part the spacing and the training system to be used. As shown in Table 5, trees may range in size from 5 to 130 % of standard vigorous trees. The very dwarfed trees must be planted at high densities in order to achieve maximum yield. Poorly anchored but good rootstocks require trellises or other support systems to assume good performance. A study of Table 5 will indicate which stocks have the right combination of traits for a given situation.

Pear decline is caused by a mycoplasma organism which is transmitted to the tree by pear psylla insects feeding on the leaves. The disease at present seems to be limited to western North America and parts of Europe, but it could spread to any region where the psylla are found. The expression of the disease requires, in addition to the pathogenic organism and the insect, a tolerant cultivar and a susceptible rootstock. The insect injects the mycoplasma into the leaf of the tolerant cultivar, where it multiplies and migrates in the phloem conducting system to the root. As the organism moves across the graft union from tolerant to susceptible tissue, the rootstock phloem is killed, resulting in the lower trunk being girdled. The root starves and dies and the top then declines or dies. Partially susceptible rootstocks result in a partial girdling which causes the root to be weakened. The top then declines and is weak but does not die. Such trees may be brough back to health by annual injections of tetracycline. This is an expensive alternative to the use of resistant rootstocks. As shown in Table 3, many of the rootstocks are resistant to decline. Most seedling stocks, however, will have a proportion of susceptible individuals. Selected clonal stocks are uniformly resistant. The most susceptible rootstocks are *P. ussuriensis* and *P. pyrifolia*, so they are not recommended where decline occurs or where it might occur in the future. However, trees on *P. betulifolia* are the least susceptible.

Uptake of mineral nutrients from the soil is a function of both soil factors and rootstocks genetics. The genetic effects are shown in Table 6 in a slightly acid sandy loam soil in western Oregon, with 'Williams' as the scion cultivar (10). In different soils and climates, nutrient uptake would be different. For example, *P. pashia*, *P. fauriei*, *P. betulifolia*, and *P. calleryana* are not tolerant to high-pH soils and become chlorotic (Table 3), while *P. elaeagrifolia* and *P. amygdaliformis* are healthy and thrive in soils with pH 7.5 to 8.0. Rootstocks such as *Sorbus* species (mountain ash) and *Amelanchier* species, (service berry) which take up relatively large amounts of Mn, could cause Mn toxicity in low-pH soils with abundant available Mn.

TABLE 5. Pear Tree Performance as Influenced by Rootstocks[a]

	Management			Tree Size		Yield		Fruit Quality			
	Root-sprouts[b]	Anchor-age[c]	Compat-ibility[d]	% of Standard[e]	Uni-formity[f]	Pre-cocity[g]	Effi-ciency[h]	Size[i]	Black End[j]	Cork Spot[k]	Over-all[l]
European											
P. caucasica	2	5	5	80	3	3	4	3	No		4
P. communis											
seedling											
Kirschensaller	1	5	5	90	4	3	4	3	No		
Williams (Bartlett)	2	5	5	90	4	3	4	3	No	3	4
Winter Nelis	2	5	5	90	4	3	4	3	No	3	4
clonal											
Anjou	0	5	5	90	5	3	5	3	No		4
Bartlett	0	5	5	50	4	3	4	3	No		4
Old Home (OH)	5	5	5	100	5	1	1	3	No	5	3
OH × Farmingdale (F)											
OH × F 51	1	5	5	60	5	3	4	3	No	4	4
OH × F 34	1	5	5	70	5	3	4	3	No	4	4
OH × F 69	1	5	5	70	5	3	5	3	No	4	4
OH × F 87	1	5	5	70	5	3	5	3	No	4	4
OH × F 230	1	5	5	70	5	3	5	3	No	4	4
OH × F 333	1	5	5	70	5	3	5	3	No	4	4
OH × F 217	1	5	5	80	5	3	5	3	No	4	4
OH × F 267	1	5	5	80	5	3	5	3	No	4	4
OH × F 361	1	5	5	80	5	3	4	3	No	4	4
OH × F 18	1	5	5	100	5	3	4	3	No	4	4
OH × F 97	1	5	5	100	5	3	4	3	No	4	4
OH × F 112	1	5	5	100	5	3	4	3	No	4	4
OH × F 198	1	5	5	100	5	3	4	3	No	4	4
Oregon 1	1	5	5	100	5	3	5	5	No	4	4
P. cordata	1	4	5	70	3	3	3	3	No		3
P. nivalis	1	5	5	90	3	3	3	3	No	3	3

(continued)

163

TABLE 5. (Continued)

	Management			Tree Size		Yield		Fruit Quality			
	Root-sprouts[b]	Anchor-age[c]	Compat-ibility[d]	% of Standard[e]	Uni-formity[f]	Pre-cocity[g]	Effi-ciency[h]	Size[i]	Black End[j]	Cork Spot[k]	Over-all[l]
Circum-Mediterranean											
P. amygdaliformis	1	5	5	80	2	3	3	4	No		
P. elaeagrifolia	1	5	5	100	3	3	4	4	No		
P. gharbiana											
P. longipes		5	5	70	4	3	3	3			
P. mamorensis		5	5								
P. syriaca	3	5	5	30	1	3	2	3			
Mid Asia											
P. pashia	2	5	5	90	3	3	3	4			
P. regelii											
P. salicifolia											
East Asia											
P. betulifolia											
seedling											
Talent (Oregon)	0	5	5	130	5	3	3	5	No	2	4
Italian	0	5	5	130	5	3	3	5	No	2	4
clonal											
Oregon 260, 261, 264	0	5	5	110	5	3	4	5			4
P. calleryana											
seedling	1	5	5	90	3	3	4	4	No	3	4
clonal											
Oregon 211	0	4	5	15	4	3	4	4			
Oregon 249	0	4	5	25	4	3	4	4			
D-6	0	3	4	70	4	3	4	4	No		5
P. dimorphophylla	1	5	5	90	3	3	4	4			
P. fauriei	4	2	5	40	4	3	2	4	No		
P. hondoensis											
P. pseudopashia											

	[b]	[c]	[d]	[e]	[f]	[g]	[h]	[i]	[j]	[l]
P. kawakamii	0	5	0	70	3	3	4	4	Yes	3
P. pyrifolia	0	5	5	100	4	3	4	4	Yes	3
P. ussuriensis		5	5			3	4	4		
Related Genera										
Amelanchier species	0	1	2	15	4	5	5	4	No	4
Crataegus species	2	2	2	45	3	4	4	3	No	4
OH/Crataegus	3	3	4	90	3	4	4	3	No	4
Cydonia oblonga (Q)										
Provence Q Lapage C	3	1	2	55	5	4	5	4	No	5
PQBA 29	3	1	2	55	5	4	5	4	No	5
QA	4	1	2	50	5	4	5	3	No	5
QC	3	1	2	50	5	4	5	3	No	5
Adams Q		1	2	50	5	4	5			5
Pillnitz Q	3	1	2	50	5	4	5	3		5
Palestine Q	3	1	2	50	5	4	5	3		5
Caucasian Q	3	1	2	50	5	4	5	3		5
Fontenay Q	3	1	2	50	5	4	5	3		5
Hardy/Q	4	2	3	50	5	4	5	3	No	5
OH/Q	3	3	3	60	5	3	4	3	No	5
Malus species										
Winter Banana/M26	2	2	3	5	3	3	2	4		4
Sorbus species	2	3	2	15	3	2	1	2	No	4

[a] Rating 0-5. Based on data from research in the United States on several *P. communis* cultivars (13, 20–22, 34, 39–42).

[b] Rootsprouts, sprouts from the ground: 0 = none, 5 = many.

[c] Anchorage: requires support is rated 1, support suggested is rated 2.

[d] Compatibility: requires compatible interstem is rated 1, interstem suggested is rated 2.

[e] Standard tree: 100%.

[f] Uniformity: tree growth variability.

[g] Precocity, time required to come into bearing: more than 10 years = 1, less than 5 years = 5.

[h] Yield efficiency: yield per surface area of mature tree, based on trunk cross-sectional area.

[i] Fruit size: in diameter, relative size within a cultivar.

[j] Black end: physiological disorder of calyx end.

[k] Cork spot: Ca-physiological disorder causing fruit pitting similar to bitter pit.

[l] Overall quality: flavor, texture, storage and shelf life of fruit.

TABLE 6. Relative Nutrient Uptake of the Scion Cultivar by Various Pear Rootstocks[a]

Nutrient	High Uptake	Low Uptake
N	*P. amygdaliformis* *P. eleaegrifolia*	*Crataegus* spp. *Cydonia oblonga* *P. kawakamii*
P	*P. amygdaliformis* *P. eleaegrifolia* *P. pashia*	*Crataegus* spp. *P. kawakamii*
K	*P. pashia* *Sorbus* spp.	*P. amygdaliformis*
Ca	Old Home OH x F clones *P. betulifolia*	*P. cordata* *P. calleryana*
Mg	*Cydonia oblonga* *Sorbus* spp.	
Mn	*Sorbus* spp. *P. fauriei* *Amelanchier* spp.	
Fe	*P. amygdaliformis* *P. eleaegrifolia* *P. syriaca* *P. pashia*	*P. kawakamii* *Cydonia oblonga* *Crataegus* spp. *P. calleryana* *P. dimorphophylla*
B	*P. betulifolia* *P. fauriei* *P. pashia* *P. ussuriensis* *P. eleaegrifolia*	*P. amygdaliformis* *P. syriaca* *P. cordata* *Cydonia oblonga*
Zn	*P. eleaegrifolia* *P. betulifolia* *P. communis*	*P. cordata* *P. kawakamii*

[a]Based on data from Chaplin and Westwood, (10).

Effect of mineral uptake on pear fruit quality in respect to rootstock has received nominal study (14). Cork spot of 'Anjou' pear (similar to bitter pit of apple) is greatly reduced with rootstocks such as 'Old Home', which result in high levels of Ca in the scion cultivar (21).

The use of interstem with pear may be used to protect against fire blight or to increase the hardiness of the lower trunk. But usually pear interstems are used to provide graft compatibility with quince stocks. 'Abbe Fetel,' 'Anjou,' 'Comice,' 'Passe Crassane,' 'Hardy,' 'Clapp Favorite,' 'Flemish Beauty,' and 'Maxine,' and to a lesser degree 'B. P. Moretini,' 'B. Esperen,' 'Conference,' 'P. Trevoux,' and 'Santa Maria' are common cultivars compatible with quince. 'Williams,' 'Bosc,' 'B. Gifford,' 'Coscia,' 'Dr. J. Guyot,' 'Laxton's Superb,' 'Starkrimson,' 'Winter Nelis,' 'El Dorado,' and 'Forelle' are at least partly incompatible with quince and thus need an interstem (3). Also, most Asian pears are incompatible with quince. 'Old Home' and 'Hardy' are the most commonly used compatible interstems for both European and Asian cultivars. 'Old Home' also can be used as an interstem for *crataegus* (hawthorn) rootstocks, but it is incompatible with mountain ash. (34).

There is evidence that virus content can affect compatibility of the scion on the quince root (28). Viruses affecting the compatibility on either Provence or Angers quince are pear vein yellows, pear ring pattern mosaic, and quince sooty ring spot. These viruses also decreased tree growth, but yield efficiency of the 'Williams' was similar to the virus-free trees.

Limited tests (34) on compatibility of pear on apple rootstocks indicate some potential for producing very small trees. The most compatible combinations for 'Williams' and 'Comice' scions have been M26 with a 'Winter Banana' interstem. Because of variations among seedling populations in both tree size and compatibility, it is best to select specific rootstock clones when genera other than *Pyrus* are used as pear stocks.

Precocity and ultimate yield and yield efficiency are usually correlated (Table 5) and these factors are related both to rootstock and to cultivar. Dwarfing stocks are not necessarily related to precocity and high yield efficiency, nor are vigorous stocks related to lack of precocity and low efficiency. For example, *P. fauriei* is dwarfing but has a low yield efficiency, while *P. betulifolia* is vigorous yet has a high efficiency. The selection of clonal stocks should take precocity and yield potential into account, along with other desirable traits. One of the first clonal pear rootstocks used in the United States was 'Old Home.' It was known to be blight resistant, easy to root from cuttings, and resistant to pear decline. It was discontinued as a clonal stock, however, because fruit productivity was consistently low (21).

Management programs for pear rootstocks vary with the weaknesses and needs of each rootstock. Poorly anchored stocks do best in supported tree walls. Stocks that tend to have root sprouting do best with nontillage culture. Irrigation and soil moisture should be matched with the tolerance of the stock to either wet or dry conditions. Fertilizers and liming should be adjusted to the pH and nutrient needs of the stock.

4. ROOTSTOCK CHARACTERIZATION

The general descriptions of pear rootstocks are given in Tables 7, 8, and 9.

Botanical descriptions are available for each species, but these are not detailed enough to describe specific clones. It is important to develop identification of clones from vegetative traits. Chemical "fingerprinting" would be useful but as yet has not been developed for pear. Authentic labeling and

TABLE 7. Leaf Characteristics of Pyrus Species for Rootstock Identification[a]

	Adult Leaf Margins	Juvenile Leaves	Adult Shoot Leaves		
			Length (cm)	Length/ Width Ratio	Length/ Petiole Ratio
Europe					
P. caucasica	Crenate	Simple	4.81	1.28	2.28
P. communis	Crenate	Simple	4.95	1.57	2.40
P. cordata	Crenate	Simple	4.37	1.40	3.06
P. nivalis	Entire or Crenate	Simple	7.87	1.79	3.93
Circum-Mediterranean					
P. amygdaliformis	Entire or sl. crenate	1-Lobed	6.52	2.88	3.20
P. elaeagrifolia	Entire or sl. crenate	Simple	5.70	2.24	4.01
P. gharbiana	Crenate	Simple	5.15	1.51	2.35
P. longipes	Crenate	Simple	3.77	1.30	1.92
P. mamorensis	Crenate	Simple	4.58	1.13	2.08
P. syriaca	Crenate	Simple	4.78	2.27	4.47
Mid-Asia					
P. pashia	Crenate	2-Lobed	7.99	1.55	2.87
P. regelii	Crenate	Laciniate	5.80	0.90	2.30
P. salicifolia	Entire	Simple	5.10	4.50	8.70
East Asia					
P. betulifolia	Coarse, serrate	1-Lobed	5.74	1.53	2.58
P. calleryana	Crenate	Simple	6.60	1.50	2.73
P. dimorphophylla	Crenate, serrate	2-Lobed	7.16	1.85	3.60
P. fauriei	Crenate	Simple	4.35	1.38	2.90
P. hondoensis	Fine serrate, setose	Simple	7.57	1.45	2.79
P. pseudopashia	Crenate	1-Lobed	6.42	2.04	2.77
P. kawakamii	Crenate	3-Lobed	6.14	1.47	3.07
P. pyrifolia	Fine serrate, setose	Simple	10.65	1.84	3.70
P. ussuriensis	Coarse serrate, setose	Simple	8.23	1.61	3.02

[a]From Challice and Westwood (9).

trueness to name are of utmost importance for clonal stocks. The penalty for mislabeling rootstocks can be as great as that of mislabeling the cultivar.

Certification of nursery trees should include virus status as well as cultivar and rootstock identity. The most important viruses of pear are vein yellows

TABLE 8. Fruit characteristics of Pyrus Species for Rootstock Identification[a]

	Skin	Stone Cells	Calyx Type	Fruit Diam. (cm)	Carpel No.	Seed Wt. (mg/seed)
Europe						
P. caucasica	Smooth	+	Persistent	4.70	5	35
P. communis	Smooth (some russet)	+	Persistent	3.80	5	26
P. cordata	Russet	+	Deciduous	1.50	2–3	12
P. nivalis	Smooth	+	Persistent	5.60	5	47
Circum-Mediterranean						
P. amygdaliformis	Smooth	+	Persistent	2.60	5	43
P. elaeagrifolia	Smooth	+	Persistent	2.40	5	38
P. gharbiana						
P. longipes	Russet	+	Deciduous (some persistent)	1.70	3–4	21
P. mamorensis						87
P. syriaca			Persistent		5	100
Mid-Asia						
P. pashia	Russet	++	Deciduous	2.39	4–5	17
P. regelii	Smooth	+	Persistent	2.0–3.0	5	168
P. salicifolia	Smooth	+	Persistent	1.90	5	32
East Asia						
P. betulifolia	Russet	0	Deciduous	0.88	2 (3)	11
P. calleryana	Russet	++	Deciduous	1.06	2	18
P. dimorphophylla	Russet	0	Deciduous	1.24	2	13
P. fauriei	Russet	0	Deciduous	1.34	2	17
P. hondoensis	Smooth	+	Persistent	2.80	5	28
P. pseudopashia	Smooth	+	Persistent	4.26	5	
P. kawakamii	Russet	++	Deciduous	1.04	4 (3)	13
P. pyrifolia	Russet	+	Deciduous	4.10	5	40
P. ussuriensis	Smooth	+	Persistent	3.75	5	48

[a]From Challice and Westwood (9).

169

TABLE 9. Phenological and Stem Characteristics of Pyrus Species[a]

	Tree Size (sdlg. tree in natural habit) (height)	Bloom Period (relative to 'Williams')	Late Season Growth	Xylem/Phloem Ratio 1st Year	2nd Year
Europe					
P. caucasica	Large, 5–6 m	Same–variable	None		
P. communis	Large, 5–6 m	Same–variable	None	1.35	2.44
P. cordata	Small–Med., 2–3 m	Sl. later	None	1.12	2.62
P. nivalis	Med., 3–4 m	Later	None	0.79	2.13
Circum-Mediterranean					
P. amygdaliformis	Small, 1-2 m (or shrub)	Same	Active	1.03	1.77
P. elaeagrifolia	Med., 3–4 m	Same	None	1.08	1.83
P. gharbiana			Some		
P. longipes	Med., 3–4 m	2 Wks. after	Some	1.73	3.25
P. mamorensis			Some		
P. syriaca	Small, 1–2 m		Active	0.83	3.21
Mid-Asia					
P. pashia	Med., 3–4 m	Same	Active	1.22	2.70
P. regelii	Small, 1–2 m (or shrub)	1–2 Wks. after	Some		1.65
P. salicifolia	Small, 1–2 m	Same	None	0.85	1.78
East Asia					
P. betulifolia	Large, 5–6 m	Same or sl. later	Some	1.57	2.64
P. calleryana	Med., 3–5 m	2–3 Wks. before	Active	1.63	3.18
P. dimorphophylla	Med., 3–4 m	10–14 Days before	Some	1.93	4.00
P. fauriei	Small, 1–2 m (or shrub)	Same	None	1.60	2.40
P. hondoensis	Med., 3–4 m	Same	Some	1.54	3.40
P. pseudopashia	Med., 3–4 m	Same	Active	1.06	2.75
P. kawakamii	Sm.–Med., 1–3 m	2–3 Wks. before	Active	1.43	1.80
P. pyrifolia	Med–large, 3–5 m	1–5 Days before	Some	1.45	2.45
P. ussuriensis	Sm.–med., 1–3 m	3 Wks. before	None	1.44	2.82

[a]From Challice and Westwood (9).

(PVYV) and stony pit (PSPV). Both are transmitted only by budding and grafting.

Vein yellows can be detected by bud inoculation into the indicator B-13 or other sensitive cultivars. PVYV appears to cause little reduction in performance of pear trees except in specific cultivars growing under favorable conditions (16). Fridlund states that performance of some rootstocks apparently is affected by PVYV, so it should be eliminated from all scion and rootstock cultivars, particularly in the case of quince, as noted earlier. Stony pit is detected by fruit symptoms on the 'Bosc' pear, which is used as an indicator. Rootstocks may carry these viruses even though they do not shown symptoms.

Other viruses reported in Europe (16) are: pear ring pattern mosaic virus (PRPMV), which does not affect yield but may cause dark rings on fruit of sensitive cultivars; apple stem grooving virus (ASGV), which is infrequently found and may not affect yields; and apple proliferation (APV) and apple rubbery wood virus (ARWV), the former causing witches broom syndrome in pear and unkown yield effect while the latter causes no symptoms in pear. Pear bark measles or pear blister canker in both North America and Europe (16) have been reported to be associated with 'Old Home' interstocks in California.

Pear rootstock propagation involves three standard techniques: (a) sexual seed germination, (b) trench or mound layerage for quince, and (c) hardwood stem cuttage (15). In certain cases root cuttings or root suckers can be used for difficult-to-root types. Leafy cuttings can be rooted under mist after an IBA treatment.

Seed preparation of most *Pyrus* for germination requires thorough washing of a water soak of one day and stratification for chilling. Chilling requirements for several *Pyrus* species and related genera are listed in Table 10. Germination may take 5 to 30 days at 20°C.

TABLE 10. Chilling Requirements of Several *Pyrus* and Related Genera for Seedling Rootstock Propagation[a]

Species	Seed Size (no. per g)	Chilling Req. (days)	Optimal Chilling Temperature (°C)
P. amygdaliformis	24	25	7
P. betulifolia	90	55	5
P. calleryana	55	30	7
P. communis	22	90	5
P. dimorphophylla	77	65	5
P. elaeagrifolia	22	90	5
P. fauriei	57	35	7
P. gharbiana	99	60	5
P. mamorensis	11	50	5
P. nivalis	18	110	5
P. pashia	55	15	10
P. pyrifolia	26	120	5
P. ussuriensis	20	80	7
Amelanchier alnifolia	112–250	180	3
Crataegus mollis	–	90	5
Malus sylvestris	21–35	100	4
Sorbus americana	185–520	90	5
Sorbus aucuparia	230–375	90	1
Sorbus decora	280	90	5

[a]From Westwood (33) and Westwood and Bjornstad (36).

Hardwood stem cuttage has been used successfully for various *P. communis* stock such as 'Old Home' and the OH × F clones. Cuttings should be taken in the fall prior to the end of rest (1) and the basal portion treated with an IBA soak of 25–250 ppm for 24 hr (17) or a 5-sec dip of IBA of 1000–5000 ppm (37). Bottom heat in the propagation bed can improve rooting.

For quick propagation of a stock, root cuttings of about .7–1.5 cm diameter and 10–15 cm long can be either cleft grafted with or without a scion variety, basally treated with an IBA dip, and planted in the nursery for further rooting. Scion rooting of a cultivar can be made with a hardwood cutting or a leafy cutting. Certain species such as *P. calleryana* and *P. betulifolia* can be best scion rooted with a leafy cutting taken in midsummer.

Rooting under mist was more successful than hardwood cutting propagation for the Circum-Mediterranean, Mid-Asian, and East Asian species (31). Hartmann et al. (17) rooted 'Old Home' and 'Bartlett' softwood cuttings successfully with relatively high concentration of IBA (6000–10,000 ppm). Hardwood cuttings of 'Old Home' had to be taken in late October and soaked with IBA at 100 ppm for 24 hr followed by storage in wet moss 21°C for 3 weeks before rooting in a nursery row. Propagation of OH × F 51 by tissue culture technique has been accomplished by Cheng (11).

5. NEW AND FUTURE ROOTSTOCKS

The *Pyrus* species and related genera posses the genetics for resistance to the major pests and tolerance to soil and temperature conditions. Also, there are genetic traits among these as rootstocks to influence chilling requirement of the bud, regulate bloom date, control tree size, fruit size and quality, precocity, yield efficiency, and nutrient uptake. Because of this variability, it is possible to breed and develop a variety of rootstocks for use in special situations or for general usage.

The use of clonal stock has generally reduced tree-to-tree variability. Some believe that because of the uniformity, there lies a danger in widespread loss of trees on a susceptible rootstock to a new disease such as pear decline. However, one should not put all hope on a universal rootstock but should consider developing several clonal stocks that can serve special purposes. The extra cost of clonal stock because of additional propagation expenses can, though, be a detriment to its development.

Present research plans at various centers are involved with the evaluation of the effect of rootstock on tree performance, for example, size control and yield and/or their development of quince clones in Europe; of *P. communis* in France, Netherlands, the United Kingdom, Canada, New Zealand, South Africa, and the United States; of *P. calleryana* in Brazil, Chile, South Africa, India, and the United States; of *P. betulifolia* in Israel, the United States, Chile, and South Africa; of *P. pyrifolia* in Japan, Korea, and New Zealand; and of other species or related genera in the United States and India (Table 11).

TABLE 11. Pear Rootstock Research and Goals

Region and Country	Principal Research and Goals	Personnel and Location
Africa		
Egypt	Selection of 'Le Conte' seedlings for fire blight–resistant clones	A-Talaat El-Wakeel, Horticulture Research, Institute, Cairo (Giza)
South Africa	Release of *P. communis* clones in 1973 BP-1 (semidwarf, precocious) patented BP-2 (standard, precocious, virus infected) BP-3 (standard, precocious) patented Comments: Difficult to root, tolerant to waterlogging, no black end fruit Goal: (1) Evaluate self-rooted trees, *P. communis*, *P. calleryana*, and *P. betualifolia*, OH × F; (2) improve rooting techniques; (3) determine planting densities	P. van Huyssteen, researcher H. J. van Zyl J. D. Stadler Fruit and Fruit Technology Research Institute Private Bag X5013 Stellenbosch 7600
Asia		
China	Breeding dwarf rootstock	Wang Yu-lin Zhengzhou Fruit Research Institute Chinese Academy of Agricultural Science Zengzhou, Honan People's Republic of China
India	Compare performance of *P. calleryana*, *P. pashia*, var. *Kumaonii*, *P. pyrifolia*. Develop dwarfing rootstocks resistant to powdery mildew and fire blight	S. S. Randhawa, researcher Indian Agricutural Research Institute Regional Station, Amartara Cart Road, Simla-4
Japan	Breeding and selection of clonal stock for dwarfing, hard-end resistance, and compatibility to Japanese pears	Katsumi Kumashiro, Fac. of Ag. Shinshu Univ., Ina, Nagano-Ken Osamu Kishimoto, Fac. of Ag. Utsunomiya Univ. Utsunomiya 321
Korea	Breeding and selection for dwarfing and compatibility with cvs. of *P. pyrifolia* and *P. ussuriensis*	Hyun-Mo Cho, Yong-Uk Shin, Beong-Woo Yae, Jung-Ho Kim, researchers Horticultural Exp. Station Office of Rural Development Suweon 170

173

TABLE 11. (*Continued*)

Region and Country	Principal Research and Goals	Personnel and Location
Asia Minor		
Israel	Selection of *P. betulifolia* clones from successful mature trees. *P. betulifolia* performs better than QA; evaluation of QA, QC, QC.1.7, Q 16-L-2, and *P. betualifolia*	R. Assaf, researcher Agricultural Research Organization Regional Experiment Station Newe-Ya'ar 31999 P. Speigel-Roy The Volcani Center, Inst. of Horticulture Bet-Dagan
Europe		
Belgium	Evaluation of QSYDO (Angers), QA, QC, Q Adams on tree performance.	W. Porreye, Deckers Research Station of Gorsem Brede Akker 3 B-3800 Sint-Truiden
France	Evaluation of clones from seed-lings of perry pears, 'Hardy,' 'Old Home,' 'Kirchensaller' vs. PQBA 29 and QSYDO for performance and tolerance to chlorosis Propagation and evaluation of clone RV 139 Release clones of P2277 and P2278 (semidwarf *P. communis*) for research	B. Thibault, I. Pochon, researchers INRA, Station de Recerches d'Arboriculture Fruitière Beaucouze, 49000 Anger
Greece	Develop quince clones compatible with cultivars and interstocks Evaluate rootstocks for nutrition, pear quality	M. Vasilakakis, B. Tsirakoglou, J. Therias, E. Sfakiotakis, D. Gerasopoulus, O. Ntinopoulps, researchers Univ. of Thessaloniki, Thessaloniki
Hungary	Develop wild pear seedling for clonal rootstocks	J. Nyeki, Kerteszeti Egyetem, Postafiok: 53, Budapest
Italy	Evaluate quince clones QA, QC, PQBA 29, Q Laspare A, PQ 98/IV, Q 212, Q 214 on several interstocks and cultivars for compatibility, virus content, clay and wet soils, replanting and high-lime soils Release of Q 212, Univ. of Pisa Develop dwarfing seedling rootstock	S. Sansavini, G. Costa, researchers Istituto di Coltivazioni Arboree, Univ. di Bologna 40126 Bologna

TABLE 11. (*Continued*)

Region and Country	Principal Research and Goals	Personnel and Location
Netherlands	Evaluate different clones of Q Adams, with QA, QC, PQBA 29, Q SYDO, C132, OH × F51, & BP-1 for dwarfing, performance Study methods to overcome incompatibility with quince rootstocks Increase hardiness of QC in nursery	R. Wertheim, H. S. van Oosten, researchers Proefstation voor de Fruitteelt 4475 An Wilhelminadorp
Poland	Evaluate *P. caucasica* and quince for winter hardiness and precocity Evaluate 10 *Sorbus* interstem for dwarfing, winter hardiness and precocity	A. Czynczyk and W. Dzieciol Research Institute of Pomology and Floriculture 96-100 SKIERNIEWICE
Romania	Released qiunces Q GS-4-62, Q GB-1-60, and BN-70 which are winter hardy and better compatibility with 'Williams' Evaluate interstock for compatibility between various cultivars and QA Develop Harbuzesti seedling stock for dwarfing on certain cultivars	V. Cociu, N. Braniste, P. Parnia, M. Straulea Research Institute for Fruit Production Pitesti-Maracineni (12)
Spain	Evaluate quince rootstocks for chlorosis resistance, compatibility, and performance Propagation of quince and self-rooting of cultivars	M. Carrera, Q. Felipe, researchers S.I.A.−D.G.A., Apartado 727 50012 ZARAGOZA
Switzerland	Evaluate rootstocks for chlorosis resistance, suitability to cultivars, size control	W. Pfammatter, G. Perraudin State Federal de Recherchers Agronomique de Changins 1964 Conthey
United Kingdom	Development by breeding of quince and *Pyrus* rootstocks for precocity, yield, fruit quality, easy rooting and compatibility Evaluate rootstocks for performance	F. Alston, R. Smith, researchers East Malling Research Station East Malling, Maidstone Kent, ME19 6BJ
Yugoslavia	Release Q CA$_7$, and Q CA$_{15}$ clones from Fruit Research Institute in Cacak; Q CA$_7$ is more compatible	P. Misic, Agroekonomskog Institute PKB, 11306 Gocka B. Risteuski, Zemfodelski fakultet, 91000 Skapje

TABLE 11. (*Continued*)

Region and Country	Principal Research and Goals	Personnel and Location
		A. Stancevic, Agronomski fakultet, 32000 Cacak
		D. Stankovic, Sanje Zivanovica 12, 11000 Beograd
Oceania		
New Zealand	Evaluation of rootstocks for performance such as BP series, Angers *P. communis* clones, and new quince	A. White, S. Tustin, researchers DSIR Research Orchards Goddards Lane, Havelock North
	Develop clonal *P. communis* rootstocks from high-performing orchard trees	
	Develop clonal *P. pyrifolia* rootstock from seedlings for Japanese cultivars	
	Evaluation for precocity, yield, size control compatibility, propagation ease, and freedom of fire blight and mildew	
North America		
Canada	Evaluate and develop rootstocks for ease of rooting, precocity, vigor control and cold hardiness, particularly clones of OH × F, QS-1, QS-2, QS-3 (Poland) and clones of *P. communis* (U.S.-309, U.S.-342), and *P. faurieri* (24)	H. Quamme, researcher Agriculture Canada Summerland, B.C. VOH IZO
The United States	Evaluation of rootstocks for winter hardiness, and performance of clones of OH × F, seedlings of *P. betulifolia*, and 'Williams' (20)	F. Larsen, Dept. of Horticulture, Washington St. Univ., Pullman, WA 99164
	Evaluation of rootstocks for performance	J. Beutel, Dept. of Pomology, Univ. of California, Davis, CA 95616
	Evaluation of rootstocks for resistance and performance on European and Asian cultivars	P. Lombard, M. Westwood Dept. of Horticulture, Oregon St. Univ., Corvallis, OR 97331
	Develop clonal rootstocks of *P. betulifolia*, *P. calleryana*, *P. communis*, and *Amelanchier*	

TABLE 11. (*Continued*)

Region and Country	Principal Research and Goals	Personnel and Location
South America		
Argentina	Evaluate and develop interstocks, rootstocks for dwarfing, and compatibility	H. Castro, R. Rodriquez, Instituto Nacional de Tecnologia Agropecuaria, EERA, Alto Valle de Rio Negro CC 52—General Roca 83332, Rio Negro
Brazil	Evaluation of rootstocks for performance and compatibility with quince, *P. communis*, and *P. calleryana*	N. Finardi, researcher UEPAE de Cascata, Pelotas, R.S.

Most European rootstock research goals were aimed at tolerance to lime-induced chlorosis and improved compatibility, since quince is widely used for its precocity and size control (6). French researchers have approached the problem by breeding and developing *P. communis* clones. However, using *P. amygdaliformis* and *P. elaeagrifolia* as a gene source would improve tolerance to chlorosis considerably (Table 8). Quince as a rootstock is limited in wide adaptation by susceptibility to lime-induced chlorosis, incompatibility with many cultivars, and a lack of winter hardiness. Also, it has shown less favorable adaptation in warm areas such as Israel, Brazil, Argentina, and South Africa. Therefore its general use is limited geographically.

The species *P. communis*, *P. betulifolia*, and *P. calleryana* offer the most useful genetic traits for a general-use rootstock. Although less easy to propagate clonally than quince, they offer winter hardiness and hot-climate tolerance, compatibility for all scion cultivars, tolerance to wet soils, reduced susceptibility to lime-induced chlorosis, and resistance to many disease, particularly pear decline and fire blight.

Oregon researchers have been successful in locating several clonal rootstocks selected from seedling rootstocks of high-performing trees such as Oregon Pear Rootstock 1. The New Zealand rootstock program will use the same method for identifying superior clonal rootstocks.

Rootstock breeding has been successful in the development of the OH × F clonal series by Brooks (5) and for the perry pear hybrid rootstocks by Brossier (7, 8). Beginning with 516 seedlings from an open-pollinated, but isolated block of 'Old Home' and ('Farmingdale') trees, Brooks has selected 13 OH × F clonal rootstocks for commercial propagation after extensive screening and field evaluation by Westwood et al. (41). These OH × F clonal rootstocks give a range of growth control from dwarf to vigorous as indicated in Table 5. Brossier began testing 80 seedlings from perry pear varieties in western France in 1962 and from these he selected 12 clonal rootstocks for field tests with 'Williams' scion. In comparison with trees on QA, 10 clonal rootstocks produced smaller trees; one clone, P2267, is 25 percent of the tree size of QA (23). Also, all 10 clonal rootstocks had greater yield efficiency than trees on QA. These clonal rootstocks could serve as a genetic source for dwarfing and high-yield efficiency since pest resistance is unknown with these.

Screening for resistance or tolerance to pests and to specific conditions is necessary in a rootstock breeding program. These techniques have been developed for fruit breeding programs and they would be useful for rootstocks.

However, field testing is a necessary part of the rootstock development, but the field evaluation should take only 10 years to test tree performance. Any field plot should use standard rootstocks as a comparison such as 'Williams' seedling, QA, etc. Appropriate annual performance data taken should be appropriate trunk cross-sectional area for comparative tree size, flower initiation or bloom density for precocity, and yield efficiency as yield per unit trunk cross-sectional area (kg/cm^2). Other tree performance data can be taken on specific years such as fruit set, fruit size, fruit quality, tree canopy volume, and other observations. Trials in commercial orchards are often valuable for studying long-term effects or evaluating the resistance to a new pest such as pear decline.

INTERNATIONAL CATALOG OF ROOTSTOCK BREEDING PROGRAMS

Argentina
Instituto Nacional de Tecnologia Agropecuaria, EERA
Alto Valle de Rio Negro
CC 52 - General Roca
83332 Rio Negro, Argentina
H. Castro

Canada
Agriculture Canada
Summerland, B. C., Canada V0H 1Z0
H. Quamme

Harrow Research Station
Agriculture Canada
Harrow, Ontario, Canada N0R 1G0
Frank Kappel

Egypt
Horticulture Research Institute
Cairo (Giza), Egypt
A-Talaat El-Wakeel

England
   East Malling Research Station
   East Malling, Maidstone, Kent
   ME19 6BJ, England
   F. Alston

France
   Station de Recherches d'Arboriculture Fruitière
   Beaucouze
   49000 Angers, France
   J. Huet, I. Pochon, B. Thibault

Greece
   Pomology Lab
   University of Thessaloniki
   Thessaloniki, Greece
   M. Vasilakakis

India
   Indian Agricultural Research Institute
   Regional Station
   Amartara
   Cartroad, Simla-4, India
   S.S. Randhawa

Israel
   Regional Experiment Station
   Newe-Ya'ar 31999, Israel
   R. Assaf

Japan
   Faculty of Agriculture
   Shinshu University
   INA Nagano-ken, Japan
   Katsumi Kumashiro

   Faculty of Agriculture
   Utsunomiya University
   Utsunomiya, 321 Japan
   Osamu Kishimoto

Korea
   Horticultural Experimentat Station
   Office of Rural Development
   Suweon 170, Korea
   Hyun-Mo Cho

New Zealand

DSIR Research Orchards
Goddards Lane, Havelock North,
New Zealand
A. White, S. Tustin

Peoples Republic of China

Department of Horticulture
Zhejiang Agricultural University
Hangshou, Zhesiang, PRC

Shen De-xu

Fruit Research Institute
Jinling Academy of Agriculture
Sci-Gongzhuling Jilling, PRC

Gu-Mo

Zhengzhou Fruit Research Institute
Zhengzhou Honan, PRC

Wang Yu-Lin

Romania

Research Institute for Fruit Production
Pitesti-Maracineni (12), Romania
V. Cociu

South Africa

Fruit and Fruit Technology Research Institute
Private Bag X5013
Stellenbosch, South Africa 7600

P. van Huyssteen

United States

Department of Horticulture
Oregon State University
Corvallis, OR 97331

P.B. Lombard

REFERENCES

1. Ali, C.N., and M. N. Westwood (1966). Rooting of pear cuttings as related to carbohydrates, nitrogen, and rest period, *Proc. Amer. Soc. Hort. Sci.*, **88**, 145–150.
2. Andersen, P. C., P. B. Lombard, and M. N. Westwood (1984). Leaf conductance,

growth, and survival of willow and deciduous fruit tree species under flooded soil conditions, *J. Amer. Soc. Hort. Sci.,* **109**(2), 132–138.

3. Baldini, E., G. Costa, and S. Sansavini (1977). A twelve year survey on various interstocks on Beurre Bosc, Beurre Anjou, Clapp's Favorite, and William pear trees on Quince A, *Acta Hort.,* **69**, 105–112.

4. Batjer, L. P., and H. Schneider (1960). Relation of pear decline to rootstock sieve tube necrosis, *Proc. Amer. Soc. Hort. Sci.,* **76**, 85–97.

5. Brooks, L. (1984). History of the Old Home × Farmingdale pear rootstocks, *Fruit Var. J.,* **38**, 126–128.

6. Brossier, J. (1961). Les Porte-greffes du Poirier, *Pomologie Francaise.*

7. Brossier, J. (1972). Etat actuel des travaux d'amélioration des porte-greffes du poirier à la station INRA d'Angers Compte Rendu du Symposium "Culture du Poirier," pp. 229–251.

8. Brossier, J. (1977). La recherche de nouveaux porte-greffes du poirier dans le genre *Pyrus communis* L., *Acta Hort.,* **69**, 41–47.

9. Challice, J. S., and M. N. Westwood (1973). Numerical taxonomic studies of the genus *Pyrus* using both chemical and botanical characters, *Bot. J. Linn. Soc.,* **67**, 121–148.

10. Chaplin, M. H., and M. N. Westwood (1980). Effects of *Pyrus* species and related genera rootstocks on mineral uptake in 'Bartlett' pear, *J. Plant Nutr.,* **2**(3), 335–346.

11. Cheng, T. Y. (1979). Microporpagation of clonal fruit tree understock: Apple, pear, and plum, *Compact Fruit Tree,* **12**, 27.

12. Cociu, V., N. Bransite, P. Pornia, and M. Straulea (1982). "The pear cultivars to marketing. Pear culture in Romania," in *The Pear,* T. van der Zwet and N. Childers, Eds., Horticultural Pub., Gainesville, FL, pp. 69–76.

13. Fallahi, E., and F. E. Larsen (1981). Rootstock influence on 'Bartlett' and 'd'Anjou' pear fruit quality at harvest and after storage, *Hort. Science* **16**(5), 650–651.

14. Fallahi, E., and F. E. Larsen (1983). Rootstock influence on leaf and fruit mineral status of 'Bartlett' and 'd'Anjou' pear, *Scientia Horticulturae,* **23**, 41–49.

15. Frecon, J. L. (1982). "Commercial production of pear trees," In *The Pear,* T. van der Zwet and N. Childers, Eds., Horticultural Pub., Gainesville, FL, pp. 215–238.

16. Fridlund, P. R. (1982). Current status of pear virus diseases in North America, *Acta Hort.,* **124**, 67–73.

17. Hartmann, H. T., W. H. Griggs, and C. J. Hansen (1963). Propagation of own-rooted Old Home and Bartlett pears to produce trees resistant to pear decline, *Proc. Amer. Soc. Hort. Sci.,* **82**, 92–101.

18. Hedrick, U. P. (1921). *The Pears of New York,* Report of the N.Y. Agr. Exp. Sta., 636 pp.

19. Kikuchi, A. (1946). Speciation and taxonomy of Chinese pears [in Japanese], *Col. Records of Hort. Res. No. 3.*

20. Larsen, F. E. (1982). "Effect of pear rootstock and interstock on tree nutrient levels, tree growth, fruiting, and certain disorders," in *The Pear,* T. van der Zwet and N. Childers, Eds., Horticultural Pub., Gainesville, FL, pp. 239–252.

21. Lombard, P. B., and M. N. Westwood (1976). Performance of six pear cultivars on clonal Old Home, double rooted, and seedling rootstocks, *J. Amer. Soc. Hort. Sci.*, **101**, 214–261.

22. Lombard, P. B., M. N. Westwood, and R. L. Stebbins (1984). Related genera and *Pyrus* species for pear rootstock to control size and yield, *Acta Hort.*, **146**, 197–202.

23. Michelen, J. C. (1980). Sélection et expérimentation des porte greffes du poirier à la Station d'Angers, *Rivista della Ortoflora frutticoltura Italiana*, **64**, 93–104.

24. Quamme, H. A. (1982). "Breeding and testing pear rootstocks,"in *The Pear*, T. van der Zwet and N. Childers, Eds., Horticultural Pub., Gainesville, FL, pp. 253–256.

25. Rajashekar, C., M. N. Westwood, and M. J. Burke (1982). Deep supercooling and cold hardiness in genus *Pyrus*, *J. Amer. Soc. Hort. Sci.*, **107**, 968–972.

26. Reimer, F. C. (1925). *Blight Resistance in Pears and Characteristics of Pear Species and Stocks*, Oregon Agr. Exp. Sta. Bul. 214, 99 pp.

27. Reimer, F. C. 1950. *Development of Blight Resistant French Pear Rootstocks*, Oregon. Agr. Exp. Sta. Bul. 485, 24 pp.

28. Sansavini, S., G. Costa, R. Credi, M. Grandi, V. Bindi, and C. Monti (1980). Influenza dello stato sanitario e du nuovi portinnesti clonali sul pero "William," *Rivista della Ortoflora frutticoltura Italiana*, **64**, 563–578.

29. Western Region W-130 Technical Committee (1978). *Phenology and Plant Species Adaptation to Climates of the Western United States*, Oregon Agr. Exp. Sta. Bul. 632, 15 pp.

30. Westigard, P. H., M. N. Westwood, and P. B. Lombard (1970). Host preference and resistance of *Pyrus* species to the pear psylla, *Psylla pyricola* Foerster, *J. Amer. Soc. Hort. Sci.*, **95**, 34–36.

31. Westwood, M. N. (1965). Clonal propagation of several *Pyrus* species, *Oregon Ornamental and Nursery Digest*, **9**(1), 3–4.

32. Westwood, M. N. (1976). Inheritance of pear decline resistance, *Fruit Var. J.*, **30**, 63–64.

33. Westwood, M. N. (1978). *Temperate-Zone Pomology*, "Rootstocks," Freeman, New York, pp. 77–107.

34. Westwood, M. N. (1981). Graft compatibilities of pear with several related rootstock genera, *Compact Fruit Tree*, **14**, 151–152.

35. Westwood, M. N. (1982). Pear germplasm of the new national clonal repository: It's evaluation and uses, *Acta Hort.*, **124**, 57–65.

36. Westwood, M. N., and H. O. Bjornstad (1968). Chilling requirements of dormant seeds of 14 pear species as related to their climatic adaptation, *Proc. Amer. Soc. Hort. Sci.*, **92**, 141–149.

37. Westwood, M. N., and L. A. Brooks (1963). Propagation of hardwood pear cuttings, *Proc. Int. Plant Prop. Soc.*, **13**, 261–268.

38. Westwood, M. N., and H. R. Cameron (1978). Environment-induced remission of pear decline symptoms, *Pl. Dis. Rptr.*, **62**, 176–178.

39. Westwood, M. N., and P. B. Lombard, (1977). Pear rootstock and *Pyrus* research in Oregon, *Acta Hort.*, **69**, 117–122.

40. Westwood, M. N., and P. B. Lombard (1982). Rootstocks for pear, *Proc. Oregon Hort. Soc.*, **73**, 64–79.

41. Westwood, M. N., and P. B. Lombard (1983). Pear rootstocks: Present and future, *Fruit Var. J.*, **37**(1), 24–28.

42. Westwood, M. N., P. B. Lombard, and H. O. Bjornstad (1976). Performance of 'Bartlett' pear on standard and Old Home × Farmingdale clonal rootstocks, *J. Amer. Soc. Hort. Sci.*, **101**, 161–164.

6

PEACH ROOTSTOCKS

Richard E. C. Layne
Research Station, Agriculture Canada
Harrow, Ontario, Canada

1. ORIGIN AND HISTORY OF CULTURE

The peach [*Prunus persica* (L.) Batsch] belongs to section *Euamygdalus* Schneid., subgenus *Amygdalus* (L.) Focke, subfamily Prunoideae and family Rosaceae. It has a basic chromosome number of $x = 8$. Seventeen species belong to the subgenus *Amygdalus*, but only peach and almond (*P. amygdalus* Batsch) are widely cultivated (24, 28, 50).

The peach originated in China, where its culture predates written history (24, 28). Peach culture spread from China to Persia following the trade routes. It spread from Persia to the Mediterranean basin and was introduced to Greek culture between 400 and 300 B.C. It was introduced to Roman culture shortly after the beginning of Christianity. The Romans subsequently spread the peach throughout their realm and as far north as England (24, 28).

Peach culture spread from the Mediterranean basin to North America, South America, southern Africa, Australia, and New Zealand during the era of exploration and colonization in the sixteenth, seventeenth and eighteenth centuries (24). The Portuguese introduced peaches to South America and the Spaniards introduced them to Mexico and Florida (24). The Indians of North America, followed by European settlers, spread the peach throughout the continent almost to the limits of its present-day culture (24).

It is almost certain that the spread of peach culture from its ancestral home in China to the rest of the world was accomplished by seeds, since this would have been the simplest and easiest manner of transport over land and water. In most countries peach culture was initially based on the growing of seedlings. Budded trees in North America were not available until the late 1700s and not commonplace until the 1800s (24). Only in this century have budded trees become the norm in commercial peach culture. Interest in finding suitable rootstocks for peach developed with the shift from the commercial culture of seedlings to that of budded trees, and has grown steadily since.

Peaches are now grown around the world, usually between latitudes of 30 and 40° N and S (63). At higher latitudes minimum winter temperatures and spring frosts are the usual limiting factors (25, 30, 37, 48). At lower latitudes insufficient chilling is the most important limiting factor adversely affecting dormancy requirements, normal foliation, bloom, and fruit production (24, 59).

According to the FAO Yearbook, world production of peaches in 1982 was 7,107,000 metric tons (t). Europe was the largest producer (3,185,000 t), followed by North and Central America (1,425,000 t), Asia (1,187,000 t), Africa (244,000 t), and Oceania (94,000 t). Of the 48 peach-producing countries listed, the top 10 in descending order of importance included: Italy (1,400,000 t), the United States (1,197,000 t), Spain (462,000 t), Greece (453,000 t), France (438,000 t), the Soviet Union (420,000 t), China (406,000 t), Turkey (252,000 t), Japan (228,000 t), and Argentina (213,000 t).

2. ROOTSTOCK USAGE

2.1. Peach Seedlings

Peach seedlings have been and still are the principal rootstock source for peach on a worldwide basis (53). The seed sources can be conveniently divided into three groups: those from wild types, those from commercial cultivars, and those from special rootstock selections (53). The wild types are usually obtained from peach trees that have escaped cultivation and are found growing wild or in a nearly wild state. In North America, the Tennesse Naturals or Indian Peaches are examples. Other countries also utilize wild-type peaches originating there or imported from elsewhere. Of 25 countries recently surveyed, 10 indicated that wild peaches were the most commonly used rootstocks for peach (53). A major problem with wild peaches is their genetic variability and general lack of uniformity in the nursery and the orchard (2, 3, 6, 53).

Commercial cultivars, usually those used for commercial processing or drying, are second in importance as rootstock seed sources for peach (13, 28, 30, 31, 34, 44, 53, 59, 63). Thus 'Halford' and 'Lovell' are of great importance as peach seedling rootstocks in Canada, the United States, and Mexico. Similarly, 'Polara' and 'Sims' are important in Argentina; 'Golden Queen,' 'Elberta,' and 'Wright' in Australia: 'Baladi' in Israel; 'Kakamas' and 'Du Plessi' in South Africa and Zimbabwe; 'Cape de Bosq' and 'Conserva' in Brazil; and 'Balc Elita' in Romania(53). This group provides more genetically uniform rootstocks both in the nursery and orchard than those from the wild and their performance is therefore more predictable.

The third group consists of selections that have little or no commercial value other than as rootstock seed sources for peach and nectarine. Examples of this group include 'Siberian C' and 'Harrow Blood' in Canada; 'Bailey,' 'Rutgers Red Leaf,' and 'Nemared' in the United States; GF 305 and 'Rubira' in France and southern Europe; BVA 1, 2, 3 and 4 in Czechoslovakia; and 'Ohatsumomo' in Japan (1−3, 6, 7, 13, 30, 34, 40, 42, 49, 52−54). Theses stocks have special attributes that led to their introduction, and have therefore assumed some local importance in the regions or countries where they originated. However, they have not assumed any worldwide importance. This is usually because the special traits they have (e.g., cold hardiness), while important in the region where they were selected, may not be as important in many other regions of peach culture.

2.2. Interspecific Peach Hybrids

Interspecific hybrids of peach × almond (GF 556, GF 677) are of importance in France and other countries of southern Europe, especially in the Mediterranean basin. They are clonally propagated and are especially useful on alkaline soils because they are resistant to chlorosis, have some tolerance to wet and dry soils, and are especially useful in replant situations (2−4, 6, 15, 28, 53, 54, 56−58). 'Nemaguard,' a seed-propagated interspecific hybrid (*P. persica* × *P. davidiana*), is used extensively in California and Florida. Resistance to root knot nematodes (*Meloidogyne* spp.) is the primary reason for its wide acceptance in regions with mild winters. However, its adverse effect on peach scions with respect to cold hardiness and bacterial canker (*Pseudomonas syringae* pv. *syringae* van Hall) limit its use farther north (9, 10, 12, 30, 31, 41, 44, 62).

2.3. Plums and Plum Hybrids

Plums are the most commonly used rootstock for peach other than peach seedlings and peach interspecific hybrids (2, 3, 5, 6, 15−17, 28, 53). They are better adapted to wet, waterlogged soils than other *Prunus* spp. and have been used primarily for that reason (2, 3, 5, 6, 15−17, 28, 55−58). Included in this group are those derived from *P. insititia* (St. Julien d'Orleans, St. Julien Hybrid No. 1, St. Julien Hybrid No. 2, St. Julien GF 655.2), *P. cerasifera* Ehrh. (Myrobalans), *P. domestica* (GF 43), *P. domestica* × *P. munsoniana* (Marianna, GF 8-1), and *P. domestica* × *P. spinosa* (Damas GF 1869).

2.4. Other Prunus Species

Other *Prunus* spp. are also used as rootstocks for peach and include *P. davidiana* (Carr.) Franch, *P. amygdalus* Batsch, *P. armeniaca* L., *P. insititia* L., *P. salicina* Lindl., *P. tomentosa* Thunb., and *P. besseyi* Bailey (28, 53). However, none of these has had major worldwide acceptance.

2.5. Other Considerations

As production, labor, and land costs escalate, peach growers are seeking methods of greatly increasing their production per hectare to offset rising costs. This is leading to more intensive systems of orchard management including higher tree densities, new pruning and training systems, and greater use of irrigation. Such changes in production practices require rootstocks that are better adapted to more intensive culture. Hightened interest in peach root- stocks is leading to an expansion and intensification of such research in many peach-producing countries and should lead to better rootstocks for peaches and nectarines in the future (6, 15, 16, 30, 31, 33, 34, 37, 52–54, 59). The shift to high-density culture is also leading to alternative approaches in order to reduce cost of nursery trees. Much research is now in progress to grow peach cultivars on their own roots, thus doing away with rootstocks altogether. While this approach has some advantages and has shown some commercial potential, it is beyond the scope of this chapter.

3. ROOTSTOCK ADAPTATIONS

3.1. Temperature

Peaches thrive best where the summers are hot and the winters are not subject to temperature extremes below -25°C except occasionally (24, 25, 34, 35, 37, 47, 48, 50, 59, 63).

Recently it has been shown that most cultivated *Prunus* spp. have evolved a freeze-avoidance mechanism associated with cold-sensitive tissues that super- cool, including flower primordia and xylem ray parenchyma (37, 47, 48). This mechanism, while protective at temperatures usually above -20°C, is usually harmful at colder temperatures and severely limits the northern distribution of commercial culture. Supercooled tissue water in the absence of ice nucleators remains in an unfrozen, supercooled state in fully acclimated shoots and buds to quite low temperatures, $\geq -20°C$. However, at temperatures usually be- tween -25 and $-45°C$ this water will suddenly freeze explosively, causing death of these tissues. Freezing injury of this type results in death of flower buds, injury or death of shoot and trunk xylem, and may result in ultimate death of whole trees.

A recent survey was made of several *Prunus* species and interspecific *Prunus* hybrids (36) and showed the limitations of some and the potential of others as cold-hardy rootstock sources for peach. Most of the hardier species and inter- specific hybrids (Table 1) have not been adequately studied as rootstocks for peach in order to be recommended for commercial use. Therefore, such studies are indicated.

Cold hardiness, at least in the northerly areas of production, is required both for seed-propagated and for clonally propagated peach rootstocks. Seed-pro- pagated stocks must possess adequate flower bud and wood hardiness to ensure tree survival as well as consistency and continuity of seed production from mother trees, especially following severe winters (30, 31, 33, 34, 37). Clonally propagated rootstocks require sufficient vegetative bud and wood hardiness to furnish an adequate supply of rooted cuttings for use as nursery plants upon

TABLE 1. Cold Hardiness of Flower Buds and Shoot Xylem of *Prunus* Taxa Preconditioned to Obtain Maximum Cold Hardiness[a]

| Species and Interspecific Hybrids | Flower Buds[b] | | Shoot Xylem[c] |
	T_{20} (°C)	T_{90} (°C)	ILTE (°C)
P. persica	−22.0	−27.8	< −34.6
(P. tenella × P. persica) × P. persica	−36.0	−44.0	−42.0
P. armeniaca	−21.7	−29.3	−36.0
P. armenica var. mandshurica	−21.0	−34.0	−40.0
P. armenica var. mandshurica × P. armeniaca	−26.0	−33.5	−39.6
P. salicina	−29.0	−37.0	−40.3
P. domestica	−30.0	−39.0	−35.0
P. nigra	−38.0	−46.5	−47.5
P. americana × P. salicina	−36.0	−46.0	−44.0
(P. americana × P. salicina) × P. americana	−37.0	−46.0	−43.0
P. besseyi	−37.8	−44.5	−41.3
P. besseyi × P. persica	−23.0	−34.0	−40.0
P. besseyi × P. armeniaca	−31.5	−36.5	−40.5
P. besseyi × P. salicina	−36.5	−45.0	−42.0
(P. besseyi × P. salicina) × P. armeniaca var. mandshrica	−33.0	−36.0	−40.5
P. avium	−31.0	−35.0	−40.5
P. cerasus	−30.5	−38.5	−41.5
P. fruticosa	−38.0	−46.0	−45.0
P. maackii	−38.0	−48.0	−47.0
P. pensylvanica	−39.0	−49.0	−44.0
P. tomentosa	−31.0	−34.0	−37.0
P. tomentosa × P. besseyi	−37.0	−44.0	−42.0

[a]After Quamme, Layne, and Ronald (48).
[b]Average temperature required to kill 20 and 90% of the flower primordia.
[c]Average temperature for initiation for low-temperature exotherm in shoot xylem.

which peach and nectarine scion cultivars will be budded. However, tissue culture techniques and micropropagation may reduce some of this need in the future.

Not only must the aboveground portion of rootstock mother trees be cold hardy, but seedling or clonal rootstocks derived from them must also have sufficient cold hardiness of the root systems to survive soil temperatures below −8°C. Such temperatures are not uncommon during the winter in northerly areas, especially in the absence of snow and/or vegetation cover (30, 31, 34, 37). Root freezing occurs more frequently on light sands and sandy loam as compared to silt loam or clay loam soils, because deeper cold penetration occurs in the light soils (37). Cold hardiness of root systems, therefore, is an important requirement, especially in more northerly areas where potentially injurious temperatures occur more frequently (30, 31, 34, 35, 37).

Studies at the Harrow Research Station have shown that peach seedling rootstocks differ markedly in the cold hardiness of their root systems (30, 31, 34, 37). 'Siberian C' roots survived a minimum outdoor soil temperature stress of −13.3°C at 20 cm below the soil surface with little or no root injury (30), whereas roots of 'Elberta,' 'Nemaguard,' 'Yunan,' and 'Shalil' were severely injured or killed, those of 'Muir,' 'Rutgers Red Leaf,' and 'Halford' sustained moderate to severe injury, and those of 'Harrow Blood' were only slightly injured. Subsequent studies involving the controlled laboratory freezing of root systems from different seed lines of peach, followed by greenhouse regrowth tests, have permitted the classification of several peach seed sources into different hardiness groups (Table 2 and Figures 1 and 2).

It is noteworthy that the three hardiest rootstocks (Table 2) originated either directly ('Tzim Pee Tao,' 'Chui Lum Tao') or indirectly ('Siberian C') from northern China. A careful search of peach germ plasm from northern China, especially at the northern limits of the natural geographic range of the wild peach, may result in finding even hardier rootstock seed sources for peach in the future.

Preliminary evaluation has also been made by the author of a number of other potential peach rootstocks in terms of their flower bud and wood hardiness. Several of them show satisfactory levels of cold hardiness to justify further tests in northern areas. Those from France include one peach — S278-1 (*P. persica*); three plums — Mariana GF 8-1 (*P. cerasifera* × *P. americana* ?), Damas GF 1869 (*P. domestica* × *P. spinosa*), and GF 655.2 (*P. insititia*); and three peach × almond hybrids (*P. persica* × *P. amygdalus*) — GF 677, (S ×

TABLE 2. Classification of Cold Hardiness of Root Systems of Peach Seedling Rootstocks Determined from Controlled Freezing and Regrowth Tests[a]

Seed Source	Origin	Hardiness Class
Siberian C	China/Canada	Very hardy
Tzim Pee Tao	China	
Chui Lum Tao	China	
Bailey	United States	Medium hardy
Harrow Blood	Canada	
Yeh Hsiemtung Tao	China	Medium tender
Sinung Chui Mi	China	
Rutgers Red Leaf	United States	
Gold Drop	United States	Tender
Lemon Free	United States	
Elberta	United States	

[a]After Layne (30, 31, 34, and unpublished data).

Figure 1. Survival tests at Harrow of 'Harrow Blood' peach seedling rootstocks subjected to controlled freezing followed by greenhouse regrowth tests. Root systems include (*a*) unfrozen control, (*b*) −7°C, (*c*) −10°C, and (*d*) −13°C. Note strong growth of new shoots and new fibrous roots associated with (*a*), weaker growth of new shoots and roots with (*b*), and weak growth of shoots and roots with subsequent death of new shoots associated with (*c*) and (*d*).

R170) 11, and (S × R185). Those from Canada include Morden 800 (*P. besseyi* × *P. sibirica*) from Morden,Manitoba; two seedling selections of *P. besseyi* (HR18T21, HR19T21); and four of *P. besseyi* × *P. tomentosa* (H7520001, H7520002, H7520006, H7520007) from the Harrow Research Station in southwestern Ontario.

3.2. Soil Moisture, Texture, and Drainage

Soil moisture, texture and drainage are interrelated (7), and *Prunus* spp. and interspecific hybrids vary greatly in their tolerance to them (2, 3, 5, 6, 15−17, 28, 44, 53−58). In general, peach, apricot and almond rootstocks are among the least tolerant to wet, waterlogged soils (55−58). Peach × almond hybrids have more tolerance than peach seedlings, and plum rootstocks have the greatest tolerance of all (2, 57, 58). Effective techniques have been developed in France for the screening of candidate peach rootstocks for resistance to waterlogging (57). These tests were formerly conducted during the dormant period, outdoors, in specially designed tanks where the soil was saturated by flooding (55−58). The tests are now conducted during the growing season and

Figure 2. Survival tests at Harrow of 'Siberian C' peach seedling rootstocks subjected to controlled freezing followed by regrowth tests. Root systems include (*a*) unfrozen control, (*b*) −7°C, (*c*) −10°C, and (*d*) −13°C. Note strong growth of new shoots and abundance of new fibrous roots associated with (*a*) and (*b*), moderate shoot and root growth associated with (*c*), and absence of shoot and root growth associated with (*d*).

selection is based on the rate at which test plants develop symptoms of water-logging from flooding. The summer tests are faster and more reliable than those conducted during the dormant period (Selesse, personal communication). Such tests have permitted the grouping of a number of peach rootstocks into different classes of resistance to waterlogging (2, 6, 55–58); (a) very good resistance (Damas GF 1869), (b) good resistance (St. Julien d'Orleans, St. Julien Hybrid No. 1, St. Julien Hybrid No. 2, St. Julien GF 655.2, (c) moderate resistance (wild peach, 'Nemaguard,' GF 305, 'Brompton,' Prunier GF 43, GF 677), and (d) low resistance (most peach, apricot, and almond seedlings). It is noteworthy that only plum species and interspecific plum hybrids were represented in groups (a) (*P. domestica* × *P. spinosa*) and (b) (*P. insititia*). Included in group (c) were peach seedlings (wild peach, GF 305), a peach × *P. davidiana* hybrid ('Nemaguard'), a peach × almond hybrid (GF 677), and two plums (*P. domestica*: 'Brompton,' Prunier GF 43). Greater utilization of tank tests like those being used in France should permit better classification of other peach rootstocks for their resistance to waterlogging in other parts of the world.

Differences in resistance to waterlogging in *Prunus* have been found to be associated with the metabolism of the cyanogenic glucoside prunasin. Under anaerobic conditions, prunasin is hydrolized to hydrogen cyanide, which is autotoxic. The differential sensitivity of peach, apricot, and plum to waterlogging is closely related to their ability to hydrolize prunasin (55).

Secondary effects from anaerobic conditions may also play a role in waterlogging injury such as invasion of roots and stem pieces by soil fungi like *Phytophthora* and *Pythium* spp., which thrive in waterlogged soils (47, 59, 60, 65).

Peach rootstocks vary in their adaptation to different soil types. Peach seedlings thrive best on well-drained sandy loam soils (see Figures 3 and 4) but give satisfactory performance on other well-drained soils such as sands, gravelly loams, and silt loams (2, 3, 6, 8, 13, 15, 27, 29, 31, 34, 44, 49). They do not perform well on fine-textured, imperfectly drained soils and are usually not recommended for such soil types.

Peach × almond hybrids, although more tolerant of finer-textured soils than peach seedlings, thrive best on well-drained sandy loam soils (2).

Plum rootstocks have the best tolerance among peach rootstocks for fine-textured, imperfectly drained soils (2, 3, 5, 6, 15–17, 28, 44, 53, 55–58). In France and southern Europe, they are commonly used for such soil types and include the following: 'St. Julien d'Orleans,' St. Julien Hybrid No. 1, St. Julien Hybrid No. 2, 'Damas de Toulouse,' INRA Damas 1869, and INRA St. Julien 655.2. These stocks have not been adequately tested in North America to be recommended for commercial use.

Incompatibility of many nectarine cultivars on INRA Damas 1869 has hindered the use of this rootstock for nectarines in France (18). Incompatability of peach on plum rootstocks in North America is the main reason why plum rootstocks are seldom used for peach (13, 43). Lack of suitable plum rootstocks for peach have therefore confined peach culture to the lighter, better-drained soils on this continent.

Finer-textured soils have greater water-holding capacity than coarser-textured ones (7), and peaches are therefore less subject to drought stress on such soils. Expansion of peach culture in some regions is restricted more by soil type than by climate within climatic zones that are suitable for peach culture. For example, there is an abundance of imperfectly drained clay loam soil and a paucity of well-drained sandy loam soil in southern Ontario and land costs are substantially lower for the finer-textured soils. A strong requirement exists, therefore, for a suitable rootstock for peach that will permit utilization of such soils. Tests are under way at the Harrow Research Station to select a suitable rootstock for this purpose (34).

3.3. Drought

Peaches are grown commercially in many arid or semiarid regions of the world (63). Irrigation, at least in some regions, is essential for tree survival and/or commercial production. Drought resistance, therefore, is a desirable feature for peach rootstocks. Seedlings of peach, apricot, and almond have shown

Figure 3. Root distribution near trunk of nonirrigated 11-year-old peach tree ('Harken'/'Siberian C') in Fox sand. Note presence of peach roots (highlighted with white paint) from near top to near bottom of the trench profile where each grid is 30 × 30 cm. Photo is courtesy of Layne, Tan, and Perry (unpublished data).

Figure 4. Root distribution at drip line of nonirrigated 11-year-old peach tree ('Harken'/'Siberian C') in Fox sand. Note abundance of peach roots (highlighted with white paint) from near top to near bottom of trench profile where each grid is 30 × 30 cm. Photo is courtesy of Layne, Tan, and Perry (unpublished data).

some degree of drought tolerance and have been used for peach on droughtly soils where irrigation is not essential (24, 28, 44, 47, 53, 63, 65). Drought resistance has been reported for almond, apricot, cherry, peach, plum, and many uncultivated *Prunus* species (28). A major problem with direct utilization of some species as rootstocks for peach is that of incompatibility (28, 43). Interspecific hybridization of drought-tolerant species with peach may be an effective means of reducing the incompatibility problem. Techniques developed in the Soviet Union for selecting fruit trees for drought resistance (28) could then be employed to apply selection pressure for drought resistance among the interspecific hybrids, and could similarly be employed for classifying existing rootstocks whose degree of drought tolerance may be unknown.

3.4. Soil Reaction

Peaches perform best on slightly acid to neutral soils because more macro- and micronutrients are readily available between a soil pH of 6 and 7 than at lower or higher levels (7).

Alkaline soils are associated with low availability of iron, especially between a pH of 7.5 and 9.0, and peaches are subject to iron-induced chlorosis (2, 3, 6, 15).

In France, an alternative to pH adjustment of the soil is practiced by the judicious choice of rootstock, because they vary substantially in resistance to iron-induced chlorosis (2). The most resistant are Damas GF 1869 (*P. domestica* × *P. spinosa*) and GF 677 (*P. persica* × *P. amygdalas*), both of which are rated as very good, In the second category and rated as having good resistance are St. Julien Hybrids 1 and 2 (*P. insititia* × *P. domestica*), Brompton, and Prunier GF 43 (*P. domestica*). In the third group, rated as having moderate resistance, are wild peach and GF 305 (*P. persica*) and St. Julien GF 655.2 (*P. insititia*). 'Nemaguard' (*P. persica* × *P. davidiana*) had the lowest level of resistance of those reported. Through breeding and selection there is good potential to improve the resistance of peach rootstocks even further to alkaline soils.

Peach seedling (*P. persica*) rootstocks are quite well adapted to slightly acid (pH > 5 to 7) soils; however, pH adjustment is recommended and commonly practiced if below 5.0 by use of calcitic or dolomitic limestone to avoid nutrient imbalances and low availability of nutrients like N, P, K, Ca, Mg, S, Mn, B, Cu, Zn, and Mo (7).

3.5. Soil Fertility and Tree Vigor

Soils vary in their natural fertility and availability of nutrients for plant growth (7). Peach rootstocks also vary in their rooting habit, rooting depth, and ability to extract nutrients from the soil (29). Besides, they vary in their inherent vigor (1–6, 8, 13, 15–17, 21, 30, 31, 33, 34, 49, 51, 53) and the vigor they impart to peach scion cultivars (1–6, 12, 15–17, 21, 40, 51–54). Scion cultivars also vary in inherent vigor when propagated on the same rootstock (40). Thus obtaining the desired level of tree vigor in a peach orchard necessitates judicious manipulation of each of these variables in addition to fertilizer and pH adjustment.

In northern areas, careful regulation of tree growth is necessary to avoid excessive vigor, which delays cold acclimation and increases susceptibility to perennial canker (*Leucostoma* spp.) and winter injury (37). Rootstocks that induce a high level of scion vigor, such as 'Nemaguard,' should therefore be avoided in northern areas, especially on soils high in natural fertility. In the southeastern United States, 'Nemaguard' is also associated with increased scion vigor, winter injury, and bacterial canker (*P. syringae* pv. *syringae*) as compared with 'Lovell'; thus the latter stock is favored, especially on peach tree short life (PTSL) sites (12).

Rootstocks that induce a high level of scion vigor should also be avoided in deep, fertile soils because trees become difficult to manage, requiring excessive pruning to confine them to their assigned space. Shading of fruiting wood may take place and fruit quality and color may be reduced. However, in situations where tree vigor might be suppressed, such as in shallow soils low in natural fertility or on replant sites, vigorous rootstocks have been employed to advantage (2–4, 6, 15, 26). In France, for example, vigor-promoting rootstocks like GF 677 and 'Nemaguard' are used to counteract the vigor-reducing influence of low soil fertility. By contrast, vigor-reducing rootstocks such as St. Julien GF

655.2 are used on deep, fertile, soils to counteract the growth-promoting influence of high soil fertility (2).

4. ROOTSTOCK RELATIONS

4.1. Nematodes

Nematodes adversely affect establishment, growth, orchard uniformity, fruit production, and tree longevity of peaches in many parts of the world. Soil fumigation is often practiced to control them, but use of resistant rootstocks, when available, provides a less expensive means of control with reduced environmental impact (6, 9, 12, 15, 24, 27, 28, 30, 41, 44, 46, 49, 52−54, 59, 60, 65).

Nineteen species of nematodes have been identified in different peach-growing regions of the world that may pose a problem with peach rootstocks (53, 54), three belong to the root knot group (*Meloidogyne*), seven to the root lesion group (*Pratylenchus*), six to the ring nematode group (*Criconemoides*), and three to the dagger nematode group (*Xiphinema*). The dagger nematode group is also known to act as vectors of viruses, including tomato ringspot virus, which causes *Prunus* stem pitting (14), and peach rosette virus (45). In a recent survey of peach researchers worldwide (53), 68% of those surveyed considered nematodes to be an important problem relating to nursery and orchard performance of peach rootstocks.

More research has been conducted in breeding and selecting nematode-resistant rootstocks for peach than has been spent on developing rootstocks with resistance to any other biotic agents in the soil (30, 59). Much of the effort has been focused on developing resistance to rootknot nematodes, especially *Meloidogyne incognita* (Kofoid and White) Chitwood and *M. javanica* (Treus) Chitwood. Rootknot nematodes have been a severe problem in warmer peach-growing regions such as California, Florida, Texas, Mexico, and Israel (59). This had led to the selection of several root knot nematode−resistant rootstocks for peach including 'Nemaguard,' 'Shalil,' S-37, 'Okinawa,' 'Higama,' and 'Nemared' (53, 54).

Root lesion nematodes (*Pratylenchus* spp.) appear to be more important in cooler climatic zones such as southern Canada and the northern United States (27, 30, 46), where root knot nematodes seldom overwinter and are thus relatively unimportant. However, less emphasis has been placed on developing rootstock with resistance to root lesion nematode than on that with root knot nematode resistance. There may, nevertheless, be some resistance or tolerance to root lesion nematode associated with 'Rubira,' 'Pisa,' 'Rutgers Red Leaf,' 'Tzim Pee Tao,' and some hybrids of 'Rutgers Red Leaf' × 'Tzim Pee Tao' (6, 15, 27, 30, 31, 34, 46, 52−54).

Studies are now under way at Clemson University, South Carolina, and the USDA at Byron, Georgia, to search for resistance to ring nematode (*Criconemoides xenoplax* Raski). It appears to play an important role in the PTSL complex in the southeastern United States (12). Ring nematodes predispose peach trees to bacterial canker (*P. syringae* pv. *syringae*) and winter injury (12, 41, 53), and currently used peach rootstocks are susceptible (53).

4.2. Fungi

Nine genera of fungi have been associated with diseases of peach rootstocks (53), three have frequently been associated with causing root rots (*Phytophthora, Clitocybe, Armillaria*), and six (*Pythium, Fusarium, Fusiccocum, Rhizoctonia, Stereum, Verticillium*) have been implicated with either below- or aboveground diseases. On a worldwide basis, 56% of researchers surveyed considered soil diseases to be of concern with respect to peach rootstocks (53).

Very little research is being conducted to develop rootstocks with resistance to soil diseases caused by fungi, although *Prunus* rootstocks differ in their response to soil fungi (30, 44, 59, 60, 65). For example, some have been found to be moderately resistant to crown rot [*Phytophthora cactorum* (Zeb. & Cohn) Schroet.], including three plum rootstocks—Myrobalan seedlings (*P. cerasifera*), Myrobalan 29 (*P. cerasifera*) hardwood cuttings, and Mariana 2624 (*P. cerasifera*) × *P. munsoniana* ?) hardwood cuttings (44). The latter stock is also moderately resistant to *Armillaria* (44). Plum rootstocks have also been found to be less susceptibile to *Verticillium* wilt than peach, almond, and apricot seedlings (44). Thus there is some genetic potential in *Prunus*, especially among the plums, for improving disease resistance of peach rootstocks (28). Peach rootstocks also influence scion susceptibility to some important fungal diseases such as perennial canker (*Leucostoma* spp.), an important consideration when selecting peach rootstocks (32).

4.3. Bacteria

Two bacterial diseases, bacterial canker (*Pseudomonas syringae* pv. *syringae* van Hall) and crown gall (*Agrobacterium tumafaciens* [E. F. Sm. and Town.] Conn.), have been associated with peach rootstocks. The plum rootstocks for peach reported to be resistant to bacterial canker (6) are DAMAS GF 1869 and St. Julien, while 'Siberian C' is highly susceptible (62). Rootstocks indirectly influence susceptibility of peach scion cultivars to bacterial canker (12, 41). Those that promote susceptibility—for example, 'Siberian C'—should be avoided in regions where this disease is an important problem (12). Resistance to crown gall (30) has been found in the following species: *P. pumila, P. domestica, P. besseyi, P. insititia, P. mume,* and *P. umbellata.* Most peach seedlings, by contrast, are susceptible (Dhanvantari, personal communication), except 'Rubira,' which appears to be tolerant (6).

4.4. Viruses

Viruses are an important problem in the peach nursery and peach orchard (19, 20, 22, 23). They are readily transmitted by budding and grafting in the nursery and adversely affect rooting of clonally propagated peach rootstocks, reduce uniformity in the nursery, and greatly reduce budding success if either the budwood or the understock or both are virus infected. The production of virus-free nursery stock, therefore, is the most important means of improving propagation success in the nursery and reducing virus disease problems in peach orchards (19, 20, 22, 23, 64).

A number of public, semiprivate, and private agencies are now engaged in providing virus-free propagating materials to commercial nurseries for the production of virus-free nursery stock, such as in Canada, the United States, France, and New Zealand (1, 2, 19, 20, 22, 23, 64). Nurseries that sell finished peach trees on international markets must meet the phytosanitory requirements of importing countries. This has encouraged them to adopt procedures which guarantee the production of virus-free nursery stock (1). As the debilitating effects of virus infection in the nursery and orchard have become better documented and understood, it has increased interest and demand for virus-free nursery stock on domestic as well as international markets. The end result has been a general improvement in the quality and health of peach nursery stock on a worldwide basis.

4.5. Mycoplasmas

Some peach diseases, such as peach X-disease and Western X-disease, are caused by mycoplasmas (23). Peach seedlings are susceptible to X-disease, but *P. mahaleb* is immune (23). X-disease is spread by leafhoppers, but no feeding preference among peach cultivars was shown by the leafhopper vector species that occur in southwestern Ontario (Elliott, unpublished data). Other species of *Prunus* have not been tested to determine whether or not they are resistant to leafhoppers.

4.6. Insects

Insects are of concern not only because of injuries produced from their feeding activity, but also from the diseases they transmit. In North America, for example, the peach tree borer (*Saaninoideae exitosa* Say) and the lesser peach tree borer (*Synanthedon pictipes* Grote and Robinson) are important insect problems in the nursery and orchard (30). Apart from the serious damage they cause on peach scions and rootstocks from feeding activity, their injuries also serve as points of entry for perennial canker (*Leucostoma* spp.), a major cause of early decline and death of peach trees in Ontario, Canada, and the northern United States (37). Peach cultivars vary in their resistance to borers, with 'Goldray,' 'Dixired,' and 'Golden Jubilee' offering the best sources of borer resistance of those studied (61). *P. armeniaca* appears to be an even better source of borer resistance than *P. persica*, *P. amygdalus*, or *P. cerasifera* (9, 44).

Leafhoppers and aphids transmit a number of important virus and mycoplasma diseases of peach (23). 'Rubira' and S2678, both of which are peach seedling rootstocks from France, are reported to be resistant (42) to two aphid species (*Myzus persicae* Suiz. and *M. varians* Davids.) that transmit Sharka (plum pox), a very serious virus disease of cultivated *Prunus* species in Western and Eastern Europe (23).

Oriental fruit moth (*Grapholitha molesta* Busck) is a problem both in the nursery and in the orchard, but is readily controlled chemically. To the author's knowledge, no reports of host resistance to this insect in *Prunus* have been made.

4.7. Training Systems

Peach rootstocks must be chosen with care to ensure that they are adapted to the orchard design and training system that is to be employed (2). Some rootstocks induce dwarfing of scion cultivars while others enhance vigor. In a high-density hedgerow system, a dwarfing rootstock may have some advantages over a vigor-promoting one, whereas in a low-density, open-centered training system, a vigor-inducing rootstock might be preferred. Rootstocks that promote precocious bearing are generally preferred, especially in high-density, intensively managed systems where high, early yields are needed to offset the greater production costs.

5. ROOTSTOCK INFLUENCES

5.1. Scion Growth and Development

Peach rootstocks induce a wide range of size control of peach scion cultivars from about 50% dwarfing to about 25% vigor induction compared with standard peach seedling rootstocks. A summary of size-controlling rootstocks for peach and their degree of vigor control is given in Table 3.

In general, *P. tomentosa* and *P. besseyi* have been associated with the highest level of dwarfing (13, 21, 30, 33, 34, 53), followed by some plum rootstocks including St. Julien X and St. Julien GF 655.2, and a few peach seedling rootstocks including 'Siberian C' and 'Rubira' (2, 6, 53). The majority of peach seedling rootstocks produce standard-size trees, but a few enhance scion vigor such as 'Halford' and 'Nemared' (40, 49). Apricot and almond seedings usually induce more vigor than peach seedlings, while peach × almond hybrids (GF 677) and peach × *P. davidiana* hybrids ('Nemaguard') induce the most vigor of those tested (2–4, 6).

In Canada, insufficient cold hardiness of St. Julien X and poor compatibility associated with *P. tomentosa* and *P. besseyi* have resulted in essentially no commercial use of these dwarfing rootstocks for peach (13). Efforts are being made to obtain interspecific hybrids of *P. besseyi* and *P. tomentosa* with peach in order to reduce incompatibility while retaining at least some potential to

TABLE 3. Influence of Some *Prunus* Rootstocks on Peach Scion Vigor

Prunus Rootstocks	Level of Vigor Induction of Scion					Ref.
	Very High	High	Medium	Med. −low	Low	
P. persica						
Higama		+				2, 6
GF 305		+				2, 6
Montclar		+				2
Halford		+				12, 40
Lovell		+				12
Bailey			+			13, 40
Elberta			+			13
Harrow Blood			+			34, 40
Siberian C				+		6, 34, 40
Rubira				+		2, 6
Pisa No. 2				+		53
P. persica × *P. amygdalus*						
GF 557	+					2, 6
GF 677	+					2, 6, 53
P. persica × *P. davidiana*						
Nemaguard		+				2, 6, 53
P. insititia						
GF 655-2				+		2, 6, 53
P. insititia × *P. domestica*						
St. Julien Hybrid No. 1		+				2, 6, 53
St. Julien Hybrid No. 2			+			2, 6, 53
P. domestica						
GF 43		+				2, 6, 53
Brompton			+			2, 6, 53
P. domestica × *P. spinosa*						
Damas GF 1869			+			2, 6, 53
P. besseyi					+	33, 53
P. tomentosa					+	33, 53

control tree size (33, 34). Such interspecific hybrids are not easily made and the F_1's are usually sterile (36); thus their ease of asexual propagation and suitability for micropropagation may eventually determine the extent to which they can be used. Some dwarfing rootstocks also induce small fruit size, such as *P. tomentosa*, and therefore are undesirable for this reason (Bernhard, personal communication). Protoplast fusion techniques may have some future application in facilitating somatic hybridization where sexual hybridization is difficult or impossible.

Much interest in recent years has been shown in dwarfing rootstocks for fruit trees because of the reduced cost and greater ease associated with the spraying, pruning, thinning, and harvesting of small trees. Peach trees are usually smaller and more short-lived than other species of cultivated *Prunus*. Root competition reduces peach tree growth and vigor at high densities. Some researchers promote dwarfing rootstocks as a prerequisite for successful high-density peach culture while others believe that invigorating stocks may actually be better by offsetting the loss of vigor associated with root and tree competition at high densities. Comparative studies with dwarfing and invigorating rootstocks in various high-density systems are needed to test these hypotheses.

In regions where winters are mild and injury from low winter temperatures is rare, vigor-inducing rootstocks are often advantageous and pose little threat. Thus rootstocks like 'Nemaguard' are widely used in California, Texas, and Florida (59), GF 677 is widely used in southern France and Italy (2, 3), and GF 557 in North Africa. Vigor-inducing rootstocks often delay hardening of the scion cultivar and may result in cold injury in regions farther north. This is why 'Lovell' and 'Halford' perform better than 'Nemaguard' in colder regions (12). In northern areas, rootstock influences on tree vigor and size must therefore be studied in relation to fall acclimation and cold hardiness from fall to spring during overwintering, to ensure that winter survival is taken into account (38–40).

Interstems have received little study as alternative means of controlling tree size in peach. *P. subcordata* Benth has been studied as an interstem for peach, and several clones have been found to exert different levels of vigor control (51). It is possible that hybrids of *P. persica* × *P. besseyi* or *P. persica* × *P. tomentosa* may be worthy of testing as interstems if they fail as rootstocks. They may act as compatibility bridges between otherwise incompatible combinations of peach on other species or interspecific hybrids.

5.2. Scion Physiology

Peach cultivars on their own roots may not perform in the same manner as when propagated on different rootstocks, whether they be peach seedlings, seedlings of other species, clones of other species, or clones of interspecific hybrids. Some of the changes in scion performance induced by the rootstock are subtle and barely detectable while others are dramatic and easily detected. The graft union may serve to impede the flow of water, nutrients, photosynthates, and growth regulators (39), thereby accounting for many of the rootstock effects observed on scion performance.

For many years it was generally assumed that it mattered little which source of peach seeds was used as rootstocks for peach, provided that they were inexpensive, readily available, germinated well, and provided uniform nursery trees for budding (30, 31). Thus growers usually purchased peach trees by scion cultivar name only and showed little interest about the rootstocks on which they were budded. Similarly, nurserymen listed trees by scion cultivar name only and rarely stated the rootstock seed sources used until recently (1).

Recent studies have shown that scion performance is affected in a number of ways depending on the rootstock seed source. These effects may be manifest the second year in the nursery, such as in budding success, time of scion defoliation, and scion vigor, but they are more commonly expressed in the orchard (39, 40). A number of rootstock influences on tree growth in the orchard have been reported including those on tree height, spread, volume, trunk cross-sectional area, and crotch angle (39). Rootstocks have also been shown to influence nutrient uptake (29), xylem water potential (66), micro-spore meiosis (66), time of bloom (66, 67), fruit production (40), cropping efficiency (40), time of leaf fall (39), carbohydrate metabolism (38), cold hardiness during overwintering (38), disease response (12, 41, 62, 65), and tree survival (11, 12, 40, 65).

While most of these reports deal with peach seedling rootstocks, they likely apply to other rootstocks used for peach. This recent body of knowledge of rootstock influences on peach scion performance has led to much greater awareness by researchers, growers and nurserymen of the important part played by rootstocks in successful peach production, and has greatly stimulated interest in peach rootstock research (33, 52−54, 59).

5.3. Orchard Management

New systems of intensive peach orchard management require the successful integration of a number of production practices in order to maximize fruit production, extend tree life, and improve management efficiency. Typically, the components include optimum selection of the best scion cultivars and rootstocks at the optimum spacing for the training system to be employed, along with the type of weed control and cover crop to be grown, and the type of irrigation system that might be used. A rootstock suitable for one type of production system may not be suitable for another. Therefore studies are needed to determine which rootstocks are best suited for a given system.

6. ROOTSTOCK CHARACTERIZATION

6.1. Description

There are a large number of different types of rootstocks being used for peach on a worldwide basis (52−54). Each has particular advantages and/or disad-vantages for the regions where they are used. Some have wide areas of adaptation and are used in more than one country and geographic region, while others appear to be more narrowly adapted.

The salient features of some seed-propagated rootstocks used for peach are summarized in Table 4, and those of some clonally propagated rootstocks are summarized in Table 5. It should be emphasized that many of these stocks,

TABLE 4. Characteristics of Some Seed Propagated *Prunus* Rootstocks for Peach and Nectarine[a]

Rootstocks	Nursery Characteristics							Orchard Characteristics							Disease Resistance				Nematode Resistance			References
	Germination	Uniformity	Seedling Vigor	Bud Take	Res. to Suckering	Root Hardiness	Res. to p. Mildew	Climatic Adap.	Waterlogging	Chlorosis	Anchorage	Suckering	Compatability	Vigor Induction	Crown Gall	Oak Root	Collar Rot	Chlorosis (Fe)	Root Knot	Root Lesion	Ring	
Rootstock of *P. persica*																						
Bailey	4	4	3	4	2	3	1	(3)	1	?	3	4	4	3	?	?	?	?	(1)	2	?	11, 13, 34, *b*
Chui Lum Tao	3	4	3	3	3	4	3	?	1	?	3	4	(3)	3	?	?	?	?	1	2	?	*b*, *c*, 46
Elberta	2	3	3	3	3	1	3	2	1	?	3	4	4	3	1	?	?	?	1	1	?	13, *b*
GF 305	4	4	4	3	3	(1)	?	?	1	2	3	4	4	4	2	?	?	2	1	2	?	2, 6
Halford	3	3	3	4	3	(2)	3	3	1	?	3	4	4	3	1	?	?	?	1	1	1	12, 13, *b*
Harrow Blood	3	2	2	3	3	3	3	(3)	1	?	3	4	4	3	1	1	?	?	1	1	?	3, 31, 34, 46, *b*
Lovell	3	3	3	3	3	(2)	3	3	1	?	3	4	4	3	1	1	1	?	2	1	2	6, 12, 41, 44, *b*
Nemared	3	3	4	(3)	4	(2)	3	(2)	1	?	3	4	4	4	?	1	?	?	4	2	1	4, 7, 46, *b*
Rubira	4	4	4	(3)	3	(2)	3	(2)	1	?	3	4	(4)	4	3	?	?	?	1	2	?	2, 6, *b*
Rutgers Red Leaf	3	4	4	3	3	2	3	2	1	?	2	4	2	3	1	?	?	2	1	3	?	39, 40, 46, *b*
Siberian C	4	3	3	3	2	4	3	2	1	2	2	4	3	2	2	?	?	2	1	(2)	1	6, 13, 31, 33, 34, 46
Tzim Pee Tao	4	4	3	3	3	4	3	?	1	?	3	4	(3)	3	?	?	?	?	?	(3)	?	*b*, 46
P. persica × *P. davidiana*																						
Nemaguard	3	3	4	3	3	1	?	2	1	2	3	4	3	4	3	1	1	1	3	1	1	2, 6, 41, 44
P. insititia × *P. domestica*																						
St. Julien Hyb. No. 1	3	3	3	3	(2)	(2)	?	?	3	3	3	2	2	3	4	?	(1)	3	?	?	?	2, 6
St. Julien Hyb. No. 2	3	3	3	3	(2)	(2)	?	?	3	3	3	2	2	2	3	?	1	3	?	?	?	2, 6
P. americana	(3)	(2)	(2)	(2)	1	4	?	?	(2)	?	1	1	1	(1)	?	?	?	?	?	?	?	13
P. besseyi	3	(2)	(2)	(2)	1	4	?	?	(2)	?	(2)	1	1	1	?	?	?	?	4	?	?	13, *c*
P. tomentosa	(3)	(2)	(2)	(2)	1	4	?	?	(2)	?	(2)	1	1	1	?	?	?	?	(2)	(3)	?	13, *c*

[a]Rated on a scale where 1 = poor, 2 = fair, 3 = good, 4 = excellent, ( ) = tentative rating,
? = unknown or unreported.
[b]After Layne (unpublished data).
[c]Ramming (unpublished data).

TABLE 5. Characteristics of Some Clonally Propagated *Prunus* Rootstocks for Peach and Nectarine[a]

Rootstocks	Nursery Characteristics									Effect on Scion		Other
	Ease of Propagation		Resistance to		Length of Period for Shield Budding	Ease of Trans-planting	Resistance to			Vigor	Fruit Quality	
	Hardwood Cuttings	Softwood Cuttings	Peach Rust	Crown Gall			Water-logging	Iron-Induced Chlorosis	Suckering			
P. persica × *P. amygdalus*												
GF 677	1	2	5	3–4	5	5	2	5	5	5	3–4	
GF 557	1	2	5	3–4	5	5	1	5	5	5	3–4	Resistant to root knot
P. insititia												
St. Julien GF 655.2	4	3	2	4	1	3	3	3	3	2	5	Concentrated fruit maturity
P. domestica												
Brompton	2–3	NR	3	NR	2	3–4	3	3	3–4	3–4		Susc. to collar rot
GF 43	2	2	3	5	2	4	3	3	4	4	3–4	Susc. to CLSV
P. domestica × *P. spinosa*												
Damas GF 1869	3	4	4	4	4	1	4	4	1	3	4	Incompatible with many nectarines and some peaches

[a]After Bernhard, Grasselly, and Salesses (6), Bordeaux, France. Ratings were on a scale of 1 = poor to 5 = good. NR = not rated.

especially the clonally propagated ones, have not been widely tested in North America, although they have been tested in southern Europe. Such tests should therefore be conducted before they are recommended for commercial use in North America.

It is unlikely that there will ever be a "perfect" rootstock for peach, but clearly some are more imperfect than others; only by careful testing can the best ones be identified. An example of such a testing program is the NC-140 regional rootstock project sponsored by the USDA that was initiated in 1984 and in which identical plantings were made in 16 cooperating states in the United States and one province (Ontario) in Canada to evaluate the performance of 'Redhaven' on its own roots in comparison with its performance on 'Halford,' 'Lovell,' 'Bailey,' 'Siberian C,' GF 677, GF 655.2, Damas GF 1869, and 'Citation' rootstocks. The results of the regional S-97 peach rootstock project involving eight peach seedling rootstocks in the southeastern United States have also been recently reported (12).

6.2. Trueness to Name and Virus Status

Virus diseases are not readily controlled by chemical or physical means; thus sanitation and eradication are principal means of achieving control (10). The production of virus-free nursery stock is the best means of controlling viruses in commercial nurseries and orchards (2, 19, 20, 22, 23, 64). This is accomplished by using only virus-free sources of understock and virus-free scionwood for propagation. Routine indexing of understock and scionwood is accomplished by budding them on a range of very susceptible, herbaceous and woody indicator plants which show diagnostic symptoms for given viruses. A positive reaction on an indicator plant is evidence that the budwood from the donor plant is infected. Recently, a new technique, called enzyme-linked immunosorbent assay (ELISA), has been used successfully for detection of some stone fruit viruses (14). Heat therapy is the most commonly used technique of ridding infected plants of viruses (19, 20). This technique is used to establish the virus-free mother plants which are usually maintained in isolated, insect-proof repositories and serve as the original source of virus-free stock (19, 20).

As particular rootstocks for peach become better known and in greater demand by commercial nurseries and growers, it is important that true-to-name, virus-free sources be maintained to ensure their genetic purity and sanitary status. Several public and semiprivate agencies are now in place in various parts of the world to ensure that such stocks are so maintained and are available in commercially significant quantities to the nursery trade. Examples of such programs in Canada are the British Columbia Certified Tree Fruit and Grape Program (23) and the Western Ontario Fruit Testing Association. In the United States there are certified budwood programs in a number of states such as California, Michigan, New York, Pennsylvania, and South Carolina (20). In France (2), there is a very well organized and funded program conducted by Centre Technique Interprofessionnel des Fruits et Legumes (CTIFL). Similar programs are being developed in Eastern Europe, as, for example, in Yugoslavia (Paunovic, personal communication).

Certification programs are developing rapidly in other countries because the international movement of budwood, rootstocks, and nursery trees are regulated by the plant quarantine regulations of importing countries (20). Thus countries that wish to develop their export markets must meet the certification requirements of the importing countries.

It is impossible to ensure trueness-to-name of peach seedling rootstocks obtained from seed trees growing in the wild. Similarly, it is difficult to be certain of genetic purity of seeds obtained from cultivars used for commercial drying and canning industries because most growers use several different cultivars and deliveries to the processor may sometimes be subject to error in cultivar identification. Mixing of different seed sources at the cannery is also possible.

Isolated seed and budwood orchards, exclusively maintained for the production of propagation materials, regularly indexed for viruses, and regularly inspected for health and genetic purity, are the best means of ensuring that derived nursery stock is true to name and virus free (19, 20, 23, 64). More specialized orchards of this type are needed to ensure that the fruit industry in each country has access to healthy and genetically true-to-name peach cultivars and rootstocks, both for domestic use and for the export trade.

6.3. Propagation Method

Peach rootstocks are propagated by seeds (Table 4) or vegetatively (Table 5). Seed-propagated stocks, particularly those of *P. persica*, are obtained from open pollination which 95% of the time results in self-pollination, assuming that the seed source is male fertile (24). Genetic purity of seed-propagated *P. persica* rootstocks can be improved even further by planting trees of a given rootstock seed source in solid blocks which are isolated from other peach cultivars, thereby reducing the chances of outcrossing. Usually, this added level of genetic purity is not necessary because the low percentage of off-type plants obtained can be readily rogued out of the nursery before they are budded.

Two hybrid plum rootstocks for peach developed in France (St. Julien Hybrid No. 1, St. Julien Hybrid No. 2) are produced in isolated orchards using self-sterile cultivars to ensure cross-pollination and the production of hybrid seeds. St. Julien Hybrid No. 1 is produced by the natural hybridization of St. Julien d'Orleans × Common Mussel while St. Julien Hybrid No. 2 is the result of natural hybridization of St. Julien d'Orleans × Brompton (2, 6).

Vegetatively propagated peach rootstocks are commercially propagated by various means including hardwood, semihardwood, and softwood cuttings under mist (2, 3, 5, 15, 28, 44, 63). Some have even been propagated by leaf cuttings (5) and nodal cuttings. In Italy, the peach × almond hybrids from France (GF 557, GF 677) are now being mass propagated by micropropagation

(3). Peach rootstocks vary in their ease of propagation by each of the fore-going methods; thus it is important to choose the most appropriated method for each rootstock. Clonally propagated stocks offer the advantage over seed-propagated ones of greater uniformity in the nursery and orchard. They have the disadvantage of usually being more labor intensive and expensive to produce than seed-propagated stocks and their rooting behavior may be different.

7. NEW AND FUTURE ROOTSTOCKS

There are a number of requirements for a generally acceptable rootstock for peach. No single rootstock possesses all of the desired attributes, but a number of them possess sufficient traits to be commercially useful. Ideally, a good rootstock for peach would also be fully satisfactory for several other cultivated *Prunus* spp. It should be easy to propagate by seed and asexually and provide good early growth in the nursery. It should be cold hardy, drought tolerant, heat tolerant, and disease, nematode, and insect resistant. It should be compatible with a broad range of scion cultivars. It should be disease free, especially virus free. It should possess some degree of vigor control of scion cultivars and be adapted to a wide range of soil types, soil reaction, soil fertility, and soil moisture. It should have a low level of suckering in the nursery and the orchard. It should provide uniform understock in the nursery for budding or grafting, be easy to bud, and have a long budding season. It should produce efficient orchard trees that crop early, and produce heavily from year to year while bearing fruit of adequate size and quality. Orchard trees should also be well anchored.

It is unlikely that a single rootstock for peach will have all of these attributes. Nevertheless, it is highly desirable to incorporate as many of these traits as possible to increase usefulness and broaden areas of adaptation of new peach rootstocks. Studies to improve peach rootstocks are under way in many parts of the world including North and South America, Western and Eastern Europe, Asia, and Oceania (53). Surely they will lead to the development of better, more versatile rootstocks for peach in the future.

As we improve our ability to make interspecific *Prunus* hybrids, either through more sophisticated techniques of interspecific hybridization or proto-plast fusion (36), it is likely that future rootstocks for peach may be complex hybrids of several *Prunus* species instead of one or two, which is now the case. This will permit greater exploitation of genetic resources present in other *Prunus* species than is now possible, and will likely lead to more versatile rootstocks for peach in the future (Table 6).

It should be possible from our existing knowledge to do much of the preliminary screening of potential rootstocks in the laboratory, growth chamber, and greenhouse, especially for characters like cold hardiness (47, 48), drought tolerance (47), disease, insect, and nematode resistance (9, 10, 24, 27),

TABLE 6. Promising Peach Rootstock Candidates Requiring Further Trial[a]

P. persica	Origin	Interspecific Hybrids	Origin
1. *Selections*		1. *Peach × Almond*	
Montalar (S-2489)	France	5-3-6-65, 216-863	Czechoslovakia
Rubira (S-2605)	France	GF 557	France
Higama' (S-2543)	France	MB 1 and 4	Hungary
(S-2535)	France	Selections	Spain, Yugoslavia
Tzim Pee Tao	China	Peach × Marcona	France
Chui Lum Tao	China		
I-D-20	Greece	2. *Peach × Plum*	
I-D-37	Greece	Myran	France
Pisa No. 5	Italy		
Pisa No. 6	Italy	Citation	U.S.
R32-10	Japan		
R32-16	Japan	3. *Plum Hybrids*	
R33-1	Japan	Myrabi (P2032)	France
R33-3	Japan	P2037 (*P. bess.* × *P. cera.*)	France
Ohatsummo	Japan		
T16	Romania	4. *Plum*	
T163	Romania	Myrobalan	Australia
		Myrobalan B	Australia
2. *Hybrids*		BD-SU-I (*P. davidiana*)	Czechoslovakia
Nemared	U.S.		
115-5, 115-104,			
114-102	U.S.		
R15-2	Japan		
R17-8	Japan	5. *P. institia*	
R22-2	Japan	St. Julien A × Plum	France
		(Hybrids No. 1, No. 2)	
R26-2	Japan	St. Julien 53.7; 655.2	France
R27-1	Japan	Pollizo de Murcia	Spain

[a]After Rom (53).

ease of conventional vegetative propagation (6), and micropropagation (3). It should also be possible to select for characters like salt tolerance, resistance to iron-induced chlorosis, and waterlogging using specially constructed tanks outdoors (55–58) before going to the nursery and orchard for more advanced testing. Utilization of such techniques should greatly reduce the time involved in developing better rootstocks for peach. Regional trials of the best selections should greatly increase confidence in making sound recommendations to nurserymen and growers (12). Certification programs (20) for the production of virus-tested, true-to-name rootstocks will ensure the genetic purity and health of the newly developed stocks. Modern peach rootstock research including these facets is likely to produce the kinds of new rootstocks needed in the shortest possible time.

INTERNATIONAL CATALOG OF ROOTSTOCK BREEDING PROGRAMS

Brazil
CNPFT-EMBRAPA
CAIXA Postal 403
Pelotas, RS 86100 Brazil
Maria Bassols Raseira

Bulgaria
Fruit Research Station
Pomprie-Burgas, Bulgaria
Jordanka Gurcheva

Canada
Harrow Research Station
Agriculture Canada
Harrow, Ontario,
NOR 1GO Canada
Richard Layne

France
Institut National de la Recherche
Agronomique B. P. 131-33140
Pont-de-la-Maye, France
G. Salesses

INRA Station de Recherches
Fruitieres Mediterraneenes
BP91 Cantarel Montfavet
France
Charles Grasselly

India
Himachal Pradesh University
Solan, India
J. S. Chauhan

Italy

Istituto di Coltivazioni Arboree
Via del Borghetto 80
Pisa, Italy 56100
F. Loreti, R. Guerriero

Istituto Sperimentale de la
Fruticoltura
Ciampino Aeroporto
00040 Roma, Italy
Antonio Nicotra

Japan

Ibaraki Prefectural Horticultural Experimental Station
Yatabe, Ibaraki, Japan 305
Masao Yoshida

Mexico

Centro de Investigoiciones
Agricolas del Nozoste
APDO 1031 Hermosillo, Sonora, Mexico
J. Martinez Tellez

Prol. Zargoza 408
Jardines dela Jacienda
Queretaro, Qro 76180 Mexico
Salvador Perez

Peoples Republic of China.

Department of Horticulture
Zhejiang Agricultural University
Hangzhou, Zhesiang, PRC
Shen De-xu

United States

Department of Horticulture and Forestry
University of Arkansas
Fayetteville, Arkansas 72701
Roy Rom, James Moore

Zaiger's Nursery
Modesto, CA 95351
Floyd Zaiger

USDA/ARS
Hort Crops Research Laboratory
Fresno, CA 93727
David Ramming

Department of Fruit Crops
University of Florida
Gainsville, FL 32611
Wayne Sherman

USDA/ARS
S.E. Fruit and Tree Nut Research Laboratory
Byron, GA 31008
Dick Okie

Department of Horticulture
North Carolina State University
Raleigh, NC 27650
Dennis Werner

Department of Horticulture
Texas A&M University
College Station, TX 77843
David Byrne

USSR
Botanical Garden of Nikitski
Ualta-Crimee, USSR
N.G. Popok

Vavilov All-Union Research Institute of Plant Growing
Krymsk, USSR
G.V. Yeremin, Y.A. Gnezdilov

REFERENCES

1. Anonymous (1983). Hilltop Orchards and Nurseries, Inc. Hartford, Michigan 75th Anniversary edition, Catalogue, 56 pp.
2. Anonymous (1978). *Les porte-greffes du pecher*, CTIFL-Documents No. 59-IVe.
3. Bellini, E. (1982). "La colturo del pesco," in *Prospettive per L'agricoltura Collinare Fiorentina*, Camera di Commercio Industria, Artigianato e Agricolture di firenze, pp. 13-57.
4. Bernhard, R., and E. Germain (1975). Analyse du mode d'action de porte-greffes vigoureux: Case de hybrides amandier × pêcher porte-greffes du pêcher, *Ann. Amélior. Plantes*, **25**(3), 321–336.

5. Bernhard, R., and C. Grassely (1959). Les pruniers porte-greffes du pêcher, *Jour. Fruit. et Maraich. d'Avignon*, pp. 75—100.

6. Bernhard, R., C. Grassely, and G. Salesses (1979). Orientations des travaux de selection des porte-greffes du pêcher à la station de recherches d'arboriculture fruitière de Bordeaux, *Proc. Eurcarpia Fruit Sect. Symp. Tree Fruit Breed*, Angers, France, Sept. 3—7.

7. Brady, N. C. (1974). *The Nature and Properties of Soils*, 8th ed., Macmillan, New York, pp. 372—421.

8. Cummins, J. N., and H. S. Aldwinckle (1983). "Rootstock breeding," *in Methods in Fruit Breeding*, J. N. Moore and J. Janick Eds., Purdue University Press, W. Lafayette, IN., pp. 294—327.

9. Daubney, H. 1983). "Insect, mite, and nematode resistance," *in Methods of Fruit Breeding*, J. N. Moore and J. Janick, Eds., Purdue University Press, W. Lafayette, IN., pp. 216—241.

10. Dayton, D. F., R. L. Bell, and E. B. Williams (1983). "Disease resistance," in *Methods in Fruit Breeding*, J. N. Moore and J. Janick, Eds., Purdue University Press, W. Lafayette, IN., pp. 189—215.

11. Doud, S. L. (1980). Hardiness and survival effects of several peach seedling rootstocks, *Compact Fruit Tree*, **13**, 123—126.

12. Dozier, W. A. Jr., J. W. Knowles, C. C. Carlton, R. C. Rom, E. H. Arrington, E. J. Wehunt, U. L. Yadava, S. L. Doud, D. F., Ritchie, C. N. Clayton, E. I. Zehr, C. E. Gambrell, J. A. Britton, and D. W. Lockwood (1984). Survival, growth, and yield of peach trees affected by rootstocks, *HortScience*, **19**(1), 26—30.

13. Elfving, D. C., and G. Tehrani (1984). *Rootstocks for Fruit Trees*, Ontario Ministry of Agr. & Food. Publ. 334., 34 pp.

14. Forer, L. B., C. A. Powell, and R. F. Stouffer (1984). Transmission of tomato ringspot virus to apple cuttings, and to cherry and peach seedlins by *Xiphinema rivesi*, *Plant Disease*, **68**, 1052—1054.

15. Grasselly, C. (1981). *Rootstock Research Programs, Past and Present: Rootstock Variety Interactions Affecting Fruit Inductions and Growth*, Peach Production Short Course III, Clemson University, S. Carolina, 6 pp.

16. Grasselly, C., G. Olivier, and M. Edin (1980). Les pruniers porte-greffes du pêcher: Vingt années d'expérience du comportement des principeaux types, *Arboriculture Fruitiere*, **322**, 47—52.

17. Grassely, C., and P. Roger (1970). Comportement dans le sud-ouest des pruniers porte-greffes du pêcher, Bull. Tech. Info. No. 254, pp. 623-632.

18. Grassely, C., and R. Saunier (1968). Incompatibilité de greffage de quelques variétés de nectarines sur prunier damas, *Pomol. Francaise*, **10**(6), 1—4.

19. Fridlund, P. R. (1980). The IR-2 program for obtaining virus-free fruit trees, *Plant Disease*, **64**, 826—830.

20. Fridlund, P. R. (1983). "Certification and supporting pyhtopathology," in *Methods in Fruit Breeding*, J. N. Moore and J. Janick, Eds., Purdue University Press, W. Lafayette, IN., pp. 398—421.

21. Funt, R. C., and B. L. Goulant (1981). Performance of several peach cultivars on *Prunus tomentosa* and *Prunus besseyi* in Maryland, *Fruit Var. J.*, **35**(1), 20—23.

22. Gilmer, R. M., K. D. Brase, and K. G. Parker (1957). *Control of Virus Diseases of Stone Fruit Nursery Trees in New York*, New York State Agr. Exp. Sta. Bull. No. 779, 53 pp.

23. Hansen, A. J., H. J. O'Reilly, and J. M. Yorston (1982). *Stone Fruit Virus Diseases of British Columbia*, British Columbia Min. Agr. and Food Bull. 82-5, 23 pp.

24. Hesse, C. O. (1975) "Peaches," in *Advances in Fruit Breeding*. J. Janick and J. N. Moore Eds., Purdue University Press, W. Lafayette, In., pp. 285–335.

25. Hummel, R. (1977). A classification of hardy North American *Prunus* cultivars and native species based on hardiness zones, *Fruit Var. J.*, **31**(3), 62–69.

26. Jailloux, F., and G. Froidefond (1978). Réceptivité à l'égard du *Fusicoccum amygdali* Del. de quelques hybrides Pêcher × Armandier porte-greffes du pêcher et de l'armandier, *Ann. Phytopathol.*, **10**(2), 171–175.

27. Johnson, P. W., V. A. Dirks, and R. E. C. Layne (1978). Population studies of *Pratylenchus penetrans* and its effects on peach seedling rootstocks, *J. Amer. Soc. Hort. Sci.*, **103**(2), 169–172.

28. Knight, R. L. (1969). *Abstract Bibliography of Fruit Breeding and Genetics to 1965*, Prunus., Comm. Agr. Bur. Tech. Comm. No. 31, 649 pp.

29. Knowles, J. W., W. A. Dosier, Jr., C. E. Evans, C. C. Carlton, and J. M. McGuire. (1984). Peach rootstock influence on foliar and dormant stem nutrient content, *J. Amer. Soc. Hort. Sci.*, **109**(3), 440–444.

30. Layne, R. E. C. (1974). Breeding peach rootstocks for Canada and the northern United States, *HortScience*, **9**(4), 364–366.

31. Layne, R. E. C. (1975). New developments in peach varieties and rootstocks, *Compact Fruit Tree*, **8**, 69–77.

32. Layne, R. E. C. (1976). Influence of peach seedling rootstocks on perennial canker of peach, *HortScience*, **11**(5), 509–511.

33. Layne, R. E. C. (1978). Dwarfing rootstocks and Canadian peach varieties, *Pennsylvania Fruit News*, **17**(4), 25–29.

34. Layne, R. E. C. (1980). Prospects of new hardy peach rootstocks and cultivars for the 1980's, *Compact Fruit Tree*, **13**, 117–122.

35. Layne, R. E. C. (1982). Cold hardiness of peaches and nectarines following a test winter, *Fruit Var. J*, **36**(4), 90–98.

36. Layne, R. E. C. (1983). "Hybridization," in *Methods in Fruit Breeding*, J. N. Moore and J. Janick, Eds., Purdue University Press, W. Lafayette, IN., pp. 48–65.

37. Layne, R. E. C. (1984). Breeding peaches in North America for cold hardiness and perennial canker (*Leucostoma* spp.) resistance—review and outlook, *Fruit Var. J.*, **38**(4), 130–136.

38. Layne, R. E. C., and G. M. Ward (1978). Rootstock and seasonal influences on carbohydrate levels and cold hardiness of 'Redhaven' peach, *J. Amer. Soc. Hort. Sci.*, **103**(3), 408–413.

39. Layne, R. E. C., H. O. Jackson, and F. D. Stroud (1977). Influence of peach seedling rootstocks on defoliation and cold hardiness of peach cultivars, *J. Amer. Soc. Hort. Sci.*, **102**(1), 89–92.

40. Layne, R. E. C., G. M. Weaver, H. O. Jackson, and F. D. Stroud (1976). Influence of peach seedling rootstocks on growth, yield and survival of peach scion cultivars, *J. Amer. Soc. Hort. Sci.*, **101**(5), 568–572.

41. Lownsbery, P. F., H. English, G. R. Noel, and F. J. Schick (1977). Influence of Nemaguard and Lovell rootstocks and *Macroposthoria xenoplax* on bacterial canker of peach, *J. Nematol*, **9**(3), 221–224.

42. Massonie, G., and P. Paison (1979). Resistance de deux variétés de *Prunus persica* L. Batsch à *Myzus persicae* Suiz. et à *Myzus varians Davids: Étude préliminaire des mécanismes de la résistance, Ann. Zool. Ecol. Anim.*, **11**(3), 479–485.

43. Nelson, S. H. (1968). Incompatibility survey among horticultural plants, *Internat. Plant Prop. Soc.*, **18**, 343–407.

44. Norton, R. A., C. J. Hansen, H. J. O'Reilly, and W. H. Hart (1963). *Rootstocks for Plums and Prunes*, California Agr. Exp. Sta. Leaf. No. 158, 8 pp.

45. Powell, C. A. (1984). Comparison of enzyme-linked immunosorbent assay procedures for detection of tomato ring spot virus in woody and herbaceous plants, *Plant Diseases*, **68**(10), 908–909.

46. Potter, J. W., V. A. Dirks, P. W. Johnson, T. H. A. Olthof, R. E. C. Layne, and M. M. McDonnell (1984). Response of peach seedlings to infection by the root lesion nematode *Pratylenchus penetrans* under controlled conditions, *J. Nemat.*, **16**(3), 317–322.

47. Quamme, H. A., and C. Stushnoff (1983). "Resistance to environmental stress," in *Methods in Fruit Breeding*, J. N. Moore and J. Janick, Eds., Purdue University Press, W. Lafayette IN., pp. 242–266.

48. Quamme, H. A., R. E. C. Layne, and W. G. Ronald (1982). Relationship of supercooling to cold hardiness and the northern distribution of several cultivated and native *Prunus* species and hybrids, *Can. J. Plant Sci.*, **62**, 137–148.

49. Ramming, D. W., and O. Tanner (1983). 'Nemared' peach rootstock, *HortScience*, **18**(3), 376.

50. Rehder, A. (1967). *Manual of Cultivated Trees and Shrubs Hardy in North America*, 2nd ed., Macmillan, New York, 996 pp.

51. Roberts, A. N., and M. N. Westwood (1981). Rootstock studies with peach and *Prunus subcordata* Benth, *Fruit Var. J.*, **35**(1), 12–23.

52. Rom, R. C. (1982). A new philosophy for peach rootstock development, *Fruit Var. J.*, **36**(2), 34–37.

53. Rom, R. C. (1983). The peach rootstock situation: An international perspective, *Fruit Var. J.*, **37**(1), 3–14.

54. Rom, R. (1984). A new generation of peach rootstocks, *Proc. 43rd Nat. Peach Council Annual Convention*, pp. 59–68.

55. Salesses, G., and C. Juste (1971). Recherches sur l'asphyxie radiculaire des arbres fruitieres à noyau 11. Comportement des porte-greffes de types pêcher et prunier: Étude de leur teneur en amygdaline et des facteurs intervenaut dans l'hydrolyse de celle-ci, *Ann. Amélior. Plantes*, **21**(3), 265–280.

56. Salesses, G., H. Saunier, and A. Bonnet (1970). L'asphyxie radiculaire chez les arbres fruitieres, Bull. Tech, Info. No. 251, pp. 403–415.

57. Saunier, R. (1966). Méthode de détermination de la résistance à l'asphyxie radiculaire de certains porte-greffes d'arbres fruitiers, *Ann. Amélior. Plantes*, **16**(4), 367–384.

58. Saunier, R. (1970). *Résistance à l'asphyxie radiculaire de quelques port-greffes d'arbres fruitiere*, CTIFL-Documents No. 26.

59. Sharpe, R. (1974). Breeding peach rootstocks for the southern United States, *HortScience,* **9**(4), 362–363.

60. Traquair, J. (1984). Etiology and control of orchard replant problems: A review, *Can. J. Pl. Path.* **6**, 54–62.

61. Weaver, G. M. (1965). Preliminary evidence of host resistance to the peach tree borer, *Saaninoidea exitosa, Can. J. Plant. Sci.,* **45**, 293–294.

62. Weaver, D. J., S. L. Doud, and E. J. Wehunt (1979). Evaluation of peach seedling rootstocks for susceptibility to bacterial canker, caused by *Pseudomonas syringae, Pl. Dis. Reptr.,* **63**(5), 364–367.

63. Westwood, M. N. (1978). *Temperature Zone Pomology,* Freeman, San Francisco, 428 pp.

64. Wood, G. A. (1979). *Virus and Virus-like Deseases of Pome Fruit and Stone Fruits in New Zealand,* Dept. Sci. Ind. Res. Bull. No. 226, 87 pp.

65. Yadava, U. L., and S. L. Doud (1980). "The short life and replant problem of deciduous fruit trees," in *Horticultural Reviews,* Vol. 2, J. Janick, Ed., Avi Publ. Co., Westport, CT, pp. 1–116.

66. Young, E, and J. Houser (1980). Influence of Siberian C rootstock on peach bloom delay, water potential, and pollen meiosis, *J. Amer. Soc. Hort. Sci.,* **105**(2), 242–245.

67. Young, E., and B. Olcott-Reid (1979). Siberian C rootstock delays bloom of peach, *J. Amer. Soc. Hort. Sci.,* **104**(2), 178–181.

7

CHERRY ROOTSTOCKS

Ronald L. Perry
Michigan State University
East Lansing, Michigan

1. ORIGIN OF CHERRY

Many species of cherry exist, but only two, *Prunus avium* L. for sweet cultivars and *P. cerasus* L. for sour cherry cultivars, are commercially important (Table l). Both species are natives of south-central Europe and Asia Minor (48, 118). *Prunus avium* is the oldest species and evidence suggests that *P. cerasus* ($2n = 32$) arose from an unreduced pollen grain of *P. avium* ($2n = 16$) crossed with *P. fruticosa* Pall. ($2n = 32$) (76, 116). Cultivars known as Duke cherries, which are not commercially important, result from hybridization between *P. avium* and *P. cerasus* (5, 113).

Hedrick's (48) review in 1915 of the cherry indicates that it probably became domesticated before the rise of Greek civilization. It is likely that cultivars were introduced into Greece from Asia Minor. The cultivation of the cherry persisted through the centuries in Europe, and early colonists from England introduced it to North America in 1629. Initially, *P. cerasus* cv. Red Kentish adapted more readily in North America because of superior tolerance to insects, diseases, and winter temperatures. Today, the distribution of cherry is worldwide between 35 and 55°N and S latitudes or outside of these limits where temperature and other factors are favorably modified (118).

Geographical origin and the world distribution of cherries and production are given in Tables 1, 2, and 3.

2. ROOTSTOCK USAGE

2.1 History

The primary cherry rootstocks of use in the world are seedlings or clonal selections of *P. avium* L. and *P. mahaleb* L. Wild selections of *P. avium,* widely known as "mazzard," "wild cherry," and "gean," date back to

TABLE 1. Geographical Origin of Cherry Species[a]

Species	Distribution
North America	
P. pumila L.	E. North America
P. besseyi Bailey	E. North America
P. pennsylvanica L.	E. North America
P. virginiana L.	E. North America
P. serotina Ehrh.	North America
Europe	
P. avium L.	Europe, Western Asia
P. cerasus L.	Europe, Western Asia
P. mahaleb L.	Europe, Western Asia
Caucasia	
P. laurocerasus L.	
East Asia	
P. japonica Thumb.	China, Manchuria, N. Korea
P. macradenia Koehne	China
P. tomentosa Thunb.	China, Himalayas, Turkestan
P. glandulosa Thunb.	China, Manchuria, Korea
P. incisa Thung.	Japan
P. subhirtella Miq.	Japan, China
P. campanulata Maxim.	Japan, Taiwan
P. serrula Franch.	Yunan, China
P. serrulata Lindl.	China, Japan, Korea
P. sieboldii (Carr.) Wittmack	Japan
P. mugus Hand.-Mass.	China
P. yedoensis Matsum.	Japan
P. sargentii Rehd.	Japan
P. fruticosa Pall.	Siberia, Europe
P. pseudocerasus Lindl.	China
P. padus L.	Japan, Eurasia
P. canescens Bois.	China
P. concinna Koehne	China
P. nipponica Matsum.	Japan

[a]From references 5, 25, and 48.

use for all cherries circa 330–400 B.C. by the Greeks and Romans (48). *P. mahaleb* , also known as "St. Lucie" or "perfumed cherry," provided French horticulturists the first alternative rootstock in 1768. *Mahaleb* was the best rootstock for most sweet cherries on calcareous droughty soils in France. British writers, however, in the early 1800s found that "mahaleb" dwarfed sweet cherries but could not adapt to their soils. Thus the controversy over the use of mazzard versus mahaleb, which began in Western Europe and continues worldwide into the present day, is stimulated by differences in soil adaptability (48).

TABLE 2. World Cherry Production[a]

Soviet Union	470^b(T)	Hungary	60^b(T)
United States	275	Bulgaria	60^b(S)
West Germany	230	Portugal	45^b(S)
Italy	175	Germany	40^b(S)
France	110	Greece	30^b(S)
Turkey	105	Japan	20
Yugoslavia	100^b(T)	Canada	15
Poland	85^b(T)	Belgium-Lux.	15
Spain	75	Australia	10
Czechoslovakia	65^b(S)	New Zealand	10
Rumania	65^b(T)		

Also grown commercially in:
 Sweden
 Netherlands
 Argentina

[a]Childers (20).
[b]Thousand of metric tons, sour (T) or sweet (S) cherries predominate.

TABLE 3. Cherry Production[a] **and Principal Sweet Cherry Cultivars**[b] **in North America**

	Sweet	Tart	Cultivars[c]
Michigan	52.6	113.0	Napoleon, Schmidt, Hedelfingen, Emperor Frances
Washington	110.6	0	Bing, Napoleon, Van
Oregon	74.0	4.5	Napoleon, Bing, Van
California	74.0	0	Bing, Napoleon, Larian
New York	7.4	22.0	Windsor, Napoleon, Schmidt
Utah	8.6	14.7	Bing, Lambert, Van
Pennsylvania	1.2	6.6	Windsor, Bing, Napoleon
Wisconsin	0.	12.1	—
Montana	3.3	0	Lambert, Van
Idaho	6.1	0	Bing, Lambert, Corum
Colorado	0	1.7	Bing, Lambert, Vega
U.S. total	337.8	174.6	
B.C.[d]	16.0	1.6	Van, Bing, Lambert, Stella
Ontario	3.8	15.5	Hedelfingen, Vista, Venus, Valera

[a]Childers (20) data averaged 1979–1982 in millions of pounds.
[b]In order of importance from Andersen (2).
[c]'Montmorency' makes up 95–99% of all sour cherries produced.
[d]Andersen (2) data for 1979 in millions of pounds.

In North America, first mention of cherry rootstocks was recorded in 1730 for mazzard and 1845 for mahaleb in New York (48). By 1880 mahaleb became more popular than mazzard and, according to Hedrick (48), eventually superseded mazzard by 1915 as the stock of choice for all sweet and sour cherries. Many reasons are reported for the overwhelming acceptance of mahaleb (95%); however, ease in nursery care and propagation was primary. Mazzard seedlings are weakened in the nursery because of early defoliation caused by severe infestations of leaf spot, *Coccomyces hiemalis* Higg. However, opinions differed among horticulturists and among geographic regions at this time. For example, Bailey (4) in 1914 in his nursery book recommended the use of mazzard over mahaleb. By the 1920s, mazzard had regained its prominence for sweet cherries, while mahaleb remained the primary stock for sours. A 14-year rootstock trial reported by Howe in New York in 1927 (52) convinced horticulturists that mazzard was the best rootstock. In that study, 40 cultivars of sweet, sour, and Duke (sweet × sour hybrid) cherries had a 50% survival on mahaleb compared to close to 100% on mazzard.

Shortly after, other eastern trials echoed Howe's conclusions, which favored mazzard stocks for sweet cherries and condemned mahaleb as dwarfing and short-lived (3, 21). Phillip (81) in 1926 in California reported that of 5306 ha of sweet cherries, 61, 24, and 15% were on mazzard, mahaleb, and *P. cerasus* cv. Stockton Morello respectively. This ratio changed only slightly by 1951 to 65, 30, and 5%, respectively (28). 'Stockton Morello,' a dark-fleshed morello-type sour cherry has had limited commercial use on heavy, wet soils east of Stockton, California since 1875 (28). It must be noted that most American nurserymen obtained the bulk of mazzard and mahaleb seedlings from Western Europe (France principally) prior to 1931 when the federal government established plant quarantine procedures (22). Bryant in 1940 (16) in Colorado and Coe in 1945 in Utah (23) demonstrated that mahaleb was the superior choice for sweet and sour cherry cultivars in arid, gravelly soils. Contrary to the previous reports, trees on mahaleb were more vigorous, productive, drought tolerant, and cold hardy (16, 23).

A clonal mazzard selection developed in England, F 12/1, has seen limited use in the humid, wet districts of Oregon and western Washington since the early 1960s (84). Growers have avoided severe bacterial trunk canker infections of susceptible sweet cherry cultivars by grafting scions to scaffolds of the canker-resistant F 12/1 (17).

2.2. Present-Day Use

2.2.1. North America.
The rootstock picture in the United States is changing, principally because of the increased use of marginal soils. Mazzard is becoming the rightful choice for these sites. The arid states of Utah, Montana, Colorado, and California remain loyal to mahaleb in well-drained soils (Table 4). The scion cultivars heavily grown in these states are highly compatible with mahaleb and, concurrently, provide for tolerance to buckskin disease (Western X),

TABLE 4. Estimate of Present-Day Rootstock Usage for Cherry[a]

Geographical Areas	Sweet			Sour	
	P. avium	P. mahaleb	Other	P. avium	P. mahaleb
United States					
East					
NY	90[b]	10		70	30
PA	—	—		20	80
MI, WI, others	95	5		10	90
West					
UT, MT, CO	10	90		0	100
WA	90	5	5 Colt	10	90
OR	80	20	5 Colt	60	40
CA	30	65	5 Colt & Stn. Mor.	—	—
Canada					
BC	90 (F 12/1)	10		90 (F 12/1)	10
ONT	95	5		10	90
United Kingdom	97 (F 12/1)	3		97 (F 12/1)	3
France	30	70		3	97
Federal Republic of Germany	100	—		70	30

[a]Based on surveys taken in winter 1984–1985.
[b]Percentage of sweet cherry or sour cherry trees planted.

drought, smaller tree size, good precocity, and high productivity (23, 65). The ratio of use of mazzard and mahaleb in California has completely reversed since reported by Day in 1951 (28)(Table 4). Sweet cherries in the East and the Pacific Northwest are mostly grown on mazzard, while sour cherries in most areas of the United States and Canada continue to be grown mostly on mahaleb. 'Montmorency' on mazzard is increasingly being requested, particularly for heavier wet soils.

The accumulation of viruses has caused a steady decline in the use of 'Stockton Morello' in California (65). Virus-free 'Stockton Morello,' now available, does not exhibit the same degree of size reduction as the earlier-used virus-contaminated clone. To a small extent, 'Colt,' a new hybrid semidwarfing rootstock (*P. avium* × *pseudocerasus*), is being used in California and Washington, but more time is needed to assess its value. In Washington, some growers and the Carlton Nursery Company have experienced good results with sweet cherry scions budded on 'Montmorency'/mahaleb or mazzard (J. Ballard, personal communication). A six-year study reported by Larsen (61), however, cautions that 'Bing'/'Montmorency' unions may be weak.

2.2.2. Europe. Sweet cherry growers in the United Kingdom have depended on F 12/1 for the last 50 years (114). 'Colt' is gaining in commercial acceptance for sweet, sour, and ornamental cherries (114). In France, Italy, and Spain, mahaleb seedlings or a clonal selection, SL 64, is used for gravelly, calcareous, droughty soils and mazzard seedlings are used for heavy soils (14). Sweet and morello sour cherries ('Schattenmorelle') are grown in West Germany primarily on F 12/1 or seedlings of Harz mountains mazzard. 'Schattenmorelle' is propagated on *P. mahaleb* seedlings and clones in sandy soils (7; 111; and W. Gruppe, personal communication). Sweet and morello cherries are commonly propagated on *P. mahaleb* in Bulgaria and Hungary, while in Romania scion cultivars are grown on selections of *P. mahaleb* and *P. cerasus* (55; A. Iezzoni, personal communication).

In Yugoslavia (55), 'Oblacinska' morello cherry is grown on its own roots or used as a rootstock for other morello cultivars such as 'Schattenmorelle' and 'Nefris.' Cold-hardy *P. cerasus* stocks are recommended for cherry in the Ukraine, Russia (102).

2.2.3. Australia, New Zealand, South Africa. Mazzard, F 12/1, is used almost exclusively in Victoria (25), South Australia (8), New Zealand, and South Africa (1). Sweets in Victoria are high budded in order to avoid *Pseudomonas* trunk canker infections (8).

3. ADAPTATION

3.1. Climate

Mahaleb has long been recommended for well-drained soils in colder cherry−growing climates (16, 23, 104). Under controlled freezing experiments, Carrick (18) killed mazzard roots at −10 to −11°C and killed mahaleb roots at −15°C. *P. cerasus* is considered more cold hardy than mahaleb and mazzard and was recommended for Iowa by Price and Little in 1903 (83). In a sweet cherry rootstock trial in Michigan, 'Colt' and mazzard root systems of several four- five-year-old trees were dead following the 1984 winter (Perry, unpublished) data). Reports from Canada, Holland, and Germany indicate that 'Colt' plant tissue is very cold tender (100, 113).

3.2. Soils

Soil is one of the most important factors to successful production in cherry-growing areas (23, 30). Generally, *P. mahaleb*, *P. avium*, and *P. cerasus* adapt best in light, well-drained soils, loam to clay loam soils, and heavy clay soils, respectively (23, 110, 113). Although there is some evidence to the contrary, the literature is in general agreement that order of rootstock sensitivity to saturated soils is as follows: mahaleb > mazzard > *P. cerasus* (23, 28, 53, 87).

'Early Burlat' trees submerged in water at INRA Bordeaux began expressing symptoms of asphyxiation at 12 days on *P. mahaleb* clone SL 64, 14−15 days on mazzard seedling, and 22 days on *P. cerasus* cv. 603 (14). Chandler (19) described mahaleb as "not tolerant of wet soils" (Fig. 1) while mazzard is "moderately tolerant of poorly aerated soils similarly to peach, almond or apricot roots, but not so tolerant as myrobalan plum, apple or pear roots." Day (28) suggests that mahaleb is more sensitive to wet soils than peach. For Ontario, Upshall (110) suggests that mahaleb can only be used in deep, well-drained soils and mazzard is recommended where high water tables in the spring occur. He warns that "productivity even on mazzard on poorly drained soil is likely to be low and perhaps unprofitable."

The root system of mahaleb is characterized as being poorly branched and prone to a deep, vertical rooting habit. Conversely, roots of mazzard form a deep root system which is well branched with a dense mat of highly fibrous roots near the soil surface (23, 28, 87). A root distribution study conducted by Perry in Michigan in 1984 revealed that mahaleb roots of three-year-old 'Montmorency' trees grown in raised beds did not extend as deep into a dense clay 'B' horizon as did mazzard roots with the same scion cultivar.

Mahaleb-rooted trees are more drought tolerant and tolerant of calcareous soil than mazzard and 'Stockton Morello' (14, 16, 23, 28, 48). *Mahaleb* has long been recommended for deep, well-drained, sandy-porous soils, In contrast,

Figure 1. Two dead five-year-old 'Montmorency' trees (left center) on mahaleb seedling rootstocks in clay soil at Hartford, Michigan. Healthy trees to the right on M×M 2 (mazzard × mahaleb hybrid) and far left on M×M 14.

'Stockton Morello,' which has been recommended in California for wet clay soils, has proven to be unacceptable in sandy-porous soils (23, 28). Trees on 'Stockton Morello' in light soils are very dwarfed and require staking to correct poor anchorage (23, 28). Poor anchorage has also been reported for sweet cherries grown on *P. cerasus* clones in Britain (37) and in California (88). Day (28) and others have often suggested that the deep-rooting habit of mahaleb avoids drought stress by drawing from greater groundwater reserves. However, observations in a drought stress study conducted by Beckman and Perry (unpublished data) during the summer of 1982 suggest that an internal mechanism may be involved. In that study, three-year-old seedling trees of mahaleb and mazzard were grown in 12-L containers in full sun and regularly irrigated or nonirrigated. Nonirrigated mazzard seedlings expressed drought stress symptoms readily, while mahaleb rarely appeared to be affected. The dense development of surface roots of mazzard in contrast to absence of mahaleb surface roots subjects trees on mazzard to limb dieback as found in deeply cultivated California orchards (28).

Regardless of orchard crop and rootstock, shallow soils limited by a dense layer or shallow water table present a management problem (8, 87). Cherry rootstock recommendations for these soils depend on climate and geography. Mahaleb has performed best on shallow soils in the arid climates of Utah and California (23, 28), while mazzard is preferred in the less arid East (3, 80, 110).

3.3. Pests

There are a number of disease, insect, and rodent problems which can plague cherry rootstocks in the nursery and in the field. Table 5 describes rootstock differences when exposed to various pests. Rootstock use in many growing areas is heavily dependent upon tolerance or susceptibility characteristics to pests. For example, canker-susceptible sweet cherry scions are grown in humid districts of the North American Pacific Northwest when budded high on resistant F 12/1 (16). Conversely, F 12/1 cannot be used in Michigan, where light soils harbor crown gall.

3.3.1. Nematodes. *Pratylenchus* spp. nematodes are serious pests of cherry orchards in the East (63) and West (95). Fumigation of unthrifty cherry orchards can temporarily revive trees infested with *Pratylenchus* or *Xiphenema* spp. and avoid death in severe cases (31). Tissue of trees collected from fumigated 'Montmorency' trees was found to be more cold hardy than from nonfumigated trees in New York (31).

3.3.2. Phytophthora. A widespread problem is the susceptibility of mahaleb to various *Phytophthora* root rot pathogens (Fig. 2) (47, 69). Substantial tree losses have occurred in California orchards on mahaleb rootstock, where soils have poor internal drainage and/or where irrigation water is poorly managed (72).

TABLE 5. Susceptibility Rating of Cherry Rootstocks to Various Pests[a]

Pests	Mazz. F 12/1	Mazz. Sdlng.	Mahaleb	'Colt'	P. cerasus	Ref[b]
Bacterial canker						
Pseudomonas spp.	T	S	T	M	T	17, 62, 114
Crown and root rot						
Phytophthora spp.	T	T	S	T	T	72
Armillaria mellea	T	T	S	—	S	28
Verticilium spp.	S	S	S	—	S	62, 72
Crown gall						
Agrobacterium tumafaciens	S(high)	S	M	S(high)	M	62, 72
Nematodes						
Root knot						
Meloidogyne spp.	T	T	M	—	T	29
Lesion						
Pratylenchus vulnus	—	S	M	—	S	29
Pratylenchus penetrans	—	S	M	—	S	95
Cherry leaf spot						
Coccomyces hiemalis	S	S	T	M	M	48

[a]S = susceptible, M = moderately susceptible, and T = tolerant or resistant.
[b]References listed at the end of this chapter.

225

Figure 2. Crown rot of a seven-year-old 'Montmorency'/mahaleb tree caused by *Phytophthora* spp. in Michigan. The affected area is noted by the darkened (rotted) xylem of the mahaleb shank immediately below the union.

3.3.3. X-Disease. X-disease of cherry is caused by an organism similar to a virus, called a mycoplasma, affecting cherry (86). A similar pathogen causes Western X and buckskin disease in the West (39, 86). It has been a serious problem in some cherry-growing areas of California, Utah, and Michigan (23, 28, 57, 86). In Michigan, cherry trees on mahaleb rootstock collapse suddenly in midsummer while trees on mazzard rootstock decline slowly over several years (86). Transmission of buckskin or X-disease from mahaleb to the scion does not occur as readily as when propagated on mazzard (28). California growers attempt to avoid this disease by planting trees on mahaleb and top working cultivars high on scaffold branches. Upon infection, branches are removed before the disease spreads. This procedure also takes advantage of the fact that trunks and scaffolds of mahaleb are highly tolerant of bacterial canker or gummosis caused by *Pseudomonas syringae* pv. *syringae* and pv. *morsprunorum* (62, 73, 81).

3.3.4. Viruses. There are considerably more virus diseases which adversely affect cherry than any other stone fruit (58). Plant quarantine procedures are established to protect the U.S. cherry industry from several of the more serious virus diseases that are pollen and seed transmitted (58). All of the mazzard seed and most mahaleb seed today in the United States is being produced in the Pacific Northwest. 'Colt' and F 12/1, which are vegetatively propagated, entered the United States from nurseries in Europe which are certified by the plant protection service of that country as being free of all diseases including viruses (58).

Virus diseases can affect cherry trees in a variety of ways including leaf and fruit discoloration; foliage, fruit, and stem, deformation; reduced crop yields, reduced vigor, union separation, and subsequent death. Unfortunately, as in the case of 'Stockton Morello,' dwarfing and precocity seen early as an economical benefit (23) appear to have been virus induced. It was found to be a symptomless carrier of rusty mottle virus, crinkle leaf virus, fasciation, and many other viruses (15). The accumulation of these viruses soon rendered 'Stockton Morello' useless in California (65) until recently freed of virus. Unfortunately, the new virus-free clone is very vigorous and may not have any economic advantages over mazzard.

There are some differences in virus tolerance between *P. mahaleb* and *P. avium* cv. mazzard. However, they both will transmit, with or without symptoms, cherry rasp leaf virus (74), green ring mottle virus (77), prune dwarf virus group (41), *Prunus* ringspot virus (75), rusty mottle virus (112), yellow bud mosaic virus (91), spur cherry virus (11), and virus gummosis of 'Montmorency' (10). Spur cherry virus symptoms of sweet cherry scions appear more severe when propagated on *P. avium* than on *P. mahaleb*. Conversely, virus gummosis appears more severe on *P. mahaleb* than on *P. avium*. The prune dwarf virus group, which includes sour cherry yellows, is the most economically serious virus disease of sour cherry in the midwestern and eastern states (41). Gilmer, et al. (41) reported in 1976 that incidence of PDV in New York orchards 12−15 years age approached 100%. Since this virus is pollen and seed transmissible, seedling rootstocks are considered the primary source. Scion and rootstocks must be kept free of PDV and seed source trees isolated and protected from infected pollen. All *P. avium* is susceptible to Little Cherry Virus while *P. mahaleb* is tolerant (115).

The widespread incidence of *Prunus* stem pitting (PSP) in California was repeatedly found in sweet cherry trees propagated on mahaleb and 'Stockton Morello' by Mircetich et al. (71). No symptoms of stem pitting were found in the same orchards for trees on mazzard. Subsequent controlled inoculation tests confirmed field survey results. Tomato ringspot virus was implicated in causing PSP in Nanking cherry, but negative for peach, mahaleb, mazzard, and 'Stockton Morello' in California (71). Subsequently, tomato ringspot virus has been identified as the causal agent of PSP of peach in Pennsylvania (97) and in West Virginia (6).

Cherry growers must realize that orchard longevity and production can only be gained by the purchasing of virus-free (scion and rootstock) trees from reputable nurseries.

3.4. Training System

In most cases scion varieties on mazzard and mahaleb are not trained or pruned differently. It is recommended to Michigan growers that they train sweet and sour cherries to a modified central leader (59). In contrast, sweet cherries in California are trained to a vase shape (open center) (28). Wide crotch angles of scaffolds are gained in the modified-central-leader system, which helps to avoid bark inclusion, winter injury, and subsequent canker infections. Trees usually are slightly more vigorous on mazzard in Michigan. Narrow crotch angles are often associated with increased vigor (35). In Europe, 'Colt' is reported to induce scions with wider crotch angles (14, 114).

In California, Day (28) recommended that sweet cherry varieties on 'Stockton Morello' needed annual short pruning or heading back in order to increase branching. Scions on this stock appear to be more apically dominant. Gruppe (personal communication) in Germany has found that annual pruning of bearing 'Hedelfingen' trees grown on interspecific hybrid dwarfing rootstocks has helped to regenerate spur and branch development.

4. ROOTSTOCK/SCION RELATIONSHIP

4.1. Tree Size and Vigor

An examination of the literature will leave the reader confused regarding the effects of rootstock on scion vigor. Much of this confusion is due to differences in rootstock seedling source, soil and climate effects, virus content, presence or absence of nematodes, and specific scion/rootstock interaction. In general, the order of scion vigor for the following stocks, beginning with most vigorous, is: F 12/1 > mazzard seedling > mahaleb seedling (1, 28, 52, 92, 117). Sweet cherry may be larger at first on mahaleb but mazzard-rooted trees usually overtake and surpass them in good cherry-growing soils (116). Grown in gravelly soils of Utah (23) and Colorado (16), trees on mahaleb are more vigorous than on mazzard. Soil and moisture effects can induce or reduce growth dramatically among mazzard and mahaleb. Trees on mahaleb can be more vigorous in drought conditions than on mazzard. Conversely, trees on mazzard will be more vigorous in heavier, wet soils. Although all cherry root systems are affected, trees on mazzard and 'Stockton Morello' will grow less vigorously than trees on mahaleb in soils infested with nematodes (23).

Many studies have concluded that the source of the seedling is as influential on the productivity and vigor of the scion as is the species of the rootstock (3, 22, 35, 47, 85, 117, 119). Westwood, et al. (119) found that, after 13 years, 'Montmorency' on mahaleb clone PI 192703 was more vigorous than when propagated on *P. avium* cv. OCR-1 and 40% more vigorous than trees on mahaleb clone PI 193688. In Missouri, vigor of 19-year-old 'Montmorency' trees was greater on New York mazzard seedling than on 'Gold' seedlings, and Turkish 4 mahaleb seedling was more vigorous than other mahaleb seedling strains (68). In England, 'Merton Glory' and 'Merton Biggarreau' sweet cherry cultivars on 'Colt' were 80−90% of the trunk girth size of the same scions on SL 64 mahaleb clone and F 12/1 mazzard (79). Surprisingly, the latter two stocks were equal in vigor in this plot. However, trunk girth may be a poor indication of vigor for trees on 'Colt' and possibly some other scion/rootstock combinations because of a pronounced tapering of the scion trunk. Trees measured by height and spread of the crown on 'Colt' at the end of the rootstock trials were 47−55% the size of trees on control rootstocks (79). Sweet cherries in Washington (113) and sweets and 'Montmorency' in Michigan (Perry, unpublished data) on 'Colt' have been growing as strongly as the more invigorating stocks in the respective plots. Preliminary data from field experiments in Michigan indicate that trees on 'Colt' are greatly reduced in vigor when grown in heavy soils (Perry, unpublished data).

When allowed to form their own tops, *P. mahaleb* in general produces a large "round-topped" shrub or small tree; *P. avium* are tall, vigorous, upright−growing trees (Fig. 3); 'Stockton Morello' grows as a dwarf morello-type tree (27); and 'Colt' is upright but compact with short internodes. However, the growth habit among seedling strains of mahaleb is extremely variable (Fig. 1). Bauman (7) in Germany found dramatic differences in morphological characteristics among mahaleb clones selected from rootstock suckers in an established 20-year-old orchard. The clones were grouped into four distinguishing types.

Brase (12, 13) found that budding height appears to have a dwarfing effect on scion vigor when propagated on mahaleb. After 15 years, 'Giant' sweet cherry and 'Montmorency' budded at 66−76 cm above soil line were 50% the

Figure 3. A *P. avium* tree, 20 years old, selected originally from mountains in Poland and growing in Prof. Dr. Werner Gruppe's cherry species collection planting near Heldenbergen, West Germany. Tree height is approximately 10 m.

size of some scions budded low' on mazzard. Hutchinson and Upshall (54) reported dwarfing as high as 28% for 'Montmorency,' 'Hedelfingen,' and 'Bing' when budded at 23–38 cm above soil line compared to a 7-cm budding height on the same stock. No dwarfing is reported in the Pacific Northwest when sweet cherries are high-budded on *P. avium* (85).

The vigor of each cherry scion/rootstock combination will differ as revealed by Hutchinson and Upshall (54). In their study, 'Bing' grew faster on mahaleb than on mazzard F 12/1 during the first four years. The growth rate was greatest on F 12/1 after the sixth year. Greater cropping on mahaleb rootstock likely competed with shoot growth and influenced the change in growth rate.

'Montmorency' used as an interstem on mahaleb seedling root is reported to reduce sweet cherry scion vigor slightly while inducing good productivity in the Pacific Northwest (35).

4.2. Compatibility

Compatibility or affinity of scion and rootstock of like cherry species is predictably good, for example, sweet cherry varieties on *P. avium* or sour cherry varieties on *P. cerasus*. However, symptoms of incompatibility may occur when sweet cherry is propagated on *P. mahaleb, P. cerasus,* or other species and interspecific hybrids. These symptoms may occur rapidly, indicated by poor bud take, or they may be delayed for 6 to 10 years (14, 28, 37). In cases of incompatibility, lack of union affinity is characterized by a clean break, discontinuous but partial union, or abnormal scion behavior (50). Where death is not immediate, symptoms may include extreme precocity of flowering, small leaves and fruit (50), yellowing leaves, stunted growth, early fall senescence of leaves, scion or rootstock overgrowth, excessive rootstock suckering, excessive early fruiting, and subsequent death (105, 116). A pronounced reduction in vigor and precocity of flowering can also be associated with a compatible union and a rootstock effect on dwarfing (50, 89). Hill and Beakbane (50) distinguish compatible from incompatible dwarfing by leaf and fruit size. These symptoms can occur or are accentuated when a scion or rootstock is infected with viruses (25, 116).

Studies have been conducted to determine the cause of incompatibility in cherry. Differences in phenolic compounds and their infusion into tissues adjacent to the graft union have been associated with incompatible cherry unions (120). Schmid and Feucht (92) found that scion/rootstock combinations of greatest genetic diversity exhibited the most pronounced reduction of proteins and enzymes in the cambium, above and below the union. In that study, 'Hedelfingen' on F 12/1 mazzard synthesized the highest number of proteins and *P. fruticosa* synthesized the lowest, while *P. cerasus* clones W. 11, W. 10, and W. 13 were intermediate.

Scion or rootstock overgrowth may often appear to be an obvious indication of incompatibility. This characteristic for various scion/rootstock combinations is accentuated when scions are high-worked on mahaleb (20). However, sweet cherry scions budded on to 'Stockton Morello' unions are long-lived and strong (81). Day (28) reported examples of sweet cherries in California maintaining conspicuous scion overgrowth on mahaleb for over 50 years. In Germany, 10- and 13-year-old 'Hedelfingen' sweet cherry grafted on interspecific hybrid rootstocks possess a large scion overgrowth while the union remains strong and trees maintain good health and productivity (Gruppe, personal communication) (Fig. 4).

Figure 4. Scion overgrowth of a 10-year-old 'Hedelfingen' tree grafted on interspecific hybrid dwarfing rootstock clone #148-8 (*P. cerasus* × *P. canescens*) growing near Heldenbergen, West Germany.

Reports of sweet cherry compatibility with mahaleb are inconsistent among and within scion cultivars (Table 6). This is due more to genetic variability within mahaleb seed strains (7) than to scion cultivar variability. At INRA, Bordeaux, France, mahaleb clones have been selected which are incompatible with several sweet cherry cultivars after 7–8 years (SL 233) or compatible with all sweet cherry cultivars tested (SL 64) (14).

TABLE 6. Compatibility Rating for Cherry Scion Cultivars on Mahaleb Rootstocks[a]

Good[b]	Fair[c]	Poor[d]
Bing (23,28)	Black Tartarian (28)	Burbank (28)
Centennial (23)	Giant (81)	Chapman (28)
Elton (28)	Republican (28)	Eagle Seedling (8)
Gold[a]		Early Burlat (66)
Hedelfingen[a]		Early Rivers (37)
Knight (28)		Hedelfingen (96)
Lambert (23,28)		Merton Heart (37)
Long Stem Bing (28)		Larian (66)
Montmorency (104)		Van (66)
Napoleon (Royal Ann) (23, 28)		Williams Favorite (8)
Seneca (23)		

[a]Perry, unpublished data; compatibility of 'Hedelfingen'/mahaleb depends on mahaleb strain used.
[b]Good = Smooth unions with slight overgrowth, trees long-lived and productive.
[c]Fair = Scion or rootstock overgrowth indicating a structurally weak union.
[d]Poor = Immediate or delayed incompatibility, death within 10 years.

4.3. Precocity and Yield

In many rootstock tests reported over the years, sweet and sour cherry varieties are more precocious and produce larger crops on *P. mahaleb* than *P. avium* (Fig. 5) (3, 23, 28, 42, 68, 85, 119). Also, mahaleb seedling trees flower earlier in their life than mazzard seedlings. Sweet cherries in California on the older 'Stockton Morello' clone were more precocious and tended to outproduce sweet cherries on mahaleb and mazzard (28, 81). This characteristic was lost with heat therapy and subsequent freedom from virus. In Utah (23), sweets at the young bearing stage yielded equally well on mahaleb and 'Stockton Morello,' which was double the yield of mazzard trees. By the ninth year, order of yields was mahaleb > mazzard > 'Stockton Morello.' Lack of tree size and bearing surface area of trees on morello in the gravelly soils and arid climate of Utah allowed trees on mazzard to overtake and outproduce morello.

In England (79), sweet cherry cultivars 'Mertin Glory' and 'Mertin Biggarreau' were more precocious on 'Colt' for the first two bearing years than on F 12/1 mazzard, SL 64 mahaleb, and F 4/13 mazzard (recently released as 'Charger'). 'Charger,' F 12/1 and SL 64, in decreasing order, outproduced trees on 'Colt' in subsequent years. Reports from the Pacific Northwest indicate that sweets on interstems of 'Montmorency'/mahaleb seedling are slightly more precocious and productive than when propagated directly on mahaleb seedling (35, 61).

Influence in cherry scion productivity by rootstock is heavily dependent upon seedling or clonal source within *Prunus* species (3, 22, 35, 68, 85, 117,

Figure 5. Blossom density is highest and a more horizontal growth habit is evident for the six-year-old 'Hedelfingen' sweet cherry tree grafted on mahaleb seedling (left) than on the mazzard seedling (right). The 'Hedelfingen'/mahaleb trees growing in this rootstock trial near Traverse City, Michigan, have recently begun to express incompatibility symptoms.

119). Some mahaleb and mazzard clones have been evaluated and identified as inducing good productivity while maintaining small tree size (35, 68, 79, 85, 119).

Rootstock does not affect cherry fruit size and maturation of cherry fruits to any large extent. However, rootstocks which induce large crops such as 'Stockton Morello' influence scions to produce smaller fruits and delay maturity (28, 65). Although not well established in the literature, sweet cherry cultivars are reported to mature 3–10 days later on mazzard than on mahaleb rootstock (14). Thus mahaleb rootstock is an important choice where early sweet cherry markets, as in California, are targeted. Anthony, et al. (3) observed that 'Montmorency' bloomed 4–15 days earlier on mahaleb than mazzard roots.

4.4. Scion Maturation and Cold Hardiness

Edgerton and Parker (31) demonstrated in 1958 that 'Montmorency' scion twigs were hardier on mahaleb than mazzard root in November and February samplings. Following the severe Michigan winter of 1983–1984, winter injury and *Pseudomonas* canker infestation for three sweet cherry scion cultivars were greatest on mazzard and 'Colt' than on mahaleb, M×M 39, M×M 14, and M×M 60 rootstocks (Perry, unpublished data). Delayed scion maturation induced by rootstock effect appeared to be related to injury severity.

Delay in sweet cherry scion maturation on mazzard compared to mahaleb has also been observed in Washington (35). Although this phenomenon appears to be associated with vigor, own-rooted 'Colt' rootstock is dwarfed and compact while foliage remains lush and vegetative long after *P. mahaleb* and *P. avium* have lost leaves in the fall in Michigan. Therefore greater injury to sweet cherry scions on 'Colt' in Michigan may be more influenced by the inherent characteristic of the rootstock to remain vegetative.

4.5. Management Programs

Rootstock selection can influence management practices for cherry. Tile drainage must be provided for mahaleb where internal drainage is questionable or a high water table is anticipated. Under these conditions, additional drainage would be helpful to mazzard root growth, but it would be extremely crucial to mahaleb root development (87, 110). Supplemental irrigation must be provided in droughty, light soils for mazzard roots, whereas this provision is not as critical for mahaleb. More research is needed to determine differences in water uptake, requirements, and hydraulic conductance among cherry rootstocks.

The development of densely matted roots near the soil surface of mazzard limits tillage depth. Dieback conditions of sweet cherry on mazzard have frequently been observed in California orchards where deep cultivation is practiced. Sweet cherry in the same orchards on mahaleb were not affected (28). Excessive suckering, which can develop readily from *P. avium* and *P.*

cerasus rootstocks, must be controlled in order to reduce a devitalization effect on the scion.

Very limited research has been conducted on differential mineral uptake by cherry rootstocks. Kirkpatrick (60) found that potassium, magnesium, and phosphorus concentrations were lower in leaves of 'Montmorency' on mahaleb than on mazzard. In Oregon, mineral nutrient levels of four sweet cherry cultivars varied little among 13 mazzard × mahaleb clonal crosses, 'Stockton Morello,' and F 12/1 mazzard (19). According to a review by Westwood and Wann (118) in 1954, cherry is most frequently deficient of N and Zn. Fe, Mn, Bo, and K deficiencies have been reported in localized cherry-growing areas (119).

Mahaleb has long been known for its superior tolerance to chlorosis when grown in calcareous soils (14, 23, 28, 48). The chlorotic symptoms are most likely an expression of a micronutrient deficiency, more specifically a tie-up of iron or zinc. Trees on mahaleb are less subject to zinc deficiency (little leaf) than are those on mazzard root (14, 23, 28).

5. ROOTSTOCK CHARACTERIZATION

5.1. Botanical Description

A botanical description follows for the three *Prunus* species primarily used directly or as parents in the development of commercially used cherry rootstocks.

5.1.1. P. avium L. *Origin.* Southern and central Europe and Asia Minor.

Distribution. Mazzard is found in the wild on gravelly, well-drained, calcareous soils throughout Europe and the temperate zone of North America.

Description. Fruit are generally small, dark, with a large pit. Seedling trees are known commonly as mazzard or wild cherry and, when mature, reach 9 to 12 m (Fig. 3). Trees are vigorous, upright-spreading, semihardy; trunks are 30 cm in diameter, roughened; branches are stocky, smooth, dull ash gray, with few small lenticels; branchlets are thick, long, grayish-brown, smooth, with small, inconspicuous lenticels. Leaves are resinous at opening, 10–15 cm long, 5–7.5 cm wide, strongly conduplicate, oblong-ovate, thin, dark green, rugose on upper surface, dull green, more or less pubescent on lower surface; apex is acute, the base is more or less abrupt; margin coarsely and doubly serrate, glandular; petiole is 2.8 cm long, slender, dull red, with from one to three small, globose, reddish glands on the stalk, stipules are small, lanceolate. Buds are rather small, of medium length, pointed; leaf scars are prominent; flowers are white, 2.8 cm across, grow in clusters of two or three; pedicels are 2.5 cm long, slender, glabrous. Fruit are 2.5 cm in diameter or less, cordate; flesh is yellow, red or dark purple with colorless or colored juice (108).

Roots are well branched, fibrous, with a periderm that is reddish brown in color with large, conspicuous lenticels (108). Rooting along the rootstock shank is dense and abundant (9). New roots are thick and fleshy. Rhytidome (bark) flakes readily and, when the inner bark is exposed, oxidizes rapidly, turning an orange color (108). When submerged in water, excised root bark pieces will cause water to turn yellow-orange in color (28).

5.1.2. *P. mahaleb* L. *Origin.* Southern and central Europe and Asia Minor.

Distribution. Mahaleb is found in the wild on gravelly, well-drained, infertile soils throughout central Europe and Asia. Unlike mazzard, mahaleb trees are less conspicuous, more bush form, and slower growing.

Description. Seedling trees of this rootstock are also known as mahaleb, perfume cherry, and St. Lucie. Trees are small with slender growth, upright-spreading, open topped, often bushy; trunk is seldom more than 20 cm in diameter; branches are roughed, ash-gray over reddish brown; the numerous branchlets are slender and firm-wooded, with short internodes, dull gray, glabrous, with very numerous large raised lenticels. Leaves are 2.5 cm long, 2.8 cm wide, ovate to obovate, thick, leathery; upper surface is dark green, glossy, smooth; lower surface is light green, slightly pubescent along the midrib; apex and base are abrupt; margin is finely crenate, with reddish-brown glands; petiole is 1.2 cm long, slender, greenish, with no or with from one to three small, globose, greenish glands. Buds are small, short, obtuse; flowers appear late, after the leaves, are small, averaging 1.2 cm across, white, fragrant; borne in clusters of six to eight scattered on a main stem 2.5 cm in length, with the terminal pedicels 0.6 cm long and basal pedicels 1.2 cm long; pedicels are slender, glabrous, greenish. Fruits are very small, 0.6 cm × 0.8 cm, roundish ovate; the flesh is red-black, very astringent, sour, not edible. (108).

Roots are coarse and stringy, sparsely branched, with a straw colored periderm and a white inner bark and cambium (108). New roots are thin and long. Rootstock shanks are void of roots in comparison to mazzard (9). Upon exposure to air, inner bark and wood of root tissue does not turn orange color as mazzard, and water will remain clear when root bark pieces are submerged (28).

5.1.3. *P. cerasus* L. *Origin.* Evidence suggests that P. *cerasus* arose from an unreduced pollen grain of P. *avium* ($2n = 16$) crossed with P. *fruiticosa* Pall. ($2n = 32$) in the Caucasian region of Eastern Europe (76).

Distribution. This species is found growing wild in southeastern Asia and southeastern Europe. Grown for their fruits, sour cherries are classified into two distinct groups, the amarelles and the morellos (48).

Description. Known as sour or acid cherry, trees are small, topped with tight gray bark, glabrous growth, and suckering from the root (5). Morello cultivars are usually smaller, bushier, and more compact than amarelles. Branches of morellos are more horizontal, often drooping, less regularly arranged and more slender (48). Leaves are short-ovate to ovate-obovate, 7.5−10 cm long, abruptly short-pointed, stiff in texture, light green or gray-green, doubly serrate; petiole or base of blade with more or less undeveloped glands, glabrous or with sparse hairs on veins beneath (5). Morello leaves are more thin, small, darker green, or pendant (48). Lateral buds have small scales from which white flowers 2.5 cm in diameter arise. Flowers appear with first foliage or slightly in advance in small clusters on slender, glabrous pedicels 2.5 cm in length or longer (5). Fruit of amarelle cherries are pale, red, mildly acidic in flavor with light or colorless flesh and juice. Morello fruits are dark red, more spherical in shape, and with a more highly acidic red flesh and juice (48).

Examples of scion cultivars of amarelles include 'Montmorency,' 'Kentish,' and 'Early Richmond.' Wheras those of morellos include 'Stockton Morello,' 'Vladimir' (48), 'Schattenmorelle,' 'Oblacinska,' and 'North Star' (55).

The root characteristics are similar to *P. avium*, and water turns yellow to orange in color soon after submerging root bark pieces (28).

5.2. Rootstock Clones and Seed Sources

5.2.1. *P. mahaleb* L. This is a heterogeneous species as demonstrated by the wide variety of phenotypic characteristics (7). There are some clones available which are extremely compact and bush-form such as the clone known as 'Dwarf Mahaleb,' selected by Hansen at U.C. Davis, California (67). Conversely, there are clones collected from the wilds in Turkey and known as "Turkish" mahalebs which are extremely vigorous and upright in growth habit (35). A major portion of mahaleb seed used for propagation in the United States today, comes from a surprisingly small number of clones and trees in Washington (20). Nurseries grow seedling-liners from seed which is harvested from trees of primarily five clones introduced by Blodgett in 1946. These clones are part of a collection selected initially by D. Cation of Michigan, and are identified as numbers 902, 904, 908, and 916, which together cross-pollinate and set fruit. Seeds are blended at harvest and made available under the collective name Mahaleb 900. Groups of seed-stock source trees of the 900 series are grown in a few locations in Washington and are regularly monitored for virus status. Seeds and seedlings produced from healthy trees in these orchards are tested and declared virus free by the Washington State Agriculture Department (M. Aicheley, personal communication) (Fig. 6). Mahaleb 900 has demonstrated superior performance in rootstock tests for sweet cherry in Washington (35) and for 'Montmorency' in Missouri (68).

A clonal mahaleb selection is not yet in commercial use in North America, but one known as S.L. 64 is gaining acceptance in Europe.

5.2.2. S.L. (Sainte Lucie) (64). *Origin.* INRA, Grande Ferrade, France. Selected by M. Thomas in 1954 from *P. mahaleb* seedling population (103).

Characteristics. Successfully propagates asexually from softwood and semi-hardwood cuttings. Compatible with all sour and many sweet cherry cultivars, such as 'Early Burlat,' 'Giant Hedelfingen,' and 'Napoleon.' Adapts to a wide range of calcareous, droughty, or fertile soils but requires good internal drainage. SL 64 is available now virus free. Scion cultivars on this rootstock in

Figure 6. A Mahaleb 900 (right) and a N.Y. mazzard (left) seed source block of trees growing at Sunnyside, Washington (Pacific Coast Nursery), that has been monitored for virus status by M. Aicheley (in photo).

Bordeaux, France, are compact, precocious, and productive (14). Trees grown in trials in England have thus far been more vigorous than 'Colt' and 'Charger' and slightly less vigorous than F 12/1 mazzard (79).

5.2.3. *P. avium* L. Mazzard stocks used for budding by U.S. nurseries are seedlings of *P. avium* L. from which the cultivated sweet cherries derived. Like mahaleb, seedlings were formerly imported from Europe but are now grown from domestic selected sources found principally in the Pacific Northwest. American horticulturists in the 1930s selected wild seedlings which were found superior in cold hardiness and resistance to leaf spot (22, 64). Further propagation and selection of mazzard was conducted by Brase at the New York Agricultural Experiment Station (102). Blodgett in Washington screened selections obtained from Brase in New York and established a certified seed-production orchard of primarily a small red-fruited mazzard identified as New York mazzard (#570) and a light-red-fruited clone known as "Sayler" mazzard (M. Aicheley, personal communication). Superior cold hardiness demonstrated by New York mazzard over Sayler mazzard has made the former selection the dominant source for mazzard seed and seedlings to the American nursery industry today. New York mazzard is described as a "silver-barked" mazzard (M. Aicheley) which as a group originated in the Harz Mountains of

West Germany (37). Groups of New York mazzard trees are grown in a few locations in Washington and are regularly monitored for virus status. Seeds and seedlings produced from healthy trees on these orchards are tested and declared virus free by the Washington State Agriculture Department (M. Aicheley, personal communication) (Fig. 6).

In Europe, selections have been made of wild superior *P. avium*, including the silver-barked mazzard. Sweet cherry cultivars grafted on them have been reported to be long-lived and productive (43).

A small percentage of mazzard seed planted in the United States is in fact collected from cultivar sweet cherry trees. These seedlings are generally slightly less vigorous. During an 18-year period, 'Montmorency' was less productive and smaller in size on seedlings of 'Gold' than New York mazzard (68). Trees on 'Gold' were the same size as on a Turkish mahaleb selection, but less productive. F 12/1 is the only clonal selection of *P. avium* which is being used commercially.

5.2.4. *F 12/1 Mazzard. Origin.*   East Malling Research Station, Kent, England. Selected by N.H. Grubb among mazzard seedlings in about 1933 (43).

Characteristics. F 12/1 successfully propagates asexually by layering; it is compatible with all sweet and most sour cherry cultivars and adapts to a wide range of loam to clay-loam soils. F 12/1 is available now as virus free. Scion cultivars on this rootstock are more vigorous than on mazzard seedling. This stock is used in some humid cherry-producing areas for its trunk resistance to *Pseudomonas* canker. Roots are extremely susceptible to crown gall caused by *Agrobacterium tumifaciens*. It is not recommended in Michigan because of crown gall and because scion cultivars remain too vigorous late in the fall and it is very prone to root suckering. F 12/1 is often used in Europe to transform bush form morello cherries to more easily manageable tree forms.

5.2.5. *P. cerasus L.*   Compared to *P. avium* and *P. mahaleb*, *P. cerasus* has had minor usage as a rootstock for sweet and sour cherries. Seedlings of sour cherry are highly variable in the nursery (43) and have had little use. Clones of *P. cerasus* have been selected, propagated, and commercially used as rootstocks, interstocks, and, in Eastern Europe, as direct producing scion cultivars. Generally, *P. cerasus* rootstocks are more cold hardy and perform better in wet and heavy soils than *P. avium* and *P. mahaleb* (113). Only two clones, 'Stockton Morello' and 'Kentish,' have had commercial use as a rootstock.

5.2.6. *Kentish.*   Also known as 'Kentish Red,' 'Kentish Morello,' and 'Early Richmond,' it is an amarelle-type sour cherry (48). While little used today, it was recommended for imperfectly drained soils in Australia. It was propagated primarily by layering (42).

5.2.7. *'Stockton Morello'. Origin.*   Illinois, where it was known as "American

Morello" and introduced by Sol Runyon in the Stockton area of California (28).

Characteristics. Propagated commercially by root suckers harvested from orchard trees (27) or by softwood cuttings (46). Layerage has proven unsuccessful (27, 46). 'Stockton Morello' was recommended for heavy soils in California (27). Sweet (27) and sour (104) cherry scions were dwarfed and highly precocious on this stock. Now available as virus free. Sweet cherry scions are more vigorous than previous virus-contaminated stocks. It is less used today commercially.

5.3. Virus Status

Virus infection of cherry trees is a serious yet common problem. U.S. nurseries which produce seedling liners of *P. avium* and *P. mahaleb* cooperate with state inspection teams which monitor the virus status. Of major concern are those widespread viruses which can be pollen and, subsequently, seed transmitted (Table 7).

Gilmer and Kamalsky (40) conducted a survey for the incidence of necrotic ringspot virus (NRSV) and sour cherry yellows virus (SCYV) in 1962 of commercial mazzard and mahaleb seedlings in New York. Eighteen and 3% of the mazzard and mahaleb seedlings, respectively, were infected primarily with SCYV. Two to 3% infection of those sampled is usually considered maximum allowable to rate certification in many states. A similar survey of mahaleb seedlings in New York in 1957 produced virus incidence results of 9—18%. Sweet and sour cherry buds are less likely to unite initially with virus-contaminated rootstocks (especially mahaleb). Experience has determined that

TABLE 7. Methods of Transmission of Common Cherry Viruses

	Method of Transmission		
Virus	Seed (pollen)	Nematode	Graft
Cherry raspberry leaf (74)		X	X
Green ring mottle (77)			X
Prune dwarf virus (41) (sour cherry yellows)	X		X
Prunus ringspot (75) (necrotic ringspot)	X		X
Rusty mottle (112)			X
Spur cherry (11)			X
Yellow bud mosaic (43)		X	X
Cherry leaf roll	X	X	X

trees on mahaleb rootstock, if propagated from virus-free scionwood, almost invariably meet certification standards (40).

Tomato ringspot virus (TomRSV) has been implicated in the cause of *Prunus* stem pitting (PSP) disease in peach, plum (6, 97), and *P. mahaleb* (71). Although not proven in nematode transmission tests for cherry (95), TomRSV remains the leading suspected causal agent. Natural primary transmission of TomRSV to peach and plum is by *Xiphinema* spp. nematodes (51, 97).

Prior to development of a serological test such as enzyme-linked immuno-sorbent assay (ELISA), the detection of viruses in cherry relied upon symptoms expressed by bud- or tissue-inoculated indicator plants. Conventional tests, such as the *Prunus*-to cucumber and longer-term *Prunus*-to-*Prunus* tests (13 years), are still used as backup for the detection of many viruses. Indicator plants used include (36, 58):

P. serrulata cv. Shirofugen: prune dwarf and Prunus ringspot

P. serrulata cv. Kwanzan: green ring mottle

P. avium cv. Bing: rusty mottle

P. avium cv. Sam: little cherry, necrotic rusty mottle

Cucumber: Prunus ringspot, prune dwarf, plumpox (Sharka)

Nicotianna tabacum: cherry leaf roll virus

The ELISA test is now recognized as the most reliable and efficient method in detecting the incidence of TomRSV (51), cherry leaf roll, prune dwarf, *Prunus* necrotic ringspot, sour cherry yellows, green ring mottle, and little cherry viruses (D. Ramsdell, personal communication).

6. NEW AND FUTURE ROOTSTOCK

6.1. Rootstock Objectives

Cherry rootstocks have come about largely by screening and selection of elite seedlings of, principally, *P. mahaleb, P. avium,* and *P. cerasus* populations. While this method continues today, greater degrees of improvement are being realized and predicted for the progeny of interspecific crosses. Reports of size and vigor reductions of sweet cherry scions by as much as 50−60% account for the most significant accomplishments of interspecific hybrid rootstocks. Breeding and selection objectives differ between sweet and sour cherry and among geographical areas.

6.1.1. Sweet Cherry. Throughout the world, the one common objective is tree size reduction. A large percentage of sweet cherry fruits are highly prized, hand harvested, and sold fresh. Handpicking costs for trees, which reach 8−9 m in height, makes growing this crop unprofitable and highly inefficient. Unfortunately, a 10–20% size reduction is still inadequate in making the sweet cherry tree a viable economic unit. After (a) size reduction come

b. increased scion precocity and cropping,
c. wide range of compatibility,
d. uniformity in performance (asexual propagation),
e. cold hardiness (if in cold areas),
f. adaptation to a wide range of soils, and
g. disease and pest tolerance.

The order of objectives (a) through (g) may be altered according to geographical limitations. For example, tolerance to calcareous soils is of utmost importance in many cherry-growing areas of France (14).

6.1.2. Sour Cherry. The objectives for the improvement of rootstocks for sour cherry are less defined and less consistent. The largest tree, 'Montmorency,' rarely grows beyond 4–5 m in height, and in Eastern Europe, morellos are even smaller, with weeping bush forms. Second, most of the sour cherry crop is mechanically harvested and processed. Therefore tree size is not as prime an objective as it is for sweet cherry.
In Michigan, industry requirements for an ideal rootstock are

a. increased scion precocity and cropping,
b. uniformity in performance (asexual propagation),
c. adaptation to a wide range of soils,
d. disease and pest tolerance,
e. cold hardiness (of scion and root)
f. compatibility, and
g. tree size reduction.

Increased scion vigor, induced by the rootstock, is a benefit to morello scions in Europe. Morello scion cultivars become a more manageable form of tree when propagated on such stocks as the vigorous F 12/1 mazzard (W. Gruppe, personal communication).

It must not be overlooked in this discussion that while *P. avium* does not readily propagate asexually, *P. cerasus* generally does (34, 46). Since clones of this species are precocious, compact, cold hardy (55), and generally possess roots tolerant of wet soils and *Phytophthora* root rot, it may be possible to bypass the need for rootstocks. 'Oblacinska,' for example, is primarily grown on its own roots in Eastern Europe (55). Unfortunately, poor anchorage and excessive suckering is a common drawback among *P. cerasus* clones. Additionally, it is difficult to grow own-rooted *P. cerasus* into maiden trees in the nursery due to their inherent nature to possess a lack of apical dominance. Sour cherry scion breeding programs in Michigan, Sweden, and Eastern Europe (A. Iezzoni, personal communication) are including rooting ability as an important attribute and a step towards developing compact, high-density orchards in the future. Hutchinson (53) in 1969 suggested the growing of own-rooted 'Montmorency' trees on heavy clay soils in Ontario. Own-rooted 'Montmorency' trees are being included in contemporary rootstock trials in the United States today.

6.2. Screen Procedures

Rootstocks are of little value if uniform root-bearing propagules cannot be produced. Seedling progeny of cherry are commonly nonuniform. Therefore breeding and selection programs have long emphasized asexual propagation by layering or cuttings as the initial mass screen of a population (83, 109). Rooted layers or cuttings are then multiplied and planted in the nursery and budded with sweet or sour cherry scions. While there is some progress being made in West Germany in using protein analysis to predict incompatible combinations (92), actual field tests continues to be the most dependable method (50).

Concurrently, layers or cuttings are set aside for exposure to biotic and abiotic pressures. Researchers learn early through testing about susceptibility or tolerance levels of rootstock candidates to soilborne pathogens, low temperature, flooding, etc. (14, 44, 105). Elite candidates that show promise during these initial screens are virus tested and propagated with a limited number of scion cultivars. Finished trees are planted in the field and maintained in a commercial style, with records kept for several years.

A variance to the foregoing procedure is employed by H. Schimmelpfeng, Research Horticulturist, Technische Universitat Munchen, Weihenstephen, West Germany (Fig. 7). Mr. Schimmelpfeng has successfully selected clones from *P. cerasus* open-pollinated seedling populations (Weihroot clones) by grafting a common sweet cherry scion cultivar and evaluating its performance among thousands of trees. Root suckers and roots are collected and multiplied from promising elite candidates by conventional asexual means or in meristem culture. Propagules are then lined out in a nursery to serve as stocks for second and third test trial generations.

6.3. Cherry Rootstocks in the Future

Cherries are divided into two sections for botanical and breeding classification: Eucerasus and Pseudocerasus. The most widely used species for rootstock development within sections includes (94, 109):

Euracerus	*Pseudocerasus*
P. avium L.	*P. incisa* Thunb.
P. cerasus L.	*P. nipponica* Matsum.
P. mahaleb L.	*P. subhirtella* Miq.

Figure 7. Mr. H. Schimmelpfeng standing near an elite sweet cherry tree 'G.S. Knorpel,' 8 years old, grafted on an open-pollinated seedling of Weihroot clone W 11 (*P. cerasus*) at Weihenstephen, West Germany.

Euracerus	*Pseudocerasus*
P. fruticosa Pall.	*P. xhillieri* (Hillieri)
P. canescens Beis.	*P. xpandora* (Waterer)
	P. conccina (Koehne)
	P. pseudocerasus L.
	P. japonica Thunb.

New rootstock and interstocks will be described here as selections within Eucerasus and promising interspecific hybrids within and among the two sections. Cummins (25, 26) published two extensive reviews on the testing of numerous exotic cherry species and hybrids as rootstocks for sweet and sour cherries.

6.3.1. *Eucerasus rootstocks.* *P. avium* L. Future improvements by selecting within this species are mainly for disease resistance, precocity, and productivity, but does not include vigor control (33).

'Charger.' A clonal mazzard was recently named and released by the East Malling Research Station, Kent, England (13). This stock was selected for its resistance to bacterial canker under the selection code 1/57/4/13 (79).'Charger' propagates more easily than F 12/1 by layering or from cuttings (13). Rootstock tests in England revealed that 'Charger' produces sweet cherry trees intermediate in size between 'Colt' and F 12/1, while being more productive without adverse fruit size affects (79). Testing has only been conducted in the United Kingdom and with sweet cherry scions.

6.3.2. *P. cerasus* L. Some clones have been selected which have produced dwarf sweet cherry trees, while others will support sour cherry. Questionable compatibility (sweet scions), lack of anchorage, and excessive suckering are common problems.

CAB 6P and CAB 11E. Selections were made among wild sour cherry populations growing in the Emilia-Romagna area of Italy, Eight advanced selections were included in a 12-year rootstock trial with scion cultivars 'Durone della Marca' and 'Bigarreau Moreau.' 'Durone della Marca' performed best on CAB 6P and CAB 11E, while CAB 6P was the best stock for 'Bigarreau Moreau.' Unions were strong and smooth on these stocks and vigor was reduced by 20–30% of F 12/1 mazzard. These stocks are propagated by meristem culture or softwood cuttings (18). They were not tested with sour cherry scions.

Weihroot: W 10, W 11, W 13. Wild selections were made from *P. cerasus* Bavarian genotypes, West Germany, by H. Schimmelpfeng (90). These stocks are clonally propagated, tolerate *Phytophthora*, demonstrate good compatibility with sweet cherry scions, and reduce scion vigor by 20–30% (90). W 11 has produced a smoother union with 'Hedelfingen' than with W 10 and W 13 (92), while the same clone (W 11) was least compatible with Sam (93). Eighteen-year-old 'Schattenmorelle' trees at Technische Universitat Munchen, Weihenstephan, West Germany, propagated on the Weihroot clones have performed better on W 10 for vigor and cropping than mahaleb, F 12/1, and other Weihroot clones. Trees on W 10 are 20–30% smaller and possess a less dense canopy than when propagated on F 12/1 (Schimmelpfeng, personal communication). Some promising open-pollinated seedlings of W 11 have been selected and are under evaluation for dwarfing and precocity influence on sweet cherry cultivars at Weihenstephan by Schimmelpfeng. (Fig. 7).

'Montmorency' (Clonal). 'Merton Bigarreau' was slightly smaller and less productive on 'Charger' F 12/1 than on 'Montmorency' in a 14-year test in England (79). Anchorage was good and suckering only moderate on this stock. Trefois (106), after 8 years, reports that 'Schattenmorelle' is most productive and vigorous on 'Montmorency' at Gembloux, Belgium. In another test at Gembloux, 'Montmorency' own-rooted trees were smaller than when propagated on F 12/1 and were most productive (cumulative) after 7 years (Table 8).

Excessive suckering was observed and reported at Gembloux on *P. cerasus* clonal rootstocks GM 101,GM 103, and 'Montmorency.'

Vladimir. Vladimir is a generic name given to a group of morello cherries which originated in Russia and were introduced to the United States about 1900 (48). Ungrafted trees are weak growing with weeping branch characteristics. Ryugo and Micke (88) in California reported that 'Bing,' 'Black Tartarian' and 'Napoleon' trees on Vladimir after 13 years were small (3–4 m), precocious, and productive. However, the scions overgrew the union, trees required support, and suckering of the rootstock was excessive (Fig. 8). Vladimir has good resistance to root rot caused by *Phytophthora* species (69). This stock does not have commercial potential because of its undesirable characteristics.

Other *P. cerasus* clones which are showing potential and being evaluated are clones 603 at INRA, Bordeaux, France (14), and VV 1 at Falticeni Station, Romania (A. Iezzoni, personal communication).

TABLE 8. Effects of Rootstock on 'Montmorency' Tree Size and Fruit Production after Seven Years at Gembloux, Belgium[a]

Rootstock[b]	Crown Vol. (m^3)	Yield (kg) Cumulative 1978–1980	Yield Efficiency (kg/m^3)	Number of Fruits per Kg.	Crown Vol. (%)
GM 9	9.2	6.4	0.15	323	32
GM 8	10.5	15.4	0.38	265	36
Montmorency	23.6	26.8	0.30	252	81
GM 15	23.0	18.9	0.18	270	79
GM 101	24.2	25.8	0.25	306	83
F 12/1	29.0	20.4	0.20	263	100

[a]From R. Trefois and A. Monin, (106).
[b]Genetic background:
   GM 9: *P. incisa* × *P. serrula*
   GM 8: *P. pandora* × *P. subhirtella*
   GM 15: *P. campanulata* × *P. incisa*
   GM 101: *P. cerasus*

Figure 8. A 10-year-old 'Bing' grafted on Vladimir at U.C. Davis, California, showing excessive suckering.

6.3.3. P. mahaleb L. Clone selections of *P. mahaleb* have been evaluated as rootstocks primarily for tart cherries in Oregon (19) and Eastern Europe (55) and for sweet and tart scions in France (155).

Westwood et al. (19) published a report on the performance of 'Montmorency' on several PI (Plant Introduction) selections. After 13 years, trees on PI 192688 were 58% the size of F 12/1, while being more precocious and productive. PI 163091 was the most yield efficient, followed by PI 194098, PI 193693, and PI 193688, with F 12/1 significantly less efficient than all others. Researchers at Station d'Arboriculture Frutiere, INRA, Bordeaux, France, continue their search for clones of *P. mahaleb* selections which are compatible with sweet cherry scions and tolerant to calcareous, droughty soils. SL 63 and SL 275 are elite candidates currently under test (14).

6.3.4. Interspecific Hybrid Rootstocks. The use and crossing of exotic cherry species to produce compatible dwarfing rootstocks have been attempted with limited success in the past. However, new breeding programs are beginning to release rootstocks with great promise.

M × M Clones (P. avium × P. mahaleb). Several promising clonal rootstocks were selected from 30,000 open-pollinated seedlings of *P. mahaleb* from McGill and Son Nursery, Forest Grove, Oregon, by Lyle Brooks (99). Those seedlings which possessed large leaves and upright growth characteristics were presumed to be *P. avium × P. mahaleb* hybrids and propagated for further testing. Sweet and sour cherry scion cultivars are currently under initial test in plots throughout the United States on six of the most promising clones. Clone M × M 14 is the only dwarfing rootstock, while others are semidwarfing to vigorous (Table 9 (119).

In Oregon, Stebbins and Cameron (98) reported on the performance of 'Napoleon,' 'Bada,' and 'Corum' sweet cherry scions high-budded on M × M 2, M × M 60, M × M 97 compared to F 12/1. Trees were two-thirds and one-third the size on F 12/1 or M × M 2 when propagated on M × M 60 and M ×M 97, respectively. Scions are generally more precocious and productive on M × M clones than mazzard or F 12/1 in Oregon (98, 99) and Michigan (Perry, unpublished data). Scion performance on M × M 97 has been variable among trees (98).

None of the M × M clones are completely satisfactory as trunk or body stocks because of excessive suckering (98). Clonal stocks M × M 2, M × M 39, M × M 60, and M × M 97 were all found to be resistant to root rot caused by *Phytophthora* species (702). 'Montmorency' trees on mazzard, M × M 2, 60, 39, and 97, and 'Colt' remain healthy after five years in a clay soil near Hartford, MI (Hilltop Orchards and Nursery), while 50% of the trees on mahaleb have died and those on M × M 14 are stunted. Sweet cherry scions on M × M clones 39, 97, 14, and 60 were less injured and more canker free following a severe winter than on mazzard or 'Colt' in 1984 (Perry, unpublished

TABLE 9. General Vigor Comparison of Sweet Cherry and 'Montmorency' Trees on M×M Hybrid Rootstock[a]

Rootstock	Approx. Size (%)
Mazzard seedling	100
Mahaleb seedling	80
M×M 14	70
M×M 39	80
M×M 46	80
M×M 97	80
M×M 60	95
M×M 2	100

[a]Standard nursery height budded trees observed in Michigan and California (Perry, unpublished data.

data). These clones propagate readily from softwood cuttings (119). A limited supply of these clones is commercially available.

*OCR clones (*P. avium × P. mahaleb*).* Two hybrid clones selected by J. Milbrath (99) identified as OCR-2 and OCR-3 have been under rootstock trial in Oregon (119). Westwood et al. (119) found that 'Montmorency' on OCR-3 was slightly more vigorous and productive than on F 12/1. 'Montmorency' on OCR-2 was more yield efficient and smaller (86% of F 12/1) than F 12/1 and continued to have more commercial potential. Sweet scions on OCR-2 were vigorous but precocious in Oregon (99). This rootstock is not yet commercially available.

*Oppenheim (*P. fruticosa × P. cerasus*).* This rootstock has been under trial for sweet and sour cherry scions in West Germany. Plock (82) selected this clone from wild *P. fruticosa* seedlings at the Oppenheim station. Oppenheim appears to be a hybrid between *P. fruticosa* and *P. cerasus*. At the Oppenheim Station, 12-year-old trees of 'Hedelfingen' were two-thirds the size of F 12/1, well anchored, and productive. While trees suckered moderately, the root system appeared to be resistant to *Phytophthora* root rot (27). Oppenheim is compatible with a wide range of scion varieties (82), but incompatible with 'Sam' and 'Van' (49). Hein (49) reported that after 9 years, trees on Oppenheim were suckering excessively and required support. This clone will soon be tested in the United States, but is not yet commercially available in North America.

Belgium clones. Three clones have been identified at Gembloux, Belgium, by R. Trefois and A. Monin among many exotic cherry species and interspecific hybrids under evaluation as rootstocks for sweet and sour cherry scions. The clones are GM (Grand Manil) 9 (*P. incisa* Thunb × *P. serrula* Franch), GM

6l/1 (*P. dawyckensis* Sealy), and GM 79 (*P. canescans* Bois.). In Belgium, 'Montmorency' on GM 9 is 30% of its size on F 12/1 (Table 8). While productive, 'Montmorency' fruits on GM 9 were small compared to standard stocks (106). These clones have a wide range of compatibility with sweet cherry scion cultivars (Table 10). After 10–12 years in Belgium, sweet cherry scion vigor, compared to F 12/1 (100%), is reported to be 30, 60, and 70% on GM 9, GM 61/1, and GM 79, respectively (107). Vigor for some cultivars may be excessively reduced on GM 9 (107). These rootstocks are readily propagated by softwood cuttings or meristem culture. Oregon Rootstock, Inc., recently introduced them into the United States as virus free and, after a period of testing, they will soon be available commercially (M. Smith, personal communication).

Giessen clones. Gruppe (44) at the Justus Liebig University in Giessen, West Germany, has selected and evaluated large numbers of *Prunus* hybrids as rootstocks for sweet cherry. Many of these have excellent commercial potential, and 17 have recently been distributed for test trials in Europe and the United States. These interspecific hybrids are capable of a wide range in scion growth control (Fig. 9) and influence on precocity and yields (Table 11). Among these clones, trees on progenies of crosses between *P. cerasus* or *P. fruticosa* were most dwarfed, while trees on *P. canescans* × *P. avium* stocks produced the largest crown. The intensity of suckering was pronounced in progenies with *P. cerasus* or *P. fruticosa* parentage (44). Six-year-old trees of 'Hedelfingen' were most precocious and productive on clones originating from crosses between P. *fruticosa* × *cerasus* and P. *fruticosa* × *avium* (89). At Giessen, Roth and Gruppe (45) found several clones more resistant to waterlogging than F 12/1, 'Colt,' and *P. mahaleb* SL 64. St. Lucie 64 is the most sensitive, while clones 172/9 and 173/9 are the most resistant. Preliminary tests indicate that clones with *P. canescens* parentage are similarly sensitive to

TABLE 10. Compatibility Assessment of Sweet Cherry Scions with GM 9, 61/1, and 79 Rootstocks from Gembloux, Belgium, Rootstock Clones[a]

	GM 9	GM 61/1	GM 79
Hedelfingen	C	C	C
Early Burlat	C	C	C
Napolean	C	C	—
Van	C	—	C
Lambert	C	C	—
Schneider	C	C	C
Early Rivers	X	C	C

[a]From R. Trefois and A. Monin (105); C = compatible, X = not compatible, — = not reported.

Figure 9. 'Hedelfingen' after 10 years growing on mazzard F12/1 (left) and on interspecific hybrid clone #195-2 (*P. canescens* × *cerasus*) near Heldenbergen, West Germany. The tree at left has reached a height of 3 m while that on the right has approached 5 m.

TABLE 11. Effects of Rootstock on Hedelfingen Sweet Cherry after Eight Years in Giessen, West Germany[a]

Clone No.	Parentage	Max. Canopy Area (%)	Yield (kg/tree)	Suckers[c]	Need for Support	Rooting[b] (%)
F 12/1	P. avium	100	5.7	F	No	
148/1	P. cerasus Schattenmorelle ×	88	22.1	0	No	94
148/2	P. canescens EMRS	58	16.2	F	Yes	93
148/8		68	18.8	0	No	85
148/9		75	13.8	F	Yes	90
154/4	P. cerasus Schattenmorelle ×	56	13.2	M	No	48
154/4	P. fruticosa 64	24	7.1	F	Yes	10
154/7		47	12.3	M	No	86
172/	P. fruticosa 64 × P. avium					
172/3		81	15.4	M	No	78
172/7		41	7.8	M	Yes	90
172/9		26	4.1	F	Yes	73
173/5	P. fruticosa 64 ×	55	3.4	M	No	26
173/9	P. cerasus Schatten.	64	17.3	M	No	84
195/1	P. canescens EMRS ×	100		F	No	
195/2	P. cerasus Leitzkaner.	100		F	No	
196/4	P. canescens EMRS ×	110	11.1	F	No	
196/13	P. avium Hedelfingen	110	8.4	F	No	

[a]From W. Gruppe (45).
[b]Rooting % by softwood cuttings, avg. six years (101).
[c]Root suckers: 0 = none, F = few, M = moderate.

flooding as *P. mahaleb* (Gruppe, personal communication). Most clones can be readily propagated by softwood cuttings. Hilltop International recently introduced these stocks to the United States as virus free and, after a period of testing, stocks will soon be available commercially (W. Heuser, personal communication).

Other interspecific hybrids. Progenies of crosses between *P. avium* × *P. fruticosa* made at Geneva, New York, have shown variable results as rootstocks for sweet and sour cherry (24, 26). All clones produce many suckers and produce vigorous to fully dwarf trees. Some incompatibility has been detected with 'Emperor Francis' and 'Montmorency' (26). Commercial potential at this time is limited.

Interspecific hybrids, primarily *P. fruticosa* × *P. pseudocerasus*, *P. avium* × *P. canescens*, and *P. canescens* × *P. canescens*, are being developed and evaluated by Wagner for rootstock potential at the Research and Production Station for Pomology, Cluj-Napoca, Romania (A. Iezzoni, personal communication). Also in Romania, Parnia et al. (78) at the Fruit Research Institute, Pitesti-Maracineni have reported promising progeny of crosses between *P. avium* and *P. cerasus*. Six clones propagated readily from softwood cuttings or layering were compatible with most sour and sweet cherry scion cultivars and were tolerant of wet soils. Trees ranged in size between those grown on 'Colt' and F 12/1.

6.3.5. Interstem Trees. In an effort to find ways to reduce sweet cherry tree size more dramatically, Brase and Way (13) suggested in 1959 the testing of 'Northstar' sour cherry as an interstem. Since then, Larsen (61) has reported on results using 'Northstar,' 'Montmorency,' 'Kansas Sweet,' and 'Redrich' as interstems with sweet cherry scions and mahaleb and mazzard rootstocks. The results showed that trees on 'Northstar'/mahaleb 900 were somewhat smaller and more precocious than standard trees (56). 'Bing'/'Montmorency' unions appeared abnormal, suggesting incompatibility. However, trees with 'Bing'/'Montmorency'/mahaleb have been productive, healthy, and 20–30% smaller than standard trees for 22 years at R. E. Redman & Sons Orchard, Yakima, Washington (Fig. 10). Scaffold-grafted trees are inferior to a single graft-union (trunk) tree because of the bark inclusion problems incurred in narrow-angled crotches which is characteristic of 'Montmorency' (Fig. 10 inset) (R. Redman, personal communication). In contrast, 25-year-old trees of the same combination in The Dalles, Oregon, are large, productive trees, but yield small fruits (D. Burkhart, personal communication). Small fruit size has been the most common criticism of these interstem trees.

In California, a compact *P. mahaleb* selected at U.C. Davis and known as 'Dwarf Mahaleb' is showing promise as an interstem between sweet cherry scions and mazzard rootstock (67). Mazzard provides tolerance to *Phytophthora* and stem pitting and the *P. mahaleb* interstem is expected to give tolerance to Western X-disease. Several trees of this combination are 20 years old and slightly smaller than when propagated on *P. mahaleb* rootstock (67).

Figure 10. 'Bing' grafted on scaffolds of 'Montmorency'/mahaleb at approximately 1 m height in the field (upper-left-corner inset) at R. E. Redmond and Sons Orchard, near Yakima, Washington. Trees have been productive for over 22 years.

Smith Nursery in French Camp, California, is propagating these interstem combinations for growers in a limited basis. Interestingly, older trees of this combination may develop small lateral shoots on the interstem piece of 'Dwarf Mahaleb.' Jones and Quinland (56) in England found that tree growth was reduced by 20–30% only when small lateral shoots were allowed to develop on FB 2/58 (P. *avium* × P. *pseudocerasus*) interstem piece.

In West Germany, Hein (49) reported that two-piece trees with Oppenheim rootstock were superior in productivity and dwarfing effect to three-piece trees with reciprocal combinations of interstem/stock of F 12/1, SL 64, and 'Oppenheim.' Dwarfing and productivity of 'Hedelfingen' was greatest when 'Oppenheim' was the rootstock or interstem than other combinations after 15 years (Hein, personal communication).

INTERNATIONAL CATALOG OF ROOTSTOCK BREEDING PROGRAMS

Denmark

National Research Centre of Horticulture
Institut of Pomology
Kirstinbjergvej 12
5792 Arslev, Denmark
J. Vittrup-Christensen

France

Institut National de la Recherche Agronomique
B.P. 131-33140
Pont de lay Maye, France
G. Sallesses, Claverie

INRA Station de Recherches
Fruitières Méditerranéenes
B P 91 Cantarel Montfavet, France
Charles Grasselly

Germany

Institute for Pomology
Desden Pillnity, GDR
Brigette Walfram

Institut for Obstbau, Fachbereich
Angewandte Biologie
Ludwigstr 27
6300 Giessen-Lahn, GDR
W. Gruppe

Institute for Fruit Breeding
Bornkampsweg 31, 2070 Ahrensburg GDR
Hanna Schmidt

Technische Universität München
Lehrstuhl fur Obstbau
8050 Freising-Weihenstephan, FRG
Hermann Schimmelpfeng

Italy
Istituto Sperimental Per, La Frutticoltura
00040 Ciampino Aeroporto
Rome, Italy
Antonio Nicotra

Poland
Research Institute of Pomology and Floriculture
96-100 Skierniewice
Pomologiczna 18, Poland
Zygmunt-Stanislaw Grzyb

Romania
Facultea Horticultura
Aleea M. Sadoveanu 46
C.P. 6600 Jassy, Romania
Victor Cireasa

Research Institute for Fruit Growing
0300 Pitesti−Maracineni, Romania
Vasile Cociu

Institutul de Pomilcultura
Ro 312 Pitesti−Maracineni, Romania
G. Mladin

Soviet Union
Bulvar Pobeda, dom 5
Kwartira 11, 302029 gorod Orel, USSR
A. F. Kolessnikowa

Botanical Garden of Nikitski
Yalta-Crimee, USSR
A. Yadrof

Sweden
The Swedish University of Agricultural Science
S 230 Alnarp, Sweden
Victor Trajkovski

United Kingdom
East Malling Research Station
East Malling, Maidstone, Kent
EM 19 6BJ, England
K. R. Tobutt

United States
Department of Horticulture
Michigan State University
East Lansing, MI
Amy Iezzoni

REFERENCES

1. Anonymous (1978). Annual Report, Dept. of Agr. Tech. Serv., Republic of South Africa, 241 pp.
2. Andersen, R. L. (1981). Cherry cultivar situation, *Fruit Var. J.*, **35**(3), 83–92.
3. Anthony, R. D., R. H. Sudds, and G. E. Yerkes (1938). Orchard tests of mazzard and mahaleb cherry understocks, *Proc. Amer. Soc. Hort. Sci.*, **35**, 415–418.
4. Bailey, L. H. (1914). *The Nursery Book*, Macmillan, New York, pp. 206–208.
5. Bailey, L. H. (1924). *Manual of Cultivated Plants*, Macmillan, New York.
6. Barat, J. G., R. Scorza, and B. E. Otto (1984). Detection of Tomato Ring Spot virus in peach orchards, *Plant Dis.*, **68**(3), 198–200.
7. Bauman, G. (1977). Clonal selection in *Prunus* mahaleb rootstocks, *Acta Hort.*, **75**, 139–148.
8. Baxter, P. (1980). *Growing Sweet Cherries*, Agnote Govt. of Victoria, Australia Dept. of Agri., 3 pp.
9. Beckman, T. L., and R. L. Perry (1984). The importance of high root growth potential in cherry rootstocks, *Compact Fruit Tree*, **17**, 130–131.
10. Blodgett, E. C., and M. D. Aichele (1976). "Virus gummosis of 'Montmorency,'" in *Viruses Diseases and Noninfectious Disorders of Stone Fruits in North America*, U.S. Dept. Agric. Agric. Handbook No. 437, pp. 202–203.
11. Blodgett, E. C., and M. D. Aichele (1976). "Spur cherry," in *Virus Diseases and Noninfectious Disorders of Stone Fruits in North America*, U.S. Dept. Agric. Handbook No. 437, pp.252–255.
12. Brase, K. D. (1945). Observations on growth differences of sweet and sour cherries grated on mazzard and mahaleb body stocks, *Proc. Amer. Soc. Hort. Sci.*, 46:211–214.
13. Brase, K. D., and R. D. Way (1950). *Rootstocks and Methods Used for Dwarfing Fruit Trees*, N.Y. Agric. Exp. Sta. Bulletin 783, 50 pp.
14. Breton, S. (1980). *Le Cerissier*, Centre Technique Interprofessional Des Fruites et Legumes, Paris, France.
15. Brooks, R. M., and W. B. Hewitt (1949). Occurrence of certain diseases in sweet cherry seedlings propagated on 'Stockton Morello' rootstocks, *Proc. Amer. Soc. Hort Sci.*, **54**, 149–153.

16. Bryant, L. R. (1940). Sour cherry rootstocks, *Proc. Amer. Soc. Hort. Sci.*, **37**, 322–323.

17. Cameron, H. R. (1971). Effect of root or trunk stock on susceptibility of orchard trees to *Pseudomonas syringae*, *Plant Dis. Rep.*, **55**(5), 421–423.

18. Carrick, D. B. (1920). Resistance of the roots of some friut species to low temperature, *Cornell Univ. Agr. Exp. Sta. Mem.*, **36**, 613–661.

19. Chandler, W. H. (1957). *Deciduous Orchards*. Lea & Febiger, Philadelphia, 492 pp.

20. Childers, N. F. (1983). *Modern Fruit Science*. Horticultural Publications, Gainesville, FL 583 pp.

21. Christensen, J. V. (1966). Variety trial with sour cherries, *Saertryk Af Tidsskrift For Planteavl*, **70**, 17–21.

22. Clark, W. S., Jr, and R. D. Anthony (1946). An orchard test of mazzard and mahaleb cherry rootstocks, *Amer. Soc. Hort. Sci.*, **48**, 200–208.

23. Coe, F. M. (1945). *Cherry Rootstocks*, Utah Agr. Expt. Sta. Bul. 319, and review by *Fruit Varieties and Hort. Digest*, **1**, 18.

24. Cummins, J. N. (1972). Vegetatively propagated selections of *Prunus fruticosa* as dwarfing stock for cherry, *Fruit Var. & Hort. Dig.*, **26**(4), 85–89.

25. Cummuns, J. N. (1979). Exotic rootstocks for cherries, *Fruit Var. & Hort. Dig.*, **27**(4), 74–84.

26. Cummins, J. N. (1979). Interspecific hybrids as rootstocks for cherries, *Fruit Var. Journ.*, **33**(3), 85–89.

27. Cummins, J. N. (1984). Fruit tree rootstocks recently introduced and soon-to-be introduced, *Compact Fruit Tree*, **17**, 57–63.

28. Day, L. H. (1951). *Cherry Rootstocks in California*, Calif. Agr. Exp. Sta. Bull. 725.

29. Day, L. H., and W. P. Tufts (1944). *Nematode-resistant Rootstocks for Decidous Fruit Trees*, Calif. Agr. Exp. Sta. Cir. 359.

30. Edgerton, L. J. (1950). *Cherry Growing in New York*, Cornell Univ. Ext. Bull. No. 787, 24 pp.

31. Edgerton, L. J., and K. G. Parker (1958). Effect of nematode infestation and rootstock on cold hardiness of 'Montmorency' cherry trees, *Proc. Amer. Soc. Hort. Sci.*, **72**, 134–138.

32. Faccioli, F., C. Intrieri, and B. Marangoni (1979). New selections of cherry rootstocks: Twelve years of research, *Eucarpia*, 189–198.

33. Floor, J. (1957). Report on the selection of a dwarfing rootstock for cherries, *Euphytica*, **6**, 49–53.

34. Flore, J. A., and C. Sams (1979). Rooting softwood cuttings of sour cherry (*Prunus cerasus* L. 'Montmorency'): A preliminary report, *Compact Fruit Tree*, **12**, 85–86.

35. Fogle, H. W., E. L. Proebsting, Jr., E. C. Blodgett, and M. Aichele (1962). First-year production records from a cherry rootstock study, *Proc. Wash. State Hort. Assoc.*, **58**, 71–75.

36. Fridlund, P. R. (1976). "IR-Z, the interregional deciduous tree fruit repository," in *Virus Diseases and Noninfectious Disorders of Stone Fruits in North America*, U.S. Dept. of Agic. Agric. Handbk, No. 437, pp. 16–22.

37. Garner, R. J., and N. H. Grubb (1939). Cherry rootstocks, *Ann. Rpt. East Malling Res. Sta.*, **26**, 41.

38. Garner, R. J., and C. P. Nicoll (1957). Observations on *Prunus cerasus*, *P. mahaleb* and other species as rootstocks for sweet cherries, *Ann. Rept. E. Malling Res. Sta.*, pp. 63–72.

39. Gilmer, R. M., and E. C. Blodgett (1976). "X-Disease," in *Virus Diseases and Noninfectious Disorders of Stone Fruits in North America*, U.S. Dept. Agric. Agric. Handbk. No.437, pp. 145–155.

40. Grilmer, R. M., and L. R. Kamalsky (1962). The incidence of necrotic ring spot and sour cherry yellows virus in commercial mazzard and mahaleb cherry rootstocks, *Plant Dis. Rptr.*, **46**(8), 583–585.

41. Gilmer, R. M., G. Nyland, and J. D. Moore (1976). "Prune dwarf," in *Virus Disorders of Stone Fruits in North America*, U.S. Dept. Agric. Agric. Handbk. No. 437, pp. 179–190.

42. Grub, N. H. (1933). Cherry stocks at East Malling. I. Stocks for morello cherries, *J. Pom. and Hort. Sci.*, **11**, 276–304.

43. Grub, N. H. (1940). A resume of the cherry rootstock investigation, *Ann. Rept. East Malling Res. Sta. for 1939*, **A23**, 41–44.

44. Gruppe, W. (1979). The effects of some hybrid rootstocks on *Prunus avium* cv. 'Hedelfingen Riesenkirsche.' Preliminary results, *Proc. Eucarpia Fruit Section Symposium, Tree Fruit Breeding*, Angers, Sept. 3–7, 1979, pp. 199–221.

45. Gruppe, W. (1982). *Characteristics of some Dwarfing Cherry Hybrid Rootstocks*, Justus-Liebig Universitat, Giessen, FRG Res. Rept., 2 pp.

46. Hartmann, H. T., and R. M. Brooks (1958). Propagation of 'Stockton Morello' cherry rootstocks by softwood cuttings under mist sprays, *Amer. Soc. Hort. Sci. Proc.*, **71**, 127–134.

47. Hayes, J. E., H. S. Aldwinkle, and S. N. Jeffers (1984). Root and crown rot of cherry in New York caused by *Phytophthora megasperma* and *Phytophthora cryptogea*, *Acta Hort.* (in press).

48. Hedrick, U. P. (1915). *The Cherries of New York*, N.Y. Agri. Exp. Sta., Geneva, NY, 74 pp.

49. Hein, K. (1982). Influence of various rootstocks and interstocks on growth and yield of sweet cherries, *Proc. XXIst Intern. Hort. Cong.*, Hamburg (Abstr. 1116).

50. Hill, R., and A. B. Beakbane (1947). The application of biological observations on wild and naturalized species and varieties of fruit trees to the study of fruit tree rootstocks. A preliminary study of some *Prunus* species, *J. Pom.*, **23**, 117–133.

51. Hoy, J. W., S. M. Mircetich, and B. F. Lownsbury (1984). Differential transmission of *Prunus* tomato ringspot virus strains by *Xiphinema californicum*, *Phytopathology*, **74**, 332–335.

52. Howe, G. H. (1927). *Mazzard and Mahaleb Rootstocks for Cherries*, N.Y. State Agr. Exp. Sta. Bull. # 544 (March).

53. Hutchinson, A. (1969). *Rootstocks for Fruit Trees*, Ontario Dept. Agric. & Food, Toronto, Publ. 334, 221 pp.

54. Hutchinson, A., and W. H. Upshall (1964). Short-term trials of root and body stocks for dwarfing cherry, *Friut Var. & Hort. Dig.*, pp. 8–16.

55. Iezzoni, A. F. (1984). Sour cherry breeding in Eastern Europe, *Fruit Var. J.*, **38**(3), 121–125.

56. Jones, O. P., and J. D. Quinlan (1981). Effects of interstocks of cherry rootstock clone 15 (FB 2/58, *Prunus avium* × P. *pseudocerasus*), *J. Hort. Sci.*, **56**, 237–238.

57. Jones, A. L., and D. A. Rosenberger (1977). *X-Disease of Peach and Cherry*, Mich. State Univ. Ext. Bull. E-842.

58. Kahn, R. P. (1976). "Quarantine and the detection of stone fruit viruses in plant importations" in *Virus Diseases and Noninfectious Disorders of Stone Fruits in North America*, U.S. Dept Agric. Agric. Handbk. No. 437, pp. 23–32.

59. Kesner, C. D., and J. E. Nugent. (1984). *Training and Pruning Young Cherry Trees*, Mich. State Univ. Coop. Ext. Serv. Bull. E-1744.

60. Kirkpatrick, J. D. (1960). A study of *Prunus cerasus* cv. Montmorency in Western New York: Interrelations of rootstock, soil composition, root and soil populations of stylet-bearing nematodes, leaf composition, fruit quality, yield, and tree vigor, Ph.D. dissertation, Cornell University, Ithaca, NY.

61. Larsen F. E. (1970). A sweet cherry scion/interstock rootstock experiment, *Fruit Var. & Hort. Dig.*, **24**(2), 40–44.

62. Larsen, F. E., and R. B. Tukey (1982). *Rootstocks for Sweet Cherries*, Wash. State Univ. Coop. Ext. EB No. 1150.

63. Mai, W. F., and K. G Parker (1967). Root diseases of fruit trees in New York State. Populations of *Pratylenchus penetrans* and growth of cherry in reponse to soil treatment with nematodes, *Plant Dis. Reptr.*, **51**(5), 398–401.

64. McClintock, J. A. (1930). American mazzard cherry seedlings as rootstocks for cultivated cherries, *Proc. Amer. Soc. Hort. Sci.*, **26**, 82–85.

65. Micke, W. C., and W. R. Schreader (1978). Study of rootstocks for sweet cherries in California, *Fruit Var. & Hort. Dig.*, **32**(2), 29–31.

66. Micke, W. C., W. R. Schreader, and W. Moller (1977). *Sweet Cherries for the Home Garden*, Univ. California Coop. Ext. Serv. Lft. No. 2951, 8 pp.

67. Micke, W. C., M. Ranjit, W. R. Schreader, J. T. Yeager, and D. E. Kester (1981). *Research Evaluations of Cherry Rootstocks*, U.C. Davis Research Report Agnt. #12-14-5001-304, 18 pp.

68. Millikan, D. F., and A. D. Hibbard (1984). Increased productivity of 'Montmorency' tart cherry on WA 900 mahaleb roots, *Frt. Var. J.*, **38**(4), 143–145.

69. Mircetich, S. M., and M. E. Matheron (1976). *Phytophthora* root and collar rot of cherry trees, *Phytopathology*, **66**(5), 549–558.

70. Mircetich, S. M., and M. E. Matheron (1980). Differential resistance of various cherry rootstock to *Phytophthora* species, *Proc. Amer. Phytopath. Soc.*, **7**(2), 243.

71. Mircetich, S. M., W. J. Moller, and G. Nyland (1977). Stem pitting disease of cherries and other stone fruits, *Plant Dis. Rptr.*, **61**(11), 931–935.

72. Mircetich, S. M., W. R. Schreader, W. J. Moller, and W. C. Micke (1976). Root and crown rot of cherry trees, *California Agric.*, **30**(8), 10–11.

73. Norton, R. A., C. J. Hansen, H. J. O'Reilly, and W. H. Hart (1963). *Rootstocks for Sweet Cherries in California*, Calif. Agr. Exp. Sta. Ext. Serv. Lft. 159.

74. Nyland, G. (1976). "Cherry rasp leaf," in *Virus Diseases and Noninfectious Disorders of Stone Fruits in North America*, U.S. Dept. Agric. Agric. Handbk. No. 437, pp. 219–221.

75. Nyland, G., R. M. Gilmer, and J. D. Moore (1976). "Prunus ring spot group," in *Virus Diseases and Noninfectious Disorders of Stone Fruits in North America*, U.S. Dept. Agric. Agric. Handbk. No. 437, pp. 104–132.

76. Olden, E. J. and N. Nybom (1968). On the origin of *Prunus cerasus* L., *Hereditas*, **59**, 327–345.

77. Parker, K. G., P. R. Fridlund, and R. M. Gilmer (1976). "Green ring mottle," in *Virus Diseases and Noninfectious Disorders of Stone Fruits in North America*, U.S. Dept. Agric. Agric. Handbk. No. 437, pp. 193–199.

78. Parnia, P., G. H. Mladin, I. Dutu, and N. Stanciu (1982). New promising vegetative rootstocks for pear, sweet and sour cherry, *Proc. 21st. Intl. Hort. Cong.* (Abstr. No. 1050).

79. Pennell, D., P. B. Dodd, A. D. Webster, and P. Matthews (1983). The effects of species and hybrid rootstocks on the growth and cropping of Merton Glory and Merton Biggarreau sweet cherries (*Prunus avium* L.), *J. Hort. Sci.*, **58**, 51–61.

80. Perry, R. L. (1984). Working with soil limitations for orchard crops, *Proc. Ontario Hort. Conf.*, pp. 164–171.

81. Phillip, G.L. (1930). Cherry culture in California, *Cal. Agr. Ext. Serv. Circ.*, **46**, 20–22.

82. Plock, H. (1973). Diebedeutung der *Prunus Fruticosa* Pall. Als zwergunterterlage fur sus und sauerkirschen, *Mitteilungen*, **23**(2), 137–140.

83. Price, H. C., and E. E. Little (1903). *Cherries and Cherry Growing in Iowa*, Iowa State College Exp. Sta. Bull. No. 73, 59 pp.

84. Roberts, A. N. (1961). Current study of cherry rootstocks, *Proc. Oregon State Hort. Soc.*, pp. 92–93.

85. Roberts, A. N. (1962). Cherry rootstocks, 54th Annual Rept. Oregon Hort. Soc., pp. 95–98.

86. Rosenberger, D. A., and A. L. Jones (1977). Symptom remission in X-diseased peach trees as affected by date, method, and rate of application of Oxytetracycline HCL, *Phytopathology*, **67**, 277–282.

87. Rowe, R. N., and D. V. Beardsell (1973). Waterlogging of fruit trees, *Hort. Abstracts*, **43**(9), 534–548.

88. Ryugo, K., and W. Micke (1975). Vladimir, a promising dwarfing rootstock for sweet cherry, *HortSci.*, **10**(6), 585.

89. Schaumberg, G., and W. Gruppe (1984). Growth and fruiting habit of *Prunus avium* cv. 'Hedelfingen' on clonal cherry hybrid rootstocks, International Workshops on Improvement of Sweet and Sour Cherry Varieties and Rootstocks, Giessen, West Germany (July).

90. Schimmelpfeng, H., and G. Liebster (1979). *Prunus cerasus* als unterlage: Selektionsarbeiten, vermehrung, eignung fur sauerkirschen, *Gartenbau Wissenschaft*, **44**, 55–59.

91. Schloker, A., and J. A. Traylor (1976). "Yellow bud mosaic," in *Virus Diseases and Noninfectious Disorders of Stone Fruits in North America*, U.S. Dept. Agric. Agric. Handbk. No. 437, pp. 156–165.

92. Schmid, P. P. S., and W. Feucht (1980). Isoelectric focusing of proteins and some enzymes from secondary phloem of cherry graft combinations. I. Proteins in winter, *Sci. Hort.*, **12**, 55–61.

93. Schmid, P. P. S., and W. Feucht (1981). Differentiation of sieve tubes in compatible and incompatible *Prunus* graftings, *Sci. Hort.*, **15**, 349–354.

94. Schmidt, H. (1976). On the inheritance of the length of the juvenile period in interspecific *Prunus* hybrids, *Acta Hort.*, **56**, 229–234.

95. Serr, E. F., and L. H. Day (1949). Lesion nematode injury to California fruit and nut trees and comparative tolerance of various species of *Juglandacea, Proc. Amer. Soc. Hort. Sci.*, **53**, 134–140.

96. Simons, R. K., and R. F. Carlson (1968). Characteristics and propagation of rootstocks for deciduous fruits in the North Central region, *Hort. Sci.*, **3**, 221–224.

97. Smith, S. H., R. F. Stouffer, and D. M. Soulen (1973). Induction of stem pitting in peaches by mechanical inoculation with tomato ringspot virus, *Phytopathology*, **63**, 1404–1406.

98. Stebbins, R. L., and H. R. Cameron (1984). Performance of 3 sweet cherry, *Prunus avium* L. cultivars on 5 clonal rootstocks, *Fruit Var. J.*, **38**(1), 21–23.

99. Stebbins, R. L., J. R. Tjienes, and H. R. Cameron (1978). Performance of sweet cherry cultivars on several clonal propagated understock, *Fruit Var. J.*, **32**(2), 31–37.

100. Strauch, H., and W. Gruppe (1984). Cold hardiness of interspecific cherry hyrbrids and *Prunus* species by exosmosis, Internl. Wkshp. on Improvement of Sweet and Sour Cherry Varieties and Rootstocks, Giessen, West Germany (July).

101. Strauch, H., M. Roth, and W. Gruppe (1984). Rooting softwood cuttings of interspecific cherry hybrids and *Prunus* species by mist propagation, Internl. Wkshp. on Improvement of Sweet and Sour Cherry Varieties and Rootstocks, Giessen, West Germany (July).

102. Tarasenko, M. P. (1958). Changes in certain characters of cherry caused by the rootstock, *Agrobiologija*, **5**, 127–129.

103. Thomas, M., and J. Sarger (1965). Selection due *Prunus mahaleb* porte greffe du cerisier, *Rapport general du Congress Pomologique de Bordeaux*, pp. 175–201.

104. Toenjes, W. (1938). Mahaleb vs. morello rootstocks for early Richmond cherries, *Michigan St. Quart. Bull.*, **21**(2), 130–131.

105. Trefois, R., and A. Monin (1979). *Qualification d'unions de greffage de cultivars de cerisiers sur divers sujets porte-greffe*, Station des Cultures, Fruitieres et Maraicheres, Gembloux, Belgium.

106. Trefois, R., and A. Monin (1981). *Comportement de cultivars de cerises acides* (P. cerasus) *greffes sur de nouveaux sujets*, Station des cultures, Fruitières et Maraicheres, Gembloux, Belgium.

107. Trefois, R., and A. Monin (1981). *Developpements de cultivars de* Prunus avium *sur des nouveaux sujets porte-greffe*, Station des Cultures, Fruitières et Maraicheres, Gembloux, Belgium.

108. Tukey, H. B. (1930). *Identification of Mazzard and Mahaleb Cherry Rootstocks*, N.Y. State Agr. Exp. Sta. Cir. 117, 12 pp.

109. Tydeman, H. M., and R. J. Garner (1966). Fruit breeding: Breeding and testing

rootstocks for cherries, *E. Malling Res. Sta. Ann. Rept. for 1965–1967*, pp. 130–134.

110. Upshall, W. H., and G. H. Dickson (1950). 1949–1950 mahaleb and mazzard rootstocks for the cherry, Hort. Exp. Sta. Rept. Vineland Sta., Ontario, pp. 15–20.

111. Utermark, H. (1977). Growing and marketing fruit in the Elbe River Valley, Germany, *Compact Fruit Tree*, **10**, 38–48.

112. Wadley, B. N., and G. Nyland (1976). "Rusty mottle group," in *Virus Diseases and Noninfectious Disorders of Stone Fruits in North America*, U.S. Dept. Agric. Agric. Handbk. No. 437, pp. 242–249.

113. Webster, A. D. (1980). Dwarfing rootstocks for plums and cherries, *Acta Hort.*, **114**, 103.

114. Webster, A. D. (1984). Practical experiences with old and new plum and cherry rootstocks, *Compact Fruit Trees*, **17**, 103–117.

115. Welsh, M. F., and P. W. Cheney (1976). "Little cherry," in *Virus Diseases and Noninfectious Disorders of Stone Fruits in North America*, U.S. Dept. Agric. Agric. Handbk. No. 437, pp. 231–237.

116. Westwood, M. N. (1978). *Temperate Zone Pomology*, Freeman, San Francisco, 428 pp.

117. Westwood, M. N., and H. O. Bjornsted (1970). Cherry rootstocks for Oregon, *Oregon Hort. Soc. Ann. Rept.*, **61**, 76–79.

118. Westwood, M. N., and F. B. Wann (1954). "Cherry nutrition," in *Nutrition of Fruit Crops*, N. F. Childers, Ed., Horticultural Publications, Rutgers University, New Brunswick, NJ, pp. 158–173.

119. Westwood, M. N., A. N. Roberts, and H. O. Bjornsted (1976). Comparison of mazzard, mahaleb and hybrid rootstocks for 'Montmorency' cherry (*Prunus cerasus* L.), *J. Amer. Soc. Hort. Sci.*, **101**(3), 268–269.

120. Yu, K., and R. F. Carlson (1975). Paper chromatographic determination of phenolic compounds occurring in the leaf, bark and root of *Prunus avium* and P. *mahaleb*, *J. Amer. Soc. Hort. Sci.*, **100**(5), 536–541.

8

ALMOND ROOTSTOCKS

Dale E. Kester
University of California
Davis, California

Charles Grasselly
Station de Recherches Fruitière Méditerranéennes
Montfavet-Avignon, France

1. HISTORY

The almond originated as a cultivated crop in the Middle East in the area known as Iran. From here cultivation spread along both shores of the Mediterranean Sea to Greece, Italy, southern France, Spain, and Portugal, and also to Algeria, Tunisia, and Morocco. From these regions almond growing spread to North America, particularly California, Australia, South America, and South Africa, where Mediterranean-type climates occurred.

Originally, almond trees were grown ungrafted as seedlings, a method still utilized in Iran, Afghanistan, Turkey, and older orchards in central Sicily, Spain, the Balearic Islands, Morocco, and Greece. Two to three almond seeds would be planted in the soil in winter and the most vigorous seedling plant retained. As the trees came into bearing, those trees that produced bitter seeds were budded or grafted in the orchard using trees with better-quality nuts as sources of propagation material. In this way improved almond cultivars were probably developed in individual almond districts. The practice of field planting seeds was also considered desirable because of the deeper-rooted, longer tap—rooted (carrot-shaped) plant that was thought to be produced. These were better adapted to the dry, rocky, calcareous soil conditions in which almonds were typically planted. Even at the present time, field budding in these regions is considered desirable because trees started in this way were considered to be more resistant to dry conditions than nursery trees.

Budding came into general use about the middle of the nineteenth century and nursery propagation onto selected rootstocks became the dominant practice in producing countries where almond growing had become highly technically developed. Although almond seedlings were used initially in all growing areas, development of rootstocks from other species has been extensive and many better rootstocks adapted for almond are now available.

2. ALMOND SEEDLING ROOTSTOCKS

2.1. Use

Almond seedling rootstocks have been used primarily in Europe and other Mediterranean countries where most orchards grow on highly calcareous soil and often without irrigation (7, 14, 27). This stock has been traditionally used in Australia (1) and many parts of the world. In California, almond seedling rootstocks were used in the earlier plantings, which were often in nonirrigated hillside areas. As irrigation came to be practiced and highly fertile soils came into use, almond seedlings have ceased to be planted because of problems of survival, slow initial growth, and delayed productivity (45).

2.2. Sources

Seeds from bitter almond trees have been characteristically used by nursery-men as sources of seeds. Trees from such sources are thought to be more vigorous and generally superior to trees from sweet-kerneled seeds, although no critical evidence has been demonstrated toward this fact. It is possible that such seeds are less often eaten by rodents and thus survive better in field planting.

Seeds from known cultivars are used in many nursery practices. In Spain, the cultivars 'Atocha' and 'Garrigues' have been advocated as sources of root-stocks (14, 54) and seedlings of 'Desmayo Rojo' are also used. 'Garrigues' nursery plants have few branches are easy to graft. In California, seeds from the cultivar 'Mission' ('Texas Prolific') are typically used because of the general uniformity and vigor of the seedlings (11). In Australia, seedlings of 'Chellas-tan' and 'Nonpareil' are used (1). In Israel, several nematode-resistant almond rootstock cultivars have been introduced as 'Alnem 1,' 'Alnem 88,' and 'Alnem 201'G (15).

2.3. Growth Characteristics of Root System

Almond seedling rootstocks are vigorous, deeply rooted, and typically tap-rooted with few branches (Figs. 1 and 2). In the nursery the roots have a light color, but they become darker with age. The bark is lighter in color and thinner than that of peach rootstocks (10).

2.4. Growth Characteristics of Grafted Cultivar

Cultivars of almond grafted to almond seedling rootstocks often survive with some difficulty after transplanting because of their susceptibility to desiccation.

Figure 1. Shape of two different root systems: left, almond root; right, F1 seedling almond × *P. davidiana*. These two seedlings are from the same mother plant in same nursery conditions.

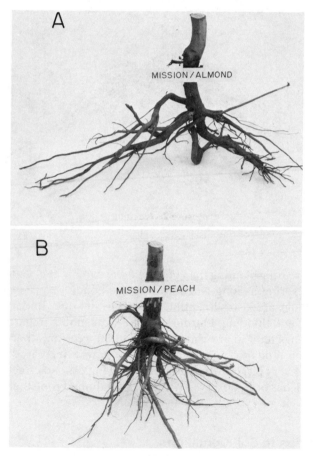

Figure 2. Root systems of three year old 'Mission' almond trees with different rootstocks. (A) almond seedling; (B) 'Lovell' peach seedling; (C) 'Marianna 2624' cutting; and (D) Peach-almond hybrid hardwood cutting ('Hansen 536').

267

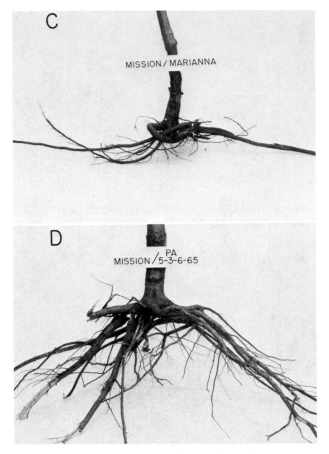

Figure 2. *(continued).*

For this reason, transplanting may best be done early in the winter rather than delayed until spring as long as good drainage conditions prevail. However, almond seedlings are also susceptible to "drowning-out" under conditions of poor drainage and flooding. During the first years in the orchard, cultivars on almond seedlings tend to grow more slowly than trees on other rootstocks so that production tends to be delayed. However, with increased age, trees on almond can be very large with great longevity (Fig. 3). No effect on blooming time or leaf fall has been observed with trees on almond rootstock as compared to others.

2.5. Adaptation to Soil Conditions

Almond seedling rootstocks are adapted to resist drought and calcareous soil conditions (26, 51, 54, 60). They appear to be able to tolerate stress better than

1

2

3

Figure 3. Rootstocks trial with 'Ferragnès' almond or almond root (1); peach root (2) and peach × almond hybrid (3)—Pictures taken from the same distance.

peach root systems, particularly as they approach permanent wilting (2). On the other hand, almond seedlings require well-drained soil and are highly susceptible to soil saturation. Even if soils are well drained, temporary soil saturation due to heavy rain or irrigation can cause losses, but the response may be complicated by susceptibility to *Phytophthora* species.

Seedlings of almond are known to be somewhat more tolerant to excess boron (28) and chlorides in the soil than peach seedlings, but are more sensitive to excess sodium (D. E. Kester and K. Uriu, unpublished data).

2.6. Disease and Insect Resistance

Almond seedling rootstocks are susceptible to various soil disease, including crown gall (*Agrobacterium tumeficiens*), oak root fungus (*Armillaria mellea*), and crown rot (*Phytophthora* species), and somewhat susceptible to verticillium wilt (*Verticillium dahliae*). Although it is possible that differences in susceptibility and resistance may exist in various species or selections, no detailed selection work has been undertaken until very recently. Reports from France as to disease resistance in seedling progeny of 'Archechoise' suggest the possibility of genetic selection (14). Variation in resistance to different species of *Phytophthora* and *Rhizoctonia* is reported in some progenies of almond (60).

Also, almond seedling rootstocks are highly susceptible to various species of nematodes including rootknot nematodes (*Meloidegyne javanica* and *M. incognita acrita*) and lesion nematodes (*Pratylenchus vulnus*) (43, 57). A source of nematode resistance to *M. javanica* and *M. arenaria* but not to *M. incognita* has been discovered in Israel (47).

In Mediterranean countries, almonds growing on almond rootstocks are susceptible to *Capnodis tenebrionis*, an insect borer, and to the larvae of *Cossus cossus*, a butterfly that lays its eggs in the "collar" or "crown" of the tree mainly when crown gall is present (24).

2.7. Propagation

Seeds from almond require 3−4 weeks' stratification in order to germinate. Gibberellic is reported to replace stratification (18). In mild climates, seeds can be planted in a nursery row at 3−4 in spacing in the late fall for germination to take place in mid- to late winter. Nursery trees can be budded in the fall at the end of the first growing season, cut back at the end of winter, and grown an additional year in the nursery before transplanting. In areas with a long growing season, as in California, seedling rootstocks can be budded in May and early June to produce nursery plants in a single growing season.

Almonds are difficult to grow from cuttings (26, 44, 46), although some varieties, such as, 'Marcona,' 'Bartre,' and 'Archechoise' (26), as well as 'Garrigues' (18), are somewhat easier than others. Micropropagation has been difficult for most cultivars of almond. Shoot tips can be established and multiplied (44). Limited success, however, has been reported with the 'Ferragnes' and 'Nonpareil' cultivars (44, 56).

3. PEACH SEEDLING ROOTSTOCKS

3.1. Use

Peach seedlings are the dominant rootstock for almond in California and in various other parts of the world where irrigation is practiced, soils are slightly acidic, and highly intensive production practices exist.

3.2. Sources

Seedlings of peach are primarily grown from seeds of either known cultivars or selected rootstock cultivars. Seeds of 'Lovell' peach have been widely used in California because seedlings are uniform, vigorous, and generally adaptable to nursery and orchard conditions. With the elimination of commercial 'Lovell' orchards, other cultivars, such as 'Halford,' have been substituted by some nurserymen. Where root knot nematodes are present, 'Nemaguard,' a cultivar released by the U.S. Department of Agriculture (USDA), has been the most widely used. It originated as a seedling hybrid of peach and *Prunus davidiana* (9). A red-leafed selection of 'Nemaguard,' known as 'Nemared,' has been introduced by the USDA (53) to facilitate rootstock identification in nursery production by the red tinge to the foliage.

3.3. Growth Characteristics of the Root System

Peach root systems tend to be somewhat shallow rooted (as compared to the almond, for instance), but with larger numbers of somewhat smaller roots (Fig. 2). Although roots will penetrate as deeply as 6 to 10 ft in uncompacted soil, a high percentage of the root system tends to be concentrated in the upper 3 ft.

3.4. Growth Characteristics of the Cultivar Top

Cultivars growing on peach seedling rootstocks are vigorous and tend to grow more than comparable trees on almond seedlings (Fig. 3). Thus trees come into bearing early and are more productive at younger ages than trees on almond root. Furthermore, survival is better after transplanting from the nursery. Recent studies have shown that trees on 'Nemaguard' rootstock may be larger and more productive than trees on 'Lovell' seedlings, at least in some locations (49).

3.5. Adaptation to Soil Conditions

Peach seedlings are sensitive to drought and calcareous soils (Figs. 4 and 5) and do best under irrigated but well-drained, slightly acidic soils, as are frequently found in the central valleys of California. Peach seedlings are more sensitive to excess boron and chlorides than are almond seedlings, but may be slightly more tolerant of excess sodium. However, peach seedlings should not be considered to have any great resistance or tolerance to any of these conditions.

3.6. Disease and Insect Resistance

Although slightly more tolerant than almond seedlings, peach seedlings are not resistant to crown gall, verticillium, oak root fungus (*Armillaria*), and crown and root rot (*Phytophthora* sp.).

Figure 4. Susceptibility to drought of almond on peach root (left) and on peach × almond (right).

Figure 5. Almond in calcreous conditions in Spain (area of Reus) on peach root (left) and peach × almond (right).

272

Peach seedlings are susceptible to most nematodes. However, selections have been made for resistance to root knot (*Meloidogyne* species), including 'Bokhara,' 'Shalil,' 'Okinawa,' and 'Nemaguard' (57, 58).

3.7. Compatibility

In general, almond cultivars have shown good compatibility to peach seedling rootstocks. Many cultivars show a distinct overgrowth at the union which has not been associated with any particular difficulty. However, one cultivar—'Milow'—produced by the University of California was found to develop a distinct brown line, some discontinuity at the union, and occasionally gum pockets. Although trees of this cultivar tend to grow more or less satisfactorily on this rootstock, some difficulty with the union has occurred in some locations.

3.8. Propagation

Peach seeds require longer (approximately three months total) stratification than almond to produce germination. In California, nurserymen maintain virus-tested seed source orchards of 'Nemaguard' and sometimes 'Lovell' or 'Halford.' These are free of *Prunus* (PRSV), which can be transmitted in low frequencies to seedlings (52). In the central valley of California, seeds are collected in the fall, cleaned carefully, and stored dry until planting. In mild winter areas, seeds are planted directly into the nursery row before the fall rains begin. Because germination percentages are sometimes low, 'Lovell' seeds are usually overplanted and then thinned to the desired spacing. 'Nemaguard' generally germinates at high temperatures but must be thoroughly soaked before field planting. Germination occurs in midwinter and seedling growth is expected to be sufficiently rapid to produce a plant large enough for budding in May and early June. Budding takes place at that time, the top is removed in several stages at about five-day intervals, and the budded plant is forced into immediate growth to produce a salable tree by the end of the first year (32).

4. PEACH–ALMOND HYBRIDS

4.1. Use and Sources

These rootstocks have been developed more recently for commercial almond orchards than the almond or peach seedling. In Europe, most interest has been in 'GF 677,' a chance seedling discovered in France in mid-1940s (Fig. 6) (5), where its value for growing peaches on calcareous soils was first noted. With the development of suitable vegetative propagation methods, this rootstock has become the dominant rootstock for almond as well as peaches in France

Figure 6. Original mother plant of peach × almond hybrid 'G.F. 677' in 1945.

since 1975. Its use has spread to other parts of the Mediterranean basin (16, 24, 51).

Selections have been made in Morocco of naturally occurring hybrids and are being tested for ease of propagation and other characteristics (3). One of these is reported to produce relatively homozygous F2 offspring. Considerable activity has been under way in France (25, 27) and the United States (36) for the selection of F1 hybrid clones from controlled crosses of peach and almond to be reproduced vegetatively. For instance, GF 557 has been developed from a cross of local almond × 'Shalil' peach (Fig. 7).

In California, hybrid rootstocks were early noted by Arthur Bright, a commercial nurseryman near LeGrand, California, who discovered their presence in seedlings grown from 'Nemaguard' source trees which were adjoining trees of almond. He developed commercial procedures for producing these rootstocks on a large scale. Hybrid rootstocks have been used in limited numbers in the San Joaquin Valley and the numbers are apt to increase. 'Titan,' a late-blooming almond cultivar, was introduced by the USDA (34) as a parent to be planted adjoining 'Nemaguard' peach in seed orchards to produce hybrid offspring. These have been referred to as 'Titan' hybrids.

Two vegetatively propagated patented clones, 'Hansen 2168' and 'Hansen 536,' have been released by the University of California as vigorous rootknot nematode—immune rootstocks (39) (Fig. 2).

Figure 7. Original seedling of peach × almond 'G.F. 557' (local almond × 'Shalil' cling).

4.2. Growth Characteristics of the Root System

The root system of hybrids is characteristically very vigorous, larger than either the peach or almond parent, deeply rooted, and well anchored. However, considerable variation has been observed in root systems of various clones of this combination, varying from those that are narrow and relatively unbranched to those that extend horizontally like a peach, for example, 'GF 677' (Fig. 8). Likewise the rootstocks of the 'Hansen 2168' and 'Hansen 536' are characteristic of each clone.

4.3. Growth Characteristics of the Top

The effect on trees grafted onto peach–almond hybrids is to combine the characteristics of the two parental species. The almond contributes the longevity, a deep root system, and adaptation to drought and calcareous soil (Figs. 3,

Figure 8. Root systems of two peach × almond clones: 'G.F. 677' (left) and 170-1, peach × 'Bartre' (right).

4, and 5). The peach provides vigorous growth, good survival after transplanting, and relatively early production. Actually, however, trees growing on the hybrid rootstock reflect the hybrid vigor of the rootstock and are more vigorous than either of the parental species (41). Because of increased vigor and size, tree yields have been greater than trees of parental almond and peach under comparable conditions (4, 16, 26, 51). Also, yield efficiency (yield per unit trunk circumference) is also greater. Blooming time of the trees grafted on hybrids are the same as trees on almond or peach rootstock, but fruit maturity tends to be delayed a few days to a week or more, reflecting the increased vigor of the plant. Defoliation of the trees in autumn also tends to be delayed, so that the trees may not develop hardiness early in the fall in frost-sensitive areas.

Almond trees growing on this rootstock under irrigation and in good soils may be too large for effective culture. Careful early training including selection of primary branches is particularly important to avoid limb crowding and later breakage.

4.4. Adaptation to Soil Conditions

The peach−almond hybrid rootstocks are best adapted for trees growing in well-drained, dry, and calcareous conditions, where trees are not irrigated (Figs 4 and 5). Under irrigated conditions, as in California, particularly in the more arid San Joaquin Valley, trees on hybrid rootstock have been observed to show less moisture stress during the long harvest period that is prevalent in multicultivar orchards. Hybrids may likewise be useful for less vigorous cultivars, particularly those that are later maturing. Anchorage is excellent in a wide variety of conditions. There has been experimental evidence that hybrid rootstocks tend to absorb less sodium than almond and less chloride than peach. Tolerance to boron excess has been intermediate with a range almond seedling progeny (D. E. Kester and K. Uriu, unpublished data). Trees on hybrid roots grow significantly better than do comparable trees on peach seedling roots when planted in replant situations after peach trees have been removed (13).

Various clones and seedlings show considerable susceptibility to wet conditions in heavy soils. However, the 'GF 677' clone in Europe has been observed to show relatively good adaptation, similar to that of the peach (11). Trees are somewhat more susceptible to waterlogging in the autumn because of their later growth, but are more tolerant in winter.

4.5. Disease and Insect Resistance

The almond × peach hybrids are equally susceptible to oak root fungus (*Armillaria mellea*), crown gall (*Agrobacterium*), and *Verticillium* as almond. Much variation in susceptibility to *Phytophthora* spp. and the associated waterlogging−crown rot complex occurs. Hybrid rootstocks in California have shown somewhat more susceptibility than peach in some field situations and equal susceptibility in others. Tests of individual clones of peach−almond hybrids by Kester and Mircetich (unpublished data) showed a range, with most selections tending towards a high degree of susceptibility to *P. syringae*. The two hybrid clones 'Hansen 536' and 'Hansen 2168' were selected as being the more tolerant but may not be sufficiently so, as has been observed in field tests. GF 557 shows considerable susceptibility. All of these are low-chilling rootstock cultivars, a characteristic possible associated with *Phytophthora* susceptibility. 'GF 677' has been found to be relatively tolerant to wet-soil−*Phytophthora* conditions, to nearly the same degree as the peach. In various orchard trials where it was compared to other clones, it was the only one to survive after water submersion (8).

Immunity to root knot nematodes (*Meloidogyne incognita acrita* and *M. javanica*) discovered in particular peach germplasm was found to be transmitted to their hybrid offspring with almond and has been used to produce the nematode-resistant hybrid clones 'Hansen 2168' and 'Hansen 536' (37). Similarly, 'GF 557' was selected from a hybrid population of 'Shalil' and an almond and is root-knot-nematode resistant (57). The seedling 'Bright Hybrids' and 'Titan' hybrids are offspring of 'Nemaguard' and almond and are largely root-knot-nematode resistant, although individuals with galls have sometimes been produced. 'GF 677' is not nematode resistant but is reported to be sufficiently vigorous that trees can withstand attack of *M. incognita* but not *M. javanica*.

‎

4.6. Propagation

Two nursery methods exist for the production of peach–almond hybrids. In one method the F1 seeds are produced by planting the parental species together in a seed orchard (33). Various almond cultivars have been used at various times. The most extensively almond cultivar used at present is 'Titan,' an introduction of the USDA in California (34). Cross-pollination with a self-incompatible almond cultivar—isolated from other cross-compatible cultivars—produces hybrid seeds. These are planted in the nursery row and almond seedlings are rogued from the other hybrid seedlings. Up to 50% hybrid seed production has been reported to occur in California nurseries, but much lower percentages were reported in Europe (25). The hybrid seedlings can be identified by their greater vigor, intermediate leaf shape, and the intermediate length of the petiole. 'Nemared' can provide good identification of the hybrid seedlings in the nursery row.

Vegetative propagation of selected hybrid clones is the second method and requires the selection of individual clones that are relatively easy to root (35, 36). Certain almond cultivars—such as 'Marcona' and 'Bartre'—were found to transmit ease of rooting, but offspring of this group tended to be susceptible to waterlogging (25). 'Hansen 2168' and 'Hansen 536' were selected from seedling populations as being relatively easy to root by hardwood cuttings (46).

Various vegetative production methods are available, but different cultivars vary in the use of method. Most hybrid clones can be readily rooted by softwood cuttings taken in the early summer (May and June) before the shoots have ceased elongation and rooted under mist (Fig. 9). 'GF 677', for instance, can be rooted from softwood cuttings about 15 cm long with two leaves, treated with indolebutyric acid, and rooted under mist (27). Between 1965 and 1980, 600,000 to 1 million cuttings have been propagated each year in France, but costs of propagation are quite high. Leaf-bud cuttings (20) also work well, but have been largely replaced by longer cuttings.

Hardwood cuttings have been used for some hybrid clones (3, 27, 35, 37, 39, 55), but poor results are achieved with others such as 'GF 677.' In general, cuttings should be collected in the fall, with November being a preferred time in mild climates when temperatures are somewhat warm, treated with a relatively high concentration of rooting hormone plus a fungicide, and planted directly into the nursery row (29) (Fig. 10). These conditions require a relatively mild winter where the soil remains somewhat warm in the fall. Likewise, 'GF 577' has relatively good rootability during mild winters.

Leafy semihardwood cuttings produce high-percentage rooting at relatively low costs (8). Cuttings are taken in the fall, with all young stems smaller 3–4 mm in diameter) than those used for hardwood or softwood cuttings (Fig. 11). These are placed in rooting medium under a plastic tunnel without mist. The new shoot grows vertically from the terminal bud in the spring and the cutting grows similarly to that of a seedling plant.

Laboratory micropropagation has been used in Italy and France to produce

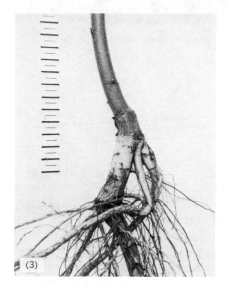

Figure 9. Different stage of leafy cuttings of peach × almond 'G.F. 677'; (1) first roots after 18–20 days (in June); (2) young stem after two months; (3) root system in winter.

about 1 to 1½ million plants per year (62). Nursery plants produced by this method have somewhat different root systems than those obtained by standard cutting procedures. However, after transplanting to the field, plants produced by various vegetative propagation methods are not different. Production costs of micropropagation are as high as those of mist propagation.

Figure 10. Semihardwood cuttings of peach × almond 18−20 days after sticking.

5. MARIANNA PLUMS

5.1. Source

The Marianna plum is believed to have originated from a cross of Myrobalan plum (*Prunus cerasifera*) × *Prunus hortulana* in the United States (11). From this hybrid various seedlings have been grown from which vegetatively propagated clones have been chosen, named, and introduced as rootstocks. Two selections used for almond in California are known as 'Marianna 2623' and 'Marianna 2624.' Although similar in most characteristics, the latter is somewhat resistant to oak root fungus and consequently is the clone used commercially. It has been used in a limited way primarily to grow almond trees in heavy, poorly drained soils or to replant in oak root fungus−infected spots. In Europe, a selection 'GF 8-1' similar to 'Marianna 2624' is used.

5.2. Growth Characteristics of the Root System

Nursery trees tend to have a flat, shallow root system, and trees grown from them tend to be shallow rooted and often have poor anchorage. To offset these characteristics, the scion cultivar may be budded relatively high, so that deeper planting can reduce this problem.

Figure 11. Rooted hardwood cutting of 'Hansen 2168' dug from nursery in April.

5.3. Growth Characteristics of the Scion Cultivar

Incompatibility with 'Marianna 2624' plum often occurs with different almond cultivars, so there is considerable range in tree size among different cultivars (Fig. 12). There are three general compatibility responses to different almond cultivars. *Compatible* almond cultivars show moderate vigor and grow about two-thirds the size of trees on almonds or peach rootstocks. Consequently, higher-density and closer plantings must be made to offset reduced yield. Such trees may have a shorter productive life than trees on other rootstocks, although little long-term experience is available.

Varieties listed in this group include the following: 'Ne Plus Ultra,' 'Peer-

Figure 12. Comparative graft union and trunks of different almond cultivars grafted to 'Marianna 2624' plum. Top row, compatible combinations (l to r.): 'Ballico', 'Thompson', 'Mission', 'Peerless', 'Merced'. Bottom row, incompatible combinations (l. to r.): 'Profuse', 'Solano', 'Nonpareil', 'Davey', '3C-29', 'Kapareil'.

less,' 'Mission' ('Texas'), 'Carmel,' 'Price,' 'Padre,' 'Harvey,' 'Carrion,' 'LeGrand,' 'Norman,' 'Ballico,' 'Thompson,' and 'Ai.'

Incompatible cultivars show a range of symptom and growth response and can be considered to produce two general groups. Completely incompatible combinations produce symptoms of incompatibility at a relatively early age (one to four years). These include failure to grow, premature defoliation, abnormalities at the graft union, early decline in seasonal growth, "dieback" of shoots, reduced size, and low yields (37). Cultivars in this group include 'Nonpareil,' 'Kapareil,' 'Milow,' 'Solano,' 'Davey,' 'Jeffries,' 'Sauret No. 2,' 'Sauret No. 1,' 'Dottie Won,' 'Ferraduel,' 'Desmayo Larguetta,' and 'Cristomorto.'

Intermediate incompatibility gives less pronounced expression of symptoms, variability in expression from tree to tree, and/or delayed appearance of symptoms. Some varieties considered to be in this group include 'Drake,' 'I.X.L.,' 'Butte,' and 'Ripon.'

Trees of the incompatible group vary in size in proportion to the degree of incompatibility from small, stunted trees to those nearly the size of compatible varieties. The union shows a breakdown of the phloem or bark region with some increase in diameter of both scion and stock. The xylem or wood area is intact, however, with no evidence of breakdown, and thus the union is mechanically strong. Externally the tissues of the stock and scion union may separate to the extent that gum exudes from it.

The pattern of symptom development of incompatibility can be explained by the following sequence: early shoot growth and cambium division begins initially in the spring. Premature cessation or blockage of the phloem creates a girdling effect and results in early defoliation and cessation of growth. During the following spring, cambium growth begins again and the cycle is repeated, but early fall symptoms may be more pronounced. As this cycle is repeated, growth becomes more and more inhibited, dieback may develop, and the tree shows signs of ill health. Tree size is reduced and loss of yield results. These conditions develop in proportion to the degree of incompatibility. Trees with intermediate incompatibility may show early defoliation in some years and under some conditions, although the trees may not be reduced in size.

Since the leading California variety—'Nonpareil'— is one of the incompatible varieties, there has been much interest in finding an interstock to allow the use of this cultivar in almond orchards. Kester (43) found that if a short (6 to 10 in.) interstock, of a compatible almond cultivar was used, incompatibility was not corrected, but some translocatable factor moved through the phloem of the interstock of the 'Marianna 2624' union, where breakdown then appeared. Some field experience suggests that if long interstocks are used, such as with high topworking, the incompatibility can be overcome or reduced. A plum clone known as 'Havens 2B' and believed to be *Prunus insititia* has been used as an interstock to overcome incompatibility (30). However, 'Nonpareil' trees grown with this interstock do not grow as well as do compatible cultivars directly on 'Marianna 2624.'

5.4. Adaptation of Soil Conditions

'Marianna 2624' rootstocks are primarily used to grow almonds where the soil is too heavy and poorly drained to grow trees on peach or almond seedlings or where the trees are periodically subject to waterlogging. The rootstocks are relatively shallow rooted and are not drought resistant. During summer in arid regions they require frequent and careful irrigation.

5.5. Disease and Insect Resistance

'Marianna 2624' rootstocks are more tolerant to oak root fungus (*Armillaria mellea*) than any other of the *Prunus* rootstocks, but they are not consistently resistant. This rootstock has been commonly used to replant in areas where losses from this disease have occurred. Most locations are pretreated with fumigants, but losses may still continue. 'Marianna 2624' is generally resistant to most *Phytophthora* species and thus can survive better in infected soils.

'Marianna 2624' is resistant to rootknot nematodes (11).

5.6. Propagation

'Marianna 2624' is readily propagated by hardwood cuttings taken during the fall or winter and planted directly into the nursery row in mild winter climates. Improvement in rooting can be obtained if the cuttings are treated with a hormone solution before planting in the nursery row. Good results can also be obtained by providing short (one to three weeks), warm precallusing treatment prior to nursery planting (31). This treatment should be done in early winter while buds are in the rest period, to avoid premature sprouting.

6. OTHER PLUMS AND PLUM HYBRIDS (Fig. 13)

6.1. European Plum

Certain varieties of European plum (*Prunus domestica*)—such as 'Brompton,' 'St. Julien,' and 'Damas'—have reasonably good compatibility to almond, but in practice have proven too weak or too susceptible to drought for conditions in Europe. These combinations have not been well tested in California.

6.2. Myrobalan Plum

Myrobalan (*Prunus cerasifera*) plum has been grown to a limited extent as a rootstock for almond in heavy, poorly drained soil conditions. In general, almond trees grown on Myrobalan have not been vigorous and have been considered unsatisfactory as an orchard tree (Fig. 14). Possibly more research should be conducted with this group.

6.3. Plum × Peach Hybrid

'Myran' is a plum hybrid *Prunus cerasifera* × *P. salicina* × 'Yunnan' peach that has been developed and tested in France (6) (Fig. 15). All cultivars that are incompatible to 'Marianna 2624,' including 'Nonpareil,' seem to be compatible with this rootstock. The root system is intermediate between those of the parents, although there is some overgrowth at the union. Almond cultivars budded to 'Myran' are reported to have a similar growth habit to that of peach rootstock and more vigorous than on 'Marianna 2624.' This stock is more tolerant to wet soil conditions than almond rootstock. It is not resistant to calcareous soil. It has shown somewhat greater tolerance to *Armillaria* (oak root fungus) than peach or the peach—almond hybrid. Because of the 'Yunnan'

Figure 13. Graft unions of almond 'Ferragnès' on different rootstock: (1) pentaploid plum; (2) Myrobalan × (Myrobalan × peach) P. 322 × 871; (3) Myrobalan 'P. 34-16'; (4) Marianna 'G.F. 8-1'; (5) Myrobalan × peach P. 322 × S. 1058; (6) Peach × Almond 'G.F. 677'.

Figure 13. *Continued.*

Figure 14. Almond on Myrobalan root ('P. 34−16'), 12 years old.

Figure 15. Myrobalan × peach hybrid 'Myran.'

parentage, it is expected to be resistant to root knot nematodes (57). It is more resistant to *Verticillium* than peach × almond hybrid. Propagation can be readily accomplished with hardwood cuttings. Softwood cuttings under mist or semihardwood cuttings in autumn also give good results.

6.4. "Pollizo" Plums

This group of plum rootstocks, apparently *Prunus insititia* of the Saint Julien type, has been traditionally utilized in the Murcia district of Spain as rootstocks of peach, apricot, and almond. This results from their adaptability to highly calcareous and compact soils in that area. Variation exists in their ease of propagation and compatibility with almond (5).

Clonal selection by local orchardists for ease of vegetative propagation by "suckering" has led to the production of a mixed group of clones that became extensively virus infected. Selection of specific clones for ease of vegetative propagation by cuttings and freed of serious viruses resulted in the selection of a new group of rootstocks being tested with promising results (17). Compatibility of almond cultivars varies (15), but selection for good growth performance has been carried out (17, 50).

6.5. Plum × Almond Hybrid

Selections with these parentages are under study in France. These show good rootability and are compatible with both almond and peach (Fig. 13).

6.6. *Prunus besseyi* × Myrobalan Plum

A selection, P2037, is being tested in France which provides semivigorous trees with good compatibility to almond. The leaves are very dark green and yield efficiency is high (Fig. 16).

6.7. *Prunus tomentosa* and *P. besseyi*

These species give very poor compatibility with almond and produce weak trees.

6.8. *P. besseyi* × Peach

A selection originating from Illinois was tested in France that gives good vigor and compatibility with almond but has poor anchorage.

7. NATIVE AMERICAN *PRUNUS*

P. fremontii is incompatible with almond. Similarly, *Prunus fasciculata*, an almondlike *Prunus* from the Mojave desert, was highly incompatible. On the other hand, E. F. Serr and H. Forde (unpublished data) produced a sterile hybrid of almond × *P. andersonii* (desert peach of Nevada) that rooted moderately well as a cutting and produced excellent trees when used as a rootstock.

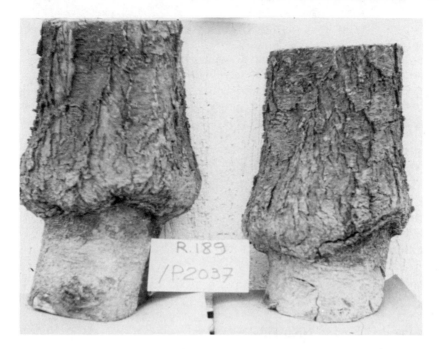

Figure 16. Graft union of 'Ai' variety on Besseyi × Myrobalan hybrid.

8. APRICOT

This rootstock often produced incompatibility the first year in the nursery including breaking off at the union.

9. MYROBALAN × (PEACH × ALMOND)

A selection of this parentage was tested in France with very good compatibility with almond and vigor and resistance to wet soil conditions; but it was difficult to propagate (6).

10. OTHER ALMOND SPECIES

Many related almond species occur in central and western Asia, extending into southeast Europe, for example, *Prunus argentea*, *P. bucharica*, *P. tangutica*, and *P. fenzliana*. Since these are closely related to the almond, it is presumed they would have similar rootstock characteristics. However, few of these have been tested for rootstock purposes but provide potential for rootstock germ-plasm (38).

Prunus webbii is a species found growing wild in southeast Europe, extending from Spain (19) to Turkey (24). Seedlings of this species are being tested as rootstocks in Yugoslavia (12) as well as being crossed with peach to produce hybrids (59). This species crosses readily with almond (38), but seedlings were found to be quite susceptible to crown and root rot conditions.

INTERNATIONAL CATALOG OF ROOTSTOCK BREEDING PROGRAMS

France

INRA Station de Recherches
Fruitières Méditerranéennes
B.P. 91 Cantarel Montfavet, France
Charles Grasselly

USSR

Botanical Garden of Nikitski
Yalta-Crimee, USSR
A. Yadrov

REFERENCES

1. Baker, B., and F. Gathercole (1977). *Commercial Almond Growing*, Dept. of Agr. and Fish. Bull. No. 9/77, S. Australia.
2. Barbara, G., and L. Fenech (1984). Effet de differents porte-greffes sur certains aspects de la physioloie de l'eau chez l'amandier, 5th Colloque G.R.E.M.P.A., *Options méditerranéennes, CIHEAM IAMZ-1984*, **2**, 63–76, Paris.
3. Barbeau, A., and El Baouami, Inram (1981). Les hybrides Amandier × Pecher naturels du Sud Morocain, 4th Colloque G.R.E.M.P.A., *Options méditerranéennes, CIHEAM IAMZ-81*, **1**, 131–134, Paris.
4. Blasco, A. B., and A. Felipe (1981). Cropping efficiency in almond: A rootstock trial under irrigation, 4th Colloque G.R.E.M.P.A., *Options méditerranéennes, CIHEAM IAMZ-81*, **1**, 137–138, Paris.
5. Bernhard, R. (1949). The peach–almond and its utilization (in French), *Rev. Hort.* (Paris), **121**, 97–101.
6. Bernhard, R. (1962). Les hybrides Prunier × Pecher et Prunier × Amandier, *Advances in Hort. Sci.*, 2.
7. Bernhard, R., and C. Grasselly (1969). Les port-greffes de l'amandier, *Bull. Tech. Inf.*, **241**, 543–549.
8. Bernhard, R., and C. Grasselly (1981). Les pechers × amandier *Arboriculture fruitiere*, **328**, 37–42.

9. Brooks, R. M., and H. P. Olmo (1972). *Register of New Fruit and Nut Varieties*, 2nd ed., University of California Press, Berkeley.

10. Damavendy Kozakonone, H. (1970). Amelioration des methodes de production de plants hybrides entre les especes Peches et Amandier, These, Université de Bordeaux.

11. Day, L. H. (1953). Rootstocks for stone fruits, *Calif. Agr. Exp. Sta.*, **736**, 1−75.

12. Dimitrovski, T., and B. Ristevski (1973). Ispitivanje pododnosti divijeg Badena Amygdalus webbii Sprach K. ao podlage, *Jugosl. Vocarstvo Brej*, **23**, 15−21.

13. Fatta del Bosco, G. Fenech, and L. Fenech (1981). Premiers resultats d'un essai de porte-greffes d'amandier sur sol vierge et fatique, 4th Colloque G.R.E.M.P.A., *Options méditerranéennes, IAMZ-84*, **2**, 75−84, Paris.

14. Felipe, A. J. (1976). "La production d'Amandes en Spagne," in *L'amandier*, *Options méditerranéennes*, CIDEAM, Montpelier, France.

15. Felipe, A. J. (1976). Compatibillidad entre cultivare de almendro y patrones circuelo "pollizo." I Congres International de Almendro y Avellana, Reus, Spain, pp. 330−336.

16. Felipe, A. J. (1978). Portainjertos para almendro, *ITEA*, **31**, 17−25.

17. Felipe, A. J. (1981). Germination acceleree d'amandes au moyen de l'acide gibberelique, *Options méditerranéennes IAMZ 1981*, ⅛**1**, 139−140.

18. Felipe, A. (1984). Bouturage ligneuxde l'amandier, 5th G.R.E.M.P.A., *Options méditerranéennes, IAMZ 1984*, **2**, 97−100.

19. Felipe, A. R., and R. Socias Company (1977). Un amandier sauvage, probablement A. webbii, non encore mentionne en espagne, 2nd Colloque G.R.E.M.P.A., *CIHEAM*, Montpelier, France.

20. Grasselly, C. (1956). La bouture de feuille. Nouvelle methode de multiplication vegetive d'un hybride Pecher-amandier, utilise come porte-greffe, *Rev. Hort. Suisse.* (April), 116−118.

21. Grasselly, C. (1972). L'Amandier: Caracteres morphologiques et physiologiques des varieties, modalite de leurs tranmissions chez les hybrides de premiere generation, These, Universtiy of Bordeaux, 156 pp.

22. Grasselly C. (1969). Etude de al compatibilite de l'espece Amandier greffee su differents Pruniers, *Ann Amel. Plantes*, p. 19.

23. Grasselly, C. (1973). *Premieres observations sur le comportement de l'hybrid Pecher × Amandier greffe sur divers Prunus*, Bull. Tech. Inf. n. 270.

24. Grasselly, C., and P. Crossa-Raynaud (1980). *L'Amandier*, Moisonneuve, Paris, 446 pp.

25. Grasselly, C., and H. Damavandy-Kozokanane (1974). Etude des possibility de production d'hybride F1 intra- et interspecific chez le sous-genre amygdalus, *Ann. Amelior. Plantes*, **24**, 405−414.

26. Grasselly, C., H. Gall, and G. Olivier (1977). Etat d'avancement des travaux sur les porte-greffes de l'Amandier, 3rd Colloque du G.R.E.M.P.A., Bari, Italy.

27. Grasselly, C., and G. Olivier (1977). Selection de nouveau clones hybrides de pecher × amandier aptes au bouturage ligneax, 2nd Coll. du G.R.E.M.P.A., Bari, Italy.

28. Hansen, C. J. (1955). Influence of the rootstock on injury from excess boron in Nonpareil almond and Elberta peach, *Proc. Amer. Soc. Hort. Sci.*, **65**, 128−132.

29. Hansen, C. J., and H. T. Hartmann (1968). The use of indolebutyric acid and captan in the propagation of clonal peach and peach–almond hybrid rootstock, *Proc. Amer. Soc. Hort. Sci.*, **91**, 135–140.

30. Hansen, C., and D. E. Kester (1955). Almond varieties on plum rootstocks, *Calif. Agri.*, **9**(9), 9.

31. Hartmann, H. T., and C. J. Hansen (1958). Effects of season of collecting, IBA, and pre-planting storage on rooting Marianna, peach and quince hardwood cuttings, *Proc. Amer. Soc. Hort. Sci.*, **71**, 57–66.

32. Hartmann, H. T., and D. E. Kester (1983). *Plant Propagation—Principles and Practices*, 4th ed., Prentice-Hall, Englewood Cliffs, NJ.

33. Jones, R. W. (1969). Selection of intercompatible almond and root knot nematode resistant peach rootstock as parents for production of hybrid rootstock seed, *J. Amer. Soc. Hort. Sci.*, **94**, 89–91.

34. Jones, R. W. (1969). Titan, a seed source for F1 almond × Nemaguard peach hybrids, *Fruit Var. Dig.*, **26**, 18–20.

35. Kester, D. E. (1969). Propagation of a peach–almond hybrid clone by hardwood cuttings, *Proc. of Intern. Plant Prop. Soc.*, **19**, 114–118.

36. Kester, D. E. (1975). Almond rootstock research in California, 2nd Colloque du G.R.E.M.P.A., Montpelier, France.

37. Kester, D. E., and R. N. Asay (1977). *Selection of almond × peach F1 hybrid clones as rootstocks of Prunus*, HortSci., **12**(4), 52 (Abstr.).

38. Kester, D. E., and R. N. Asay (1979). Hybridization among almond species, *HortSci.*, **14**(3), 408 (Abstr.).

39. Kester, D. E., and R. N. Asay (1986). 'Hansen 2168' and 'Hansen 536,' two new hybrid rootstocks, *HortSci.* **21**(2):331–332.

40. Kester, D. E., and C. Hansen (1964). Compatibility of almond varieties on 'Marianna 2624' rootstock, *Calif. Agri.*, **18**(9), 8–10.

41. Kester, D. E., and C. J. Hansen (1966). Rootstock potentialities of F1 hybrids between peach (*Prunus persica* L.) and almond (*Prunus amygdalus* Batsch), *Proc. Amer. Soc. Hort. Sci.*, **89**, 100–109.

42. Kester, D. E., C. J. Hansen, and B. F. Lownsbery (1970). Selection of F1 hybrids of peach and almond resistant and immune to root-knot nematodes, *HortSci.*, **6**(3), 32 (Abst).

43. Kester, D. E., C. J. Hansen, and C. Panetsos (1965). Effect of scion and interstock variety on incompatibility of almond on Marianna 2624 plum rootstock, *Proc. Amer. Soc. Hort. Sci.*, **86**, 169–77.

44. Kester, D. E., L. Liu, C. A. L. Fenton, and D. Durzan (1985). "Almond (*Prunus dulcis* (Miller) D. A. Webb)," in *Biotechnology of Tree Improvement*, Y. Bajaj, Ed., Springer-Verlag, Berlin.

45. Kester, D. E., and W. F. Micke (1984). The California almond industry, *Fruit Varieties Journal*, **38**, 85–94.

46. Kester, D. E., and E. Sartori (1966). Rooting of cuttings in populations of peach (*Prunus persica* L.), almond (*Prunus amygdalus* Batsch) and their F1 hybrid, *Proc. Amer. Soc. Hort. Sci.*, **88**, 219–223.

47. Kochbah, J., and P. Spiegel-Roy (1972). Resistance to rootknot nematodes in

bitter almond progenies and almond × Okinawa peach hybrids, *HortSci.*, **7**, 503−504.

48. Kochbah, J., and P. Spiegel-Roy (1976). 'Alnem 1,' 'Alnem 88' and 'Alnem 201' almonds: Nematode-resistant rootstock seed sources, *HortSci.*, **11**(3), 270.

49. Micke, W. (1984). Almond variety and rootstock trial, *Ann. Rpt. Univ. of Calif. West Side Field Station*, pp. 124−125.

50. Martinez Cutillas, A., and R. Martinez Valero (1976). El pollizo como portainjerto de almendro, I Congres Intern. de Alm. y Avel., Reus, Spain, pp. 330−336.

51. Monastra, F., and G. della Strada (1974). Portes-graffes pour l'amandier, Reunion G.R.E.M.P.A., Zaragosa, Spain.

52. Nyland, G., R. M. Gilmer, and J. Dwain Moore (1976). Prunus ringspot group, in *Virus Diseases and Noninfectious Disorders in Stone Fruits in North America*, Handbook No. 437, U.S. Govt. Printing Office, Washington, DC, pp. 104−132.

53. Ramming, D. W., and O .Tanner (1983). 'Nemared' peach rootstock, *HortSci.*, **18**(3), 376.

54. Ramos Carmona, B. (1976). Patrones francos del Almendro, 1st Congres Int. de Alm. y. Avel., Reus, Spain, pp. 377−385.

55. Rodriquez, J. (1981). Enracinement de l'hybride amandier × pecher GF 677, selection INRA, 4th Colloque G.R.E.M.P.A., *Options méditerranéennes 1981*, **1**, 135−136.

56. Rugini, E., and D. C. Verma (1983). Micropropagation of difficult-to-propagate almond (*Prunus amygdalus* Batsch) cultivar, *Plant Sci. Lett.*, **28**, 273−381.

57. Scotto La Masse, C., C. Grasselly, Minot, and R. Voisin (1984). Differential *Meloidogyne* spp. resistance in *Prunus* genus, *Revue Nematol.*, **7**(3), 265−270.

58. Sharpe, R. H., C. O. Hesse, B. F. Lownsbery, H. G. Perry, and C. J. Hansen (1969). Breeding peaches for rootknot nematode resistance, *J. Amer. Soc. Hort.Sci.*, **94**(3), 209−212.

59. Vlasic, A. (1977). L'*amygdalus webbii* Spach ed. I svsoi ibridi col pesco come pertaimnestro del amandorlo, 2nd Colloque G.R.E.M.P.A., C.I.D.A.M., Montpelier, France.

60. Stylianides, D., A. Chitzanides and I. Theochari-Athanassiou (1985). Evaluation of resistance to *Phytophthora* and *Rhizoctonia solani* in stone fruit rootstocks, 6th Colloque du G.R.E.M.P.A., *Options Méderranéemmes* IAM 2-8511:73−78.

61. Yadrov, A. A. (1970). Almond rootstocks in central Asia, *Sadovodstvo* **11**, 32.

62. Zucherelli, G. (1979). Moltiplicazione *in vitro* dei portainnesti clonali del pesco, *Frutticoltura*, **41**(2), 15−20.

9

APRICOT ROOTSTOCKS

P. Crossa-Raynaud and J.M. Audergon
Station de Recherches Fruitières Méditerranéennes
Montfavet-Avignon, France

1. INTRODUCTION

1.1. Origin of the Apricot

The cultivated apricot, *Prunus armeniaca* L., seems to have originated in northeast China between the town of Kan-Tcheou and the Russian border. That zone later extended to include central Asia, northwest China, Uzbekistan, and the town of Tashkent corresponding to Tian-Chan. Apricots are found to altitudes of 600−1000 m.

Apricot culture was practiced more than 3000 years ago in China and slowly spread through central Asia, Iran, Armenia, and Syria. From Armenia, the apricot was introduced to Europe about 70−60 B.C., through Greece and Italy, during the Roman Empire. It seems that a population of seedlings was also imported from Iran through North Africa and Spain by Arabs in the seventh century A.D.

It must be noted that great majority of trees were of seedling origin, even though the Romans knew how to graft. The use of seedlings still remains in several areas such as the upper valley of the Indus (Ladakh), northern Iran (Tabriz), oases of the mountains of Tunisia (Sbeitla), and Algeria (M'sila), as well as the valleys in the southern Atlas in Morocco (Draa). Even in Europe the extensive apricot plantations around Mount Vesuvius, near Naples, are often not grafted trees.

The use in plantations of systematically grafted trees is quite recent, and this explains the number of old varieties still existing in European countries. In spite of this variability, a characteristic of the apricot is the very strict adaptation of the old cultivated sorts to their specific local environment. The introduction of varieties from other countries has most often failed for this reason; thus world production is based on a small number of grafted cultivars.

1.2. Production and Distribution

World production comes from 30 countries, only 5 of which are in the Southern Hemisphere (Table 1). Statistical data indicate that in 1975 5 countries produced more than 100,000 metric tons: the Soviet Union, the United States, Turkey, Spain, and Italy. Two others, Hungary and France, may reach that production in favorable years. Several countries have a production reaching 60,000 metric tons: Greece, Bulgaria, Iran, Japan, and Rumania. In the Southern Hemisphere the main producing countries are Australia, South Africa, and Argentina (30,000 metric tons each). It must be noted that the production of Turkey, Iran, and Syria comes mainly from seedling orchards. In the case of Japan, a proportion of fruit is from the related species *Prunus ansu* and *Prunus mume*.

The world production of about 1,500,000 metric tons is not very high when compared to that of the apple or peach. Production in each country often varies greatly from year to year. This is partly accounted for by the small number of cultivated varieties being used and their poor adaptation to climatic variations (9, 10). The mean production per hectare in countries from which data is available is quite low: 7 /Ha in Spain, 5.5 in France, and 7.0 in Greece. It must be noted that these figures include production from a high proportion of nonirrigated orchards.

TABLE 1. Production of Apricots (In thousands of metric tons)

Country	1970	1975	1979	Country	1970	1975	1979
United States	160	166	131	Turkey	95	165	160
Canada	4,2	3,8		Israel	15	15	
Spain	162	132	151	Liban	13	18	
Italy	105	102	99	Syria	22	53	
France	74	54	54	Iran	60	?	
Greece	43	61	77	Algeria	14	14	
Bulgaria	70	?		Tunisia	19	24	
Hungary	85	?	40	Marocco	16	18	
Rumania	47	?	45	South Africa	12	23	
Czechoslovakia	34	?		Japan	68	63	
Yugoslavia	23	20		Argentina	17	19	
Austria	19	9		Chile	3,7	5	
Portugal	8	1		Australia	37	27	
Germany	5,8	1,2		New Zealand	1,2	6	
Switzerland	8	1		U.S.S.R.			290

1.3. Rootstock Use

1.3.1. Modern Orchards In modern orchards, the cultivars are grafted onto rootstocks, primarily because the cultivar scions are difficult to root successfully by means of softwood, semihardwood, or hardwood cuttings. Apricots are propagated commercially by the T-budding method, using rootstocks in the genus *Prunus*. The usual practice is to use summer and fall budding, but spring budding is practiced occasionally. The rootstocks used vary greatly from one country to another.

In North Africa and Israel, seedlings of local apricots (Balady) are used (2). In California, for 'Blenheim' and 'Tilton' cultivars, 46% of the rootstocks are apricot seedlings, usually 'Blenheim,' and 45% are peach seedlings, 'Lovell.' 'Marianna 2624,' a seedling selection of 'Marianna' (*P. cerasifera* × *P. munsoniana*?) and other rootstocks are rarely used (40).

In the main production region of Murcia, Spain, the principal rootstock for 'Bulida' is rooted suckers of a special cultivar of *Prunus insititia (domestica)* 'Pollizo.' 'Myrobalan' (*Prunus cerasifera*, cherry plum) seedlings are used in the Zaragoza region with the cultivars 'Paviot' and 'Moniqui,' while in the Valencia region apricot seedlings are rootstocks for the 'Canino' cultivar (41).

In France, in the Roussillon region, the main rootstock for the local cultivar 'Rouge de Roussillon' is rooted suckers of the 'Greengage' ('Reine Claude') plum. The principal rootstocks in the Rhone Valley for the leading cultivars 'Bergeron' and 'Polonais' are seedlings of peach or apricot and occasionally seedlings or cuttings of Myrobalan (21).

In Hungary (49), southern Czechoslovakia (26) and Germany (2) the main rootstock seems to be seedlings of apricot though research is in progress on Myrobalan.

In south Africa 90% of the trees are grafted on peach seedlings.

In Italy there does not seem to be a main rootstock as in other countries but a series depending on the quality of the soil: Apricot or peach seedlings, Myrobalan or plum cuttings.

1.3.2. Incompatibility The use of rootstocks other than *Prunus armeniaca* is restricted by problems of graft incompatibility. Studies have shown that incompatibility can have different origins and symptoms (50). Three principal categories are delienated.

1.3.2.1. Incompatibility by Translocation. Incompatibility manifest in foliar symptoms is termed incompatibility by translocation. Starch produced in the leaves will not migrate to the roots. The tree falls into a gradual decline and dies in two or three years. Herrero (24) has shown that apricots grafted on some types of Myrobalan stocks show this incompatibility. Apricots grafted on peach seedlings are sometimes compatible, but the reciprocal combination has proved to be incompatible. In this latter combination, there is little or no late summer growth, the foliage is unhealthy, there is premature leaf fall, and starch is absent in the stocks' phloem. In spite of this poor stock—scion affinity, apricot seedlings are used as rootstocks in the Tunisian oasis of Gafsa to grow local peaches because of their resistance to nematodes and salt tolerance.

1.3.2.2. Incompatibility of the Graft Union. Incompatibility at the graft union is common. In this case the rootstock is fed satisfactorily by the scion leaves, but there are anatomical problems at the point of union. A discontinuity of the bark and wood results from an involution of the cambium development. This disorder diminishes the strength of the union, and trees can break under the effect of wind. For example, Lapins (29) in British Columbia studied several apricot cultivars grafted on peach seedlings and showed that with 'Blenheim,' 'Tilton,' 'Wenatchee,' 'Sunglo,' and 'Perfection,' compatibility was good with less than 2% exhibiting structural weakness at the graft union. 'Moorpark' and 'Riland' had 2−6% structural weakness while 23% of 'Reliable' broke.

The smoothness of the break at the union and the discontinuity of wood/bark tissues were found to be reliable symptoms of incompatibility between apricot scions and peach rootstocks. Carlson (6), when comparing six rootstocks, showed that the greatest loss occurred when apricots were propagated on peach seedlings. Duquesne (12, 15) made the same observation with the cultivars 'Canino' and 'Rouge du Roussillon,' which were propagated on *P. cerasifera*, *P. domestica*, and several interspecific hybrids. Within a series of plum clones, seedlings or cuttings, it is possible to find all manifestations of graft union compatibility or incompatibility, ranging from perfect unions to those where the wood of the scion and the rootstock are separated by a thin layer of dead cells.

The apricot rootstock problem is rather difficult to solve because cultivars appear to fall into two categories with respect to graft compatibility. In the first group are cultivars compatible with a wide range of rootstocks. 'Bergeron,' 'Poizat,' 'Blenheim,' and most of the Vesuvian cultivars fall in this group. In the second group, compatibility is restricted to a small number of rootstocks such as apricot seedlings or cuttings of 'Greengage' plum or 'Pollizo.' 'Rouge du Roussillon,' 'Canino,' 'Bulida,' and most of the Spanish cultivars are in this group.

Duquesne (12) thinks that it is easier to obtain good unions by selecting cultivars with less specific rootstock requirements in breeding programs, as this characteristic seems to be easily transmitted. A more difficult approach would be to create new rootstocks to be used with difficult-to-graft noncongenial cultivars.

Graft union incompatibility is difficult to evaluate in that the developing symptoms are not readily observable. A longitudinal sawing of a three- to four-year old graft union permits examination, yet interpretation is difficult due to variation from one tree to another. 'Rouge du Roussillon' grafted on clones of 'Brompton' or 'Reine Claude' ('Greengage') plum cuttings (Fig. 1)

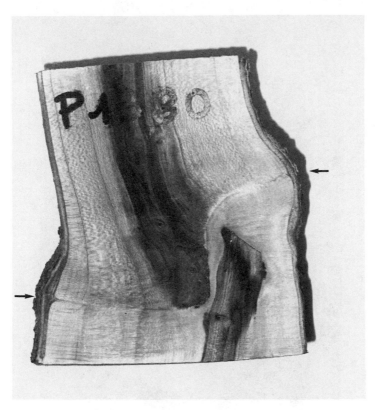

Figure 1. 'Rouge de Roussillon' cultivar grafted on 'Greengage' rootstock showing symptoms of incompatibility.

can in fact show either no sign of incompatibility or 10% or more union breakage, depending upon cultural conditions, 'Brompton' being more susceptible. Such factors as windy site conditions and rapid growth stimulated by rich soil and irrigation accelerate breaking. This happens mostly during the period of tree establishment from three to five years. Consequently, if rootstock compatibility is known to be questionable, it is necessary to moderate the growth of young trees. If the graft union is high, more than 40 cm, the tree will not break as evenly, even if the union is not strong (Fig. 2). Summer pruning also helps by slowing tree vigor and reducing any wind effect. A list of some apricot cultivars in terms of their requirements with different rootstocks is found in Table 2.

1.3.2.3. Incompatibility Caused by Virus Infection. Marenaud (35) has shown that the clorotic leaf spot virus (CLSV) can induce graft incompatibility, especially with French seedling apricot selections 'A 843,' 'Canino A 238,' and with wild

Figure 2. Highly budded 'Rouge de Roussillon' on 'Myrobalan F.G. 31.'

apricots of North Africa (Mech Mech). This grafting is sometimes impossible or, if initially successful, shoots will break off later in the nursery. The virus effect is stronger if only one component of the nursery tree, rootstock or scion is contaminated. This occurs most frequently when seedling rootstocks, which are generally virus free, are grafted with a virus-infected cultivar. This type of incompatibility can explain some of the puzzling results obtained in experimental studies using healthy seedling apricot rootstocks whose CLSV susceptibility is not known and grafting them to a virus contaminated cultivar. The CLSV seems to be very widely distributed in European orchards, and though transmission by a vector has not been proved, there probably is one.

TABLE 2. List of Some Apricot Cultivars in Terms of Their Requirements with Different Rootstocks

Compatible with All Classical Rootstocks	Problems of Compatibity with Some Rootstocks (i.e. Mariana, Myrobalan, Plums)
Ampuis	Beliana
Bergeron	Bebeco
Blenheim	Canino
Bulida	Colomer
Cafona	Dr. Mascle
Early Orange	Ferriana
Erevani	Houcall
Falca Rosic	Modesto
Hungarian Best	Moniqui
Louisette	Precoce du Portugal
Luizet	Priana
Malan Royal	Rouge de Fournes
Moorpark	Rouge du Roussillon
Paviot	Rouge de Rivesaltes
Perfection	Tardif de Bordaneil
Polonais	Screara
Precoce d'Imola	
Reale d'Imola	
Reliable	
Riland	
Sunglo	
Tilton	
Trewatt	
Wenatchee	

2. ROOTSTOCK ADAPTATION

2.1. Climate

2.1.1. Cold Temperatures Freeze damage can be observed on the trunks of apricot trees in several production areas in Eastern Europe (Rumania, Bulgaria, Moldavia, and Armenia) and Canada. Nitransky (38) has shown that cuttings of the plum cultivar 'Koslienka' (*P. insititia* (*domestica*) *rubra F. dulcis*), which has a long winter dormancy, when used as a rootstock has less damage than other rootstocks. In Rumania (1), it was observed that the plum 'Buduruz' gave hardiness to apricot scions during the winter and thus helped them to resist freeze damage during periods of oscillating temperatures in January and February. The scion budding must be high, at 80−100 cm, so as to protect the cultivar trunk from very low temperatures occuring near the soil. As a rule apricot seedlings are not particularly winter hardy. A wild apricot from the Indu-Kush mountains was tested in Germany in 1956 as a rootstock for apricots (20) and it showed a high degree of hardiness whether unworked or budded. Nitransky (39) also showed that it produced a vigorous tree. The status of this rootstock is unknown at present to our knowledge. The 'Brompton' plum has also been found to be a hardy rootstock for apricots in Germany. Lupescu (33) is of the opinion that local Myrobalan seedlings give trunks which are winter hardy. In Canada, Layne (30) has developed the apricot cultivar 'Haggith' which produced uniform seedlings with good hardiness. As rootstocks, peaches are not especially cold tolerant and some seem to be particularly sensitive (11).

Winter cold damage is not usually a problem in southern Europe, California, or Australia.

2.1.2. Heat High summer soil temperatures are not harmful to apricot seedlings in dry conditions, but can depress growth with some plum rootstocks ('Damas,' 'Brompton,' 'Saint Julien') under the same conditions.

2.2. Soils

We have seen that apricot growing is only possible in climatic conditions where winter cold is not severe and spring frost is not a hazzard. For this reason, in most European countries, the climatic zones where production is possible are restricted to the south of France, east coast of Spain, south of Italy, etc. The soils found in these zones are not always suited to apricot seedlings. Therefore it is necessary to find compatible rootstocks in the related *Prunus* species that have soil adaptability. The same holds true for most other areas of apricot production.

2.2.1. Waterlogging One of the most drastic difficulties encountered by rootstocks is waterlogging. This can occur in a wide range of soils as a result of protracted rain and/or poor interwall soil drainage. Under this condition root asphyxiation is induced. Roots are more susceptible during their growth period than when dormant. In the *Prunus* species, roots deprived of oxygen will begin to ferment and produce phytotoxic substances (43). Root asphyxiation is characterized by leaf wilting and an alcoholic smell (sour sap) emanating from the dead roots, which are brown. The appearance of symptoms varies quantitatively and qualitatively among the different *Prunus* rootstocks (Table 3). When such symptoms are present for a sufficient period, they cause not only partial or total destruction of existing roots but prevention of root regeneration. A partial destruction of the roots, if it does not induce scion death, will lessen

TABLE 3. Resistance of Different Rootstocks to Waterlogging[a]

Rootstocks[b]	Number of Days of Resistance to Waterlogging	
	In Winter	In Summer
Almond (S)	70	
Apricot	70	5–7
(Peach × almond)GF 557 (C)	75	
Peach (S)	80–85	8–10
(Peach × almond)GF 677 (C)	85	
Brompton plum (C)	120	20–25
G.F. 31 (Myrobalan × Salicina) (C)	125	20–25
Myrobalan B (S)	130	20–30
Damas plum (C)	140	40–50
Saint Julien (C)	140	
Mariana G.F. 8-1 (C)	145	50–60

[a]According to Duquesne and Bernhard.
[b]S = Seedlings; C = Cuttings.

production and fruit quality. Resistance to waterlogging could be related to the root's ability to utilize oxygen from the leaves through sap movement.

Special work has been done in France by Saunier (44) to develop a procedure for determining the resistance of fruit tree rootstocks to waterlogging. One-year-old plants, seedlings or cuttings, are subjected to controlled flooding for various periods up to 130 days. Their ability to survive or recover is then evaluated. At present, all new interspecific hybrid rootstock candidates are systematically evaluated for waterlogging characteristics, in France, using this procedure. Data presented in Table 3 are derived from this practice. A close correlation between resistance to winter and summer flooding exists.

2.2.2. Soil Lime The apricot is adapted to soils with a high lime content, although there is variability among seedlings for this characteristic. The seedlings of local North African apricots (Mech Mech) are used extensively in countries in the Mediterranian region where soils have a high lime level. Peach rootstocks are not suitable on high lime soils; apricots budded on peach seedlings and planted in high-lime soils will show typical leaf chlorosis. With Myrobalan rootstocks some resistance in apricots to lime-induced chlorosis is found. However, better results are obtained on Marianna GF8-1. The best results are obtained with the plum (Ciruelo), suckers of 'Pollizo' (*P. insititia*) of Spanish origin and peach × almond hybrids if compatibility is not a problem with these rootstocks. It must be noted that there are many Myrobalan plum types and that the behavior of one clone can be very different from another.

Duquesne (15) has shown, for example, that whereas 'Myrobalan GF 31' is susceptible to lime, another type, 'P 1245,' has good resistance.

2.2.3. Salt   Salt accumulation becomes a problem frequently in irrigated areas where rainfall is low (Middle East, North Africa), even when the irrigation water has a very low saline concentration, for example, water from the Euphrates River. Bernstein et al. (3) showed that 'Royal' apricot on 'Lovell' roots had a 66% growth reduction when planted in saline soil. Half of the reduction was attributed to chloride toxicity and half to increased osmotic concentration of the solution. They found that apricot cultivars were more sensitive to this condition than peach, plum, or prunes. 'Yunnan' (peach) rootstocks, increased the chloride accumulation and resultant toxicity in 'Royal' apricot.

It must be noted that these experiments were conducted in 3×3 plots, bordered by wood frames, in which the water was artificially salinized by using 4000 ppm of salt ($EC_2 = 5.1$ millimhos) as a 1:1 mixture of NaCl and $CaCl_2$. This concentration is usually considered unacceptable for irrigation. In North Africa orchards, saline irrigation rarely exceeds a 2000-ppm concentration ($EC_2 = 2.5$ millimhos) and the chloride ion is balanced by $CaSO_4$ and $MgSO_4$. Under these conditions, wild apricot seedlings (Mech Mech) grow vigorously and are long-lived. These trees are not grafted, however, a condition which improves their vigor and longevity (Fig. 3). When saline water is used regularly for irrigation, especially in regions of low rainfall, well-drained soils are required.

2.3. Pests and Diseases

Although the prime reason for utilizing specific rootstocks was to extend apricot culture to soils unsuited for apricot seedling stocks, a problem arose, particularly on replant orchards, with respect to pests and diseases.

2.3.1. Pests

2.3.1.1. The Peach Tree Borer.   The peach tree borer (*Aegeria opalescens*) H. Edv., according to Day (11), is confined mostly to the mountain and valleys near the coast of central California. The larvae of this moth burrow just beneath the bark at or below the soil line, although they sometimes work a considerable distance up the trunk. While the name implies that it chiefly attacks peach rootstocks, almond, myrobalan, and apricot stocks are also susceptible to attack. The graft union, particularly when it is rough, is often attacked. Heavy soil, which cracks more easily than light soil, provides an avenue for insect access to root bark. No rootstock is resistant to peach tree borer, although the apricot, which gives a smoother union with an apricot scion, has some advantage. Weaver (52) found some tolerant seedlings in populations of the peach cultivars 'Dixired,' 'Elberta,' 'Earlired,' etc.

Figure 3. Huge apricot seedlings (Mech Mech) in a Tunisian oasis.

2.3.1.2. Peach Capnodis (*Capnodis tenebrionis* L.). This large beetle is famil-
iar to growers in the Mediterranean countries. Its larvae burrow galleries in the
upper roots and at the crown of almond, plum, peach, and apricot, killing the
tree. Attacks are more common in years when trees suffer from drought. Thus
it is a particular problem in dry-land farming areas of North Africa, Italy, and
Spain. No rootstock is resistant, although some growers think that the bitter
almond stock has some resistance. The insect can be controlled by wetting or
powdering the crown and adjacent soil during the growing season with
Lindane.

2.3.1.3. Root Knot Nematodes. Three species of *Meloidogyne*—*M. arenaria*, *M.
incognita*, and *M. javanica*, are particularly injurious. The first two are
prevalant in soils of southern Europe and the Middle East, while the second
and third are more common in North Africa. These three species affect most
apricot orchards, worldwide, which are propagated on peach, almond, or plum

rootstocks. Apricot seedlings seem to be uniformily resistant to root knot nematodes. Wild peach seedlings are usually susceptible, but some selections have been selected for resistance; these include 'Nemaguard,' 'Rancho Resistant,' 'S-37,' 'Okinawa,' and 'Nemared' rootstocks. Almond seedlings, bitter or sweet, are also susceptible for the most part, but resistant selections (28) include 'Alnem 1,' '88,' and '201.' These are susceptible to *M. incognita*, which is common in North Africa. In the plums, which are mostly susceptible, 'Myrobalan B' is tolerant and 'Marianna 2624' and 'GF 8-1' are resistant to root knot species. 'GF 31', which is a *P. cerasifera* (Myrobalan) × *P. salicina* hybrid, is more or less tolerant, as are the 'French prune GF 43' and the 'Brompton' plum.

Root knot nematode resistance is mainly a hypersensitive reaction: cells around the puncture die and the nematode starves. With a high inoculum level in the soil, roots suffer numerous injuries that prove to be debilitating and production is decreased (45). Resistant rootstocks can suffer like susceptible ones if the nematode population level is increased by annual host crop growing such as tomato, lettuce, or wild hosts in volunteer vegetation.

2.3.1.4. Meadow or Lesion Nematode. *Pratylenchus vulnus* causes bark cankers, rather than beadlike knots on roots, and attacks all rootstocks. This nematode, common in Europe and California, has a major influence on tree vigor and production (32). Peach seedlings are very susceptible to *P. vulnus*; apricot, 'Myrobalan GF 31,' and 'hybrid 2038' (Myrobalan × Besseyi) seem to be more tolerant; and *P. tomentosa* is quite tolerant.

2.3.1.5. Ring Nematodes. Ring nematodes *Criconemella xenoplax* and *C. curvata*, cause severe damage to several rootstocks, including apricots, on sandy soils. These two species are very common in European orchards and vineyards.

2.3.1.6. Nematodes as Virus Vectors. The most important species in this group for apricots is *Xiphinema diversicaudatum*, which is the main vector of strawberry latent ringspot virus (SLRV). This virus is widely spread in Europe and America. Damage caused by it is increased when in association with other viruses like necrotic ringspot virus (NRSV). In many countries nurseries are subject to nematode inspection in order to reduce the spread by nursery stocks. Nematodes in orchard or nursery soils can be reduced by preplant nematode treatments, for example, dichloropropene or dibromethane fumigation. In nurseries some nonphytotoxic compounds like aldicarb, oxamly, carbofuran are used in postplant applications for nematode control. In several countries, such as France, these nematocides cannot be used in bearing orchards.

2.3.2. Diseases

2.3.2.1. Blackheart. Blackheart, or verticilliosis (*Verticillium albo-atrum*) seems to be present in different soil types throughout the world. Selection of resistant rootstocks is difficult because artificial inoculation is unsatisfactory in tests, and soils are irregularly contaminated. Irrigation, the culture of vegetables (tomato, melon, lucern, potato), and excess tree vigor all are disposing factors. Duquesne (14) has shown in a trial of several rootstocks that the symptoms of blackheart decrease each year and disappear when trees are full grown. The influence of rootstock on scions is evident. Rootstocks that induce the greatest susceptibility are: 'Myrobalan GF 31' and apricot seedlings. Peach seedlings, 'Marianna GF 8-1,' are more tolerant, and the most tolerant is 'Reine Claude GF 1380' ('Greengage'). Other observations indicate that the scion cultivar can also increase rootstock susceptibility to blackheart. Some years appear more favorable for the disease than others. Taylor (47) found one apricot cultivar, 'Zaisky Altai,' obtained from a wild population of *P. armeniaca*, to be resistant and having potential as a rootstock in infected soil. Differences in tolerance levels probably also exist in plums.

2.3.2.2. Bacterial Gummosis or Bacterial Canker. Some bacteria, *Pseudomonas syringae*, *P. morsprunorum*, and *P. viridiflava*, can induce cankers and kill apricot trees. The bacteria induce cortical lesions, which develop in mild winters and in the spring. These lesions disturb the tree's spring growth. Young trees are susceptible and may be killed in a few weeks. The disease is often associated with the sour sap condition.

Rootstocks exhibit different levels of resistance. The following is a classification in order of increasing susceptibility: peach, apricot, plum 'Greengage' and 'St. Julien,' 'Myrobalan,' 'Marianna GF 8-1,' and '26-24' (21, 40).

Day (11) suggested that the rootstock influenced the bacterial canker susceptibility of apricot scions. This has been confirmed by Duquesne (13) in several experiments, with Marianna stocks having the worst influence. The rootstock and scion have a reciprocal influence and tree death can be a consequence of cankers on the scion or any part of the rootstock above ground. The rootstock itself is not killed completely and surviving roots send up suckers. High budding on a tolerant rootstock is a good precaution but its adaptation to the soil and growing conditions are important factors in resistance (13). An illustration of this is found in that plum rootstocks induce apricot scion susceptibility when grown on dry soils but less so when on rich land. An acid soil seems to increase susceptibility.

2.3.2.3. Crown Gall (Agrobacterium tumefaciens). The bacteria causing crown gall are present in all soil types throughout the world. Trees may grow normally if the galls are only on the roots, but nursery trees are stunted and often die if attacked. Thus protection in the nursery is important and can be effectively achieved using the antagonist bacteria *A. radiobacter* K-48 strain (19).

Almond and apricot seedlings are very susceptible to crown gall, as are peach GF 305. 'Nemaguard' and peach red-leaf seedlings of 'Rutgers Red Leaf' and the new French stock 'Rubira' have good resistance. 'Marianna GF 8-1,' 'Marianna 2624,' 'Myrobalan GF 31,' 'Reine Claude,' and 'GF 1380' are resistant.

2.3.2.4. Crown Rot or Phytophthora Canker. Day (11) reported that bark canker, at or near the soil line or on roots caused by water molds of the *Phytophthora* species, often kills trees following protracted wet periods. The infection occurs mostly during the late winter or early spring when warm temperatures occur. A short period of the proper temperature plus wet conditions can result in an infection. The symptoms are similar to those induced by waterlogging, early wilting, or failure to leaf out. In waterlogging the lower roots are killed and turn dark, whereas with crown rot (collar rot) the bark is killed in a band at or near the soil surface and roots below appear normal. The disease can attack rootstocks in the nursery or orchard and is particularly devastating to young trees. Its occurrence, being weather related, is periodic. Peach and almond rootstocks are susceptible, apricot seedlings less so, and 'Marianna 26-24' is moderately resistant (11, 40). Crown rot is often called "heeling-in disease" because it attacks nursery trees after they are uprooted and temporarily "heeled in," particularly in wet soil or during protracted rain (11). As for bacterial gummosis, dried grass, weeds, leaves, or a high cover crop at the tree's crown furnish a good environment for fungal infection to occur.

2.3.2.5. Root Fungus (Armillaria mellea). The roots of all stone fruits are subject to infection by this oak root fungus. *Rosellinia necatrix* has a similar effect, but is more common in France on pome fruits. It attacks all species throughout the southern Mediterranean area. Infection occurs when the fungus penetrates large or medium-size roots by means of special mycelium formations called rhizomorphs. The only method to prevent the onset of oak rot attack is to use resistant rootstocks. Trials by Thomas in 1948, Day (11) in 1953, and, more recently, Duquesne in 1977 (16), have found some resistance in *Prunus* species (Table 4). Apricot and peach seedlings are very susceptible, with a 70% fatality rate after 7–8 years. Plums exhibit some degree of resistance. 'Reine Claude' ('Greengage') GF 1380 is susceptible, 'Myrobalan GF 31' is slightly less so; 'Marianna GF 8-1' and 'Marianna 2624' show some resistance (16). Unfortunately, compatibility between these rootstocks and several apricot cultivars is

TABLE 4. Percentage of Death by Oak Root Fungus of Apricots Grafted on Four Roostocks[a]

	% of Death by Oak Root Fungus						
Rootstocks	1970	1971	1972	1973	1974	1975	1976
Seedlings of wild apricot (Mech Mech)	1,2	6,1	24	43	51	61	73
Seedling of Peach INRA 305-1	0	2,5	30	42	56	71	71
Reine Claude G. 1380	0	0	0	1,2	2,4	5	5
Marianna G.F. 8-1	0	0	0	0	0	2.4	2.4

[a]83–85 trees of each rootstocks distributed in 8 replications; according to DUQUESNE, 1977.

unsatisfactory. Guillaumin (23) thinks that cuttings of *P. institia* (*domestica*) such as 'Mirabelle,' 'Damas,' and 'Brompton' have a satisfactory tolerance level.

Resistance to oak root fungus appears to be incomplete. In the case of Myrobalan rootstocks, the rhizomorphs penetrate the root easily, but the development of the mycelium in the root if inhibited. This explains why the tolerance level in a rootstock can decrease if it is not growing under conditions of good culture and on good soil.

2.3.2.6. Sour Sap. See discussions under waterlogging, crown rot, bacterial gummosis, and blackheart, as all these conditions are referred to by growers as sour sap.

2.4. Training Systems

Because of its growth habit and need for light, the apricot is not suitable for high-density plantings. Some attempts have been made in Italy to utilize the Baldassari palmette system. The advantages of this system, compared to the usual modified leader system for production, are not a subject for this chapter. As for rootstocks, there is presently no information available on their direct effect on production in these systems.

3. ROOTSTOCK RELATIONSHIPS

3.1. Tree Size

The influence on scion tree vigor depends upon the vigor of the rootstocks themselves. This can be evaluated when they are all grafted with the same cultivar and grown in the same suitable soil. Usually, however, trials are made in soils where some of the stocks will not be well adapted. Thus the ratings of plum and apricot rootstocks grown in a dry soil can be quite different than those obtained in an irrigated soil. This explains why classifications, as to tree size, of the usual rootstocks for apricot can differ from one trial to another (5, 7, 8, 18, 27, 31, 34, 36, 37, 39, 46, 49, 51). Good compatibility obviously plays an important role. The virus status of trees is not always reported in these studies and, when present, will also account for difference is results. If data from the listed authors and the writers' own observations are taken together, it seems that at present apricot seedlings and Myrobalan seedlings or cuttings, when used as stocks, produce the most vigorous trees. Duquesne (in 50) proposes vigor classifications in two very different French soils: a pebbly soil and a sandy loam (Table 5).

Interstems have been used in apricot culture for two reasons: to overcome problems of root—scion incompatibility and to prepare trees with a high trunk. In the first instance, the interstem must be compatible with both rootstock and

TABLE 5. Apricot Rootstock Vigor Depending on Soil Type

Rootstock Classified by Decreasing Vigor	Loess with Pebble	Sandy Loam
1	(Myrobalan × Salicina) G.F. 31	Marianna G.F. 8-1
2	Peach seedlings	G.F. 31
3	Apricot seedlings	Greengage cuttings G.F. 1380
4	Plums (*P. domestica* or *P. insititia*)	Saint Julien P. 670-3
5	*Prunus besseyi*	(Myrobalan × besseyi) P. 2038

scion cultivar. For example, with a Myrobalan rootstock and 'Canino' or 'Rouge du Roussillon,' the 'French Prune' is a good interstem. With the same cultivars, but with 'Marianna GF 8.1' rootstocks, a good interstem is 'Greengage' (Fig. 4). Hutchinson (25) studied this relationship and his observations are summarized in Table 6.

Our recent research showed that with the cultivar 'Rouge du Roussillon' there was no difference in growth on two-year-old trees having interstem lengths of 10, 20, 30, or 40 cm. But we found that when the scion and rootstock are strongly incompatible, as with 'Rouge du Roussillon' on 'Marianna GF 8.1,' 80% of the scions broke at the graft union on three-year-old trees, whereas 12.5% broke on the 'Rouge du Roussillon'/'Greengage'/'Marianna G.F. 8.1' combination and only 4% on the rather compatible 'Rouge du Roussillon'/'Greengage' association when trees were grown under the same cultural conditions.

The technique of using long interstems to give high trunks was used in the past on sites where there was a risk of winter freeze damage (i.e., the North Rhone Valley in France). Interstems for this use must grow straight and vigorously. The following combinations were used: 'Myrobalan'/'Belle de Louvain'/'Luizet,' 'Myrobalan'/'Krazinsky'/'Bergeron,' 'Marianna GF 8.1'/ Reine Claude d'Oullins/'Paviot' or 'Ampuis.'

The use of interstems is restricted by propagation difficulties and consequent increase in tree cost. The problem is complicated by the fact that all three tree components must be virus free. Since some virus diseases are exceedingly difficult to index and detect, as in the case of plum pox virus (Sharka disease) or chlorotic leaf roll mycoplasma (X disease), interstem trees are difficult to produce.

3.2. Production

The stock−scion relationship has been evaluated because of its influence on production. Apricot rootstocks appear to give more flowers and fruits (17, 46),

Figure 4. 'Rouge de Roussillon' budded on 'Marianna F.G. 8-1' with an interstem of French prune P. 707.

TABLE 6. Rootstock and Interstock Trials for Three Apricot Varieties[a]

Rootstock	Interstem	Cultivar	Observations
Myrobalan	German prune	Perfection	Satisfactory
		Tausky	Trees vigorous and productive
Myrobalan	Imperial Gage	Perfection	Satisfactory
		Early Orange	Doubtful
		Tausky	Unsatisfactory
	Grand Duke	Perfection	Unsatisfactory

[a]According to Hutchinson, 1965.

though there is some delay in coming into production, when compared to Myrobalan (7). These stocks produce fruit with more color and earlier maturity (7, 18, 50). Myrobalan, peach, and Marianna rootstocks all seem to give larger fruit. Generally speaking, a slight incompatibility will usually result in larger, sweeter, and more colorful fruit. Pollizo plum rootstock seems to give more cracking (17). Myrobalan rootstocks produces trees with fewer leaves which drop later in autumn than do apricot seedlings as rootstocks.

Understanding stock–scion relationships is as important as knowing rootstock–soil adaptability when selecting a rootstock. A good new rootstock should have the attribute of producing a smaller tree bearing larger, tastier, and more colorful fruit.

4. ROOTSTOCK CHARACTERIZATION

4.1. Apricot Seedlings

This is the rootstock most widely used for apricot on a worldwide basis. Nurserymen use, almost exclusively, seeds from the most common local cultivars: 'Blenheim' in California, 'Rouge du Roussillon' and 'Canino' in France, 'Hungarian Best' in Hungary, Czechoslovakia, Rumania, etc. These rarely give a uniform stand of seedling trees in the nursery, as there is much variation among seedlings of a given cultivar. There is also reason to believe that some cultivars like 'Rouge du Roussillon,' when selfed, produce seedlings with a delayed genetic weakness as a result of inbreeding.

In France, INRA has selected a local cultivar, 'Manicot,' which gives very vigorous and homogenous seedlings when selfed. Wild apricot trees were collected by Ballot in 1952 in southern France from which 'Manicot' and 'A-843' were selected at the Station de Recherches d'Arboriculture Fruitiere de Bordeaux. 'A-843' proved to be very susceptible to chlorotic leaf spot (CLS) virus and was used in indexing. 'Manicot' yields uniform vigorous seedlings but, like many wild apricots, is CLS-virus susceptible and thus must be budded with virus-free cultivars.

Layne (30) selected at Harrow, Canada, the 'Haggith' and introduced it as a seed source for apricot rootstocks in 1974. 'Haggith' was found on the Murray Haggith farm at Ruthven, Ontario, Canada. It is cold hardy, consistently productive, self-fertile, and disease tolerant. 'Haggith' seedlings are quite uniform in the nursery row, more vigorous than 'Manicot,' and attain buddable size in August. Scion cultivars on 'Haggith' stocks tend to have wider crotch angles and a more spreading growth than those on 'Alfred' or 'Morden 604.'

In general, apricot seedlings as rootstocks are well adapted to dry, light, pebbly soil and yield vigorous trees. They are compatible with all apricot cultivars, but often exhibit incompatibility symptoms if the bud wood is virus infected by CLSV. The usual period for T-budding is July–August. One-year-old shoots are best for propagation purposes.

4.2. Peach Seedlings

Peach seedlings are used as rootstocks on acid to neutral soils under irrigation. Seeds of unknown origin from canneries are being superseded by those of known cultivars: 'GF 305' in France, 'Lovell' and 'Nemaguard' in California. Some new rootstock cultivars such as 'Montclar,' 'Rubira' (France), and 'Nemared' (U.S.) are being tested. Peach seedlings improve scion resistance to bacterial canker and to verticilliosis, but unfortunately some widely grown apricot cultivars exhibit incompatibility symptoms on them.

4.3. Myrobalan or Cherry Plum (*P. cerasifera*)

Myrobalan plums grow wild throughout Europe. Their origin is probably in the Caucasus. Many types exist. Seeds for nursery propagation were usually gathered from the wild. They produced seedlings of diverse vigor, graft affinities, growth habit, root distribution, and other characteristics. Approximately 250 selected seedlings were evaluated at the Davis, California, Experiment Station. A few seemed superior in vigor but made no better rootstocks than selected Myrobalans propagated by cuttings (11). 'Myrobalan B' is a selection made by the East Malling Research Station propagated by cuttings and widely used by nurserymen. Difficult-to-bud cultivars are incompatible with Myrobalan seedlings. Nevertheless, Grasselly (22) has found at least one type which is compatible with peach, 'P. 2032,' though nearly all other known Myrobalan types are not. Thus it may be possible to find a Myrobalan compatible with all apricot cultivars. Research work, however, is focused on a study of natural Myrobalan hybrids. In France, 'GF 31' (*P. ceracifera* × *P. salicina*) has rather good compatibility with noncongenial apricot cultivars. Unfortunately, 'GF-31' is susceptible to Verticilliosis, while true Myrobalans are not. The term *Myrobalan* used to designate Myrobalan hybrids, is particularly confusing.

4.4. Marianna

The origin of the Marianna rootstocks is not known. It is a natural hybrid between *P. cerasifera* (Myrobalan) and an American diploid species of *Prunus* that is thought to be *P. munsonia* but could be, according to Grasselly, *P. injucunda*, based on botanical characteristics, or *P. bokhariensis*, as suggested by Day (11). Marianna is a weak tree and itself is no longer used as a rootstock. It has yielded open-pollinated seedling hybrids, some of which have been selected and propagated vegetatively. These have been selected on the basis of tree vigor and numbered '26−23,' '26−24,' etc., in order to distinguish them from the parent Marianna. 'Marianna 26−24' is presently the most widely used. 'Myrobalan 29C,' according to Grasselly, is also a Marianna seedling; Day (11) is of the same opinion. In France, an open-pollinated seedling of Marianna yielded the selection of 'GF 8-1' (3N), which in many characteristics resembles 'Marianna 2624.' Salesses (42) states that 'GF 8-1' came from a cross between a diploid 'Marianna' with an abnormal 2N ovule and *P. cerasifera*. Marianna stocks have vigorous growth, are adapted to all soil types, and show good resistance to wet soil and root fungus diseases. Terblanche et al. (48) have shown that Marianna was a better stock than peach or apricot on compacted soils because of its natural shallow-root system. Unfortunately, it is incompatible with noncongenial cultivars (Fig. 5). 'Marianna GF 8-1' has the disadvantage of being tolerant to mycoplasma diseases and to promote the multiplication of mollicutes (in its phloem). Indexing is difficult, and the only way to ensure production of healthy cuttings in infected areas is to eliminate all trees showing advanced budburst in spring and all the weak-growing cuttings.

4.5. 'Brompton' (*P. domestica*)

This plum is easily propagated by cuttings and grows vigorously in the nursery row. This allows for high budding. Compatibility with the generally noncongenial cultivars is relatively good. Trees will not break if grown slowly the first few years. 'Brompton' stocks have better tolerance to verticillium species than 'Myrobalan GF 31' and 'Greengage.' It also shows tolerance to wet soils. Scion vigor is good on 'Brompton' stocks but less than that of 'Greengage.' Scions overgrow 'Brompton' stocks.

4.6. 'Greengage' (*P. domestica*)

INRA 'Reine Claude GF 1380' ('Greengage') is an old French cultivar the suckers of which are used as rootstocks for apricots in the Roussillon region. This stock is well adapted to fertile soils. It has poor nursery characteristics, being difficult to propagate by cuttings, and carries the apricot chlorotic leaf roll.

4.7. 'Pollizo' (*P. insititia* [*domestica*])

This local plum, which produces many root suckers, is used extensively in the production area of Murcia, Spain, for the 'Bulida' when grown with irrigation (Fig. 6). 'Pollizo' is under trial in France to evaluate its performance in wet soils. It appears to have promise. 'Pollizo' is resistant to lime chlorosis and has good compatibility with peach, plum, and almond as well as apricot. A good virus-free strain has been selected by Cambra (4).

5. NEW AND FUTURE ROOTSTOCKS

The preceding consideration have shown that there is no fully adapted rootstocks for apricots. All nurserymen and growers desire a rootstock with nursery and field characteristics to overcome the limitations of those currently in use.

Figure 5. 'Marianna G.F. 8-1' compatible with 'Bergeron' cultivar (*a*) and incompatible with 'Canino' (*b*).

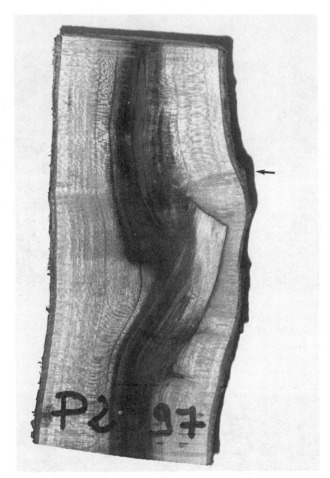

Figure 6. Very good compatibility between 'Rouge de Roussillon' cúltivar and 'Pollizo' rootstock.

Adaptability to all soils and compatibility with all cultivars is an objective that will be difficult to reach; nevertheless, the research must continue.

5.1. Selection of New Clones

5.1.1. Apricot Some years ago INRA selected the cultivar 'Manicot.' It has been used successfully as a rootstock. Selfing of 'Manicot' has yielded seedlings which are now under study for rootstock, some of which are more interesting than the 'Manicot' parent. Layne (30) is working with late-flowering selections that are also winter hardy.

5.1.2. Plums Two hundred plum cultivars (*P. domestica*) have been collected by Duquesne in France and they are being tested for compatibility with noncongenial apricot cultivars, for waterlogging resistance, propagation disposition, nursery characteristics, and facility for budding. At present, three potential rootstocks have been selected as having qualities superior to 'Greengage GF 1380' and which show three levels of vigor. They are now being compared with the Spanish *P. institia* rootstock 'Pollizo.'

5.1.3. Myrobalan It would be interesting if a Myrobalan selection could be found that is compatible with all apricot cultivars. 'Myrobalan 2032,' which is compatible with peach, is being tested in France as an apricot rootstock.

5.1.4. Peach In the United States there is interest in the newly released 'Nemared' as a rootstock. In France, the new peach rootstock selections 'Montclar,' 'Rubira,' and 'Higama,' all of which produce good seedlings, are being tested as rootstocks for apricots.

5.2. Interspecific Hybrids

A large-scale program to produce interspecific hybrids is under way in France. Several crosses between *P. cerasifera* and various apricot cultivars have resulted in 57 *P. dasycarpa* selections that are being evaluated as apricot rootstocks. These interspecific hybrids are rather easily propagated by cuttings. Some of them appear to be compatible with all apricot cultivars and show considerable differences in scion vigor.

INTERNATIONAL CATALOG OF ROOTSTOCK BREEDING PROGRAMS

France
    INRA Station de Recherches
    Fruitières Méditerranéennes
    BP81 Cantarel Montfavet,
    France

    P. Crossa-Raynaud,
    J. M. Audergon,
    J. C. Nicolas,
    J. L. Poessel

Greece
    Institute of Pomology
    Naoussa, Greece
    D. Stylianides

India
Himachal Pradesh University
Solan, India
J. S. Chauhan

REFERENCES

1. Anonymous (1974). Researches and results concerning the choice of the best rootstocks and interstocks for apricot [in Romanian], *Fruits Research Institute Annual Report* (Romania).

2. Bernhard, R., and J. Duquesne (1961). The rootstocks of apricot [in French], *Journees Nationales de l'Abricotier, Perpignan*.

3. Berstein, L., J. V. Brown, and H. E. Hayward (1956). The influence of rootstocks on growth and salt accumulation in stone fruit trees and almonds, *Proc. Amer. Soc. Hort. Sci.*, **68**, 86−95.

4. Cambra, R. (1970). Selection of Pollizos of Murcia and of other local Spanish plums [in Spanish], *Informacion Technica Economica Agraria*, **1**, 115.

5. Cambra, R. (1979). Compatibility of apricot varieties with Myrobalan and Marianna plums [in Spanish], *Annale de la Estacion Experimental de Aula Dei*, **14**, 371−375.

6. Carlson, R. F. (1965). Growth and incompatibility factors associated with apricot scion/rootstock in Michigan, *Quart. Bull. Mich. Agric. Exp. Stat.*, **48**, 23.

7. Costa, G., and M. Grandi (1975). The effect of various rootstocks on the tree behavior and fruit weight in apricot [in Italian], *Publicatione Istituto Coltivazione Arboree Bologna*, **283**, 55−62.

8. Costa, G., B. Marangoni, S. Sansavini, and R. Bordini (1974). Six years results with two apricot cultivars on twelve rootstocks (in Italian), *Riv. Ortofloro Frut.*, **58**, 286−303.

9. Crossa-Raynaud, P. H. (1961). Apricot and climate [in French], *Journees Nationales de l'Abricotier, Perpignan*, 55−58.

10. Crossa-Raynaud, P. H. (1977). Fruit varieties acclimation (in French), *Ann. Amelior. Pl*, **27**, 497−507.

11. Day, L. H. (1953). *Rootstocks for Stone Fruits*, Cal. Agric. Sci. Bull.

12. Duquesne, J. (1969). Studies of the compatibility of some cultivars of *Prunus armeniaca* (Koch) on several types of *Prunus* [in French], *Ann. Amelior. Pl.*, **19**, 419−441; **20**, 453−467.

13. Duquesne, J., H. Gall (1975). Influence of rootstocks on the susceptibility of apricot to bacteriosis [in French], *Phytoma*, 22−26.

14. Duquesne, J., J. M. Delmas, and H. Gall (1975). Verticilliosis on apricot in the Southeast and south of France [in French], *Pomologie Francaise*, **17**, 18−20.

15. Duquesne, J., K. Pomar, and J. M. Delmas (1976). Compatibility of some apricot cultivars grafted on Peach G.F. 305 and three plum rootstocks [in French], *Pomologie Francaise*, **18**, 161−171.

16. Duquesne, J., H. Gall, and J. M. Delmas (1977). New observations made on

susceptibility to root fungus disease (Armillaria) of some apricot rootstocks [in French], *Pomologie Francaise*, **19**, 95−98.

17. Egea, L., and T. Berenguer (1977). Preliminary results on the behavior of apricot cultivar 'Bulida' on different rootstocks [in Spanish], *Fruits*, **32**, 759−770.

18. Fideghelli, C. (1973). A comparative trial of four apricot rootstocks [in Italian], *Annale Istituto Sperimentale per la Frutticoltura*, **4**, 171−181.

19. Faivre-Amiot, A. (1983). Agrobacterium Radiobacter K 84 strain and its use for the biological control of crown gall [in French], 3° Colloque sur la Recherche Fruitiere, Bordeaux, pp. 225−234.

20. Friedrich, G., and P. Hoffmann (1956). Testing a wild apricot from the Indu-Kush mountain for use as a rootstock [in German], *Gartenbauvissenchaft*, **21**, 396.

21. Gautier, M. (1971). Apricot and its culture [in French], *Arboriculture*, **206**, 46−54; **207**, 46−54.

22. Grasselly, C. (1983). Possibility of using a Myrobalan clone as a rootstock for peach [in French], *Comptes Rendus Acad. Agricul. France*, pp. 346−354.

23. Guillaumin, J. J. (1982). The root fungus diseases of fruit trees [in French], 2° Colloque sur les Recherches Fruitières, Bordeaux, pp. 227−245.

24. Herrero, J. (1955). Stock scion incompatibility. I, performance of some reciprocal combinations. II, Effect of an interstock on peach−Myrobalan incompatibility [in Spanish], *Annales de la Estacion Experimental de Aula Dei*, **4**, 149−166.

25. Hutchinson, A., and O. A. Bradt (1965). Rootstock and interstock trials for three apricot varieties, *Rep. Ont. Hort. Exp. States Prod. Lab.*, **21**, 25.

26. Kalasek, J., and J. Blaha (1968). The suitability of apricot rootstocks [in German], *Mitt. Klosterneuburg*, **15**, 193−199.

27. Kapetanovic, N., and V. Pica (1976). Studies on local plums as rootstocks for plums and apricots [in Yugoslavian], *Jugosl. Vocarstvo*, **10**, 225−263.

28. Kochba, J., and P. Spiegel-Roy (1976). Alnem 1, Alnem 86, Alnem 201 almonds: Nematod-resistant rootstock seed sources, *HortScience*, **11**, 270.

29. Lapins, K. (1959). Some symptoms of stock-scion incompatibility of apricot varieties on peach seedling rootstock, *Can. J. Plant. Sci.*, **39**, 194.

30. Layne, R. E. C., and T. B. Harrison (1975). 'Haggith' apricot rootstock seed source, *HortScience*, **10**, 428.

31. Liacu, A. (1960). A contribution to the study of apricot rootstocks [in Romanian], *Lucrari Sti. Inst. Agron. Iasi.*, pp. 321−326.

32. Lownsbery, B. F., and E. F. Serr (1963). Fruit and nut tree rootstocks as host for a rootlesion nematode, *Pratylenchus vulnus*, *Proc. Amer. Soc. Hort. Sci.*, **82**, 250−254.

33. Lupescu, F. (1980). The behavior of the main apricot varieties on different rootstocks [in Romanian], *Lucrari Sti. Inst. Agron. Balescu*, pp. 299−308.

34. Manzo, P., and F. R. de Salvado (1978). Results of a twelve-year comparative study on two cultivars and four rootstocks of apricot [in Italian], *Ann. Istituto Sper. Frutt., Roma*, **9**, 33−41.

35. Marenaud, C. (1968). Manifestation of apricot species of an intraspecific incompatibilty caused by a chlorotic leaf spot virus [in French], *Etudes de Virologie, Annales des Epiphyties*, **19**, 225−245.

36. Nitransky, S. (1977). Growth and fruiting of the apricot cultivar 'Rakovsky' on different rootstocks [in Czechoslovakian], *Pol'nohospodarstvo*, **23**, 884–894.

37. Nitransky, S. (1978). Effect of rootstock on the growth and productivity of the apricot cultivar 'Paviot' [in Czechoslovakian], *Pol'nohospodarstvo*, **24**, 105–114.

38. Nitransky, S. (1983). Effect of rootstocks on termination of deep dormancy in the reproductive buds of apricot [in Czechoslovakian], *Pol'nohospodarstvo*, **29**, 457–471.

39. Nitransky, S. (1973). The growth of the apricot· cultivar 'Hungarian Best' on different rootstocks during the juvenile period [in Czechoslovakian],

40. Norton, R. C., C. J. Hansen, N. J. O'Reilly, and W. H. Hart (1963). Rootstocks for apricot in California, *Cal. Agric. Exp. Sta. Serv. Leaf*, p. 156.

41. Reig-Feliu, A., and A. Albert-Bernal (1964). Industrial handling of apricot. I, Characteristics, cold storage, freezing and preservation with SO_2 of the varieties 'Canino' and 'Bulida,' the latter on different rootstocks [in Spanish], *Bol. Inst. nac. Invest. Agrar. Madrid*, **24**, 287–325.

42. Salesses, G. (1977). Researches about the origin of two *Prunus* rootstocks, natural interspecific hybrids: An illustration of a cytological study carried out in order to create new *Prunus* rootstocks [in French], *Ann. Amelior. Pl.*, **27**, 235–243.

43. Salesses, G., R. Saunier, and A. Bonnet (1970). Root Asphyxiation of Fruit Trees [in French], *Bull. Tech. Ing. Agricoles 251*.

44. Saunier, R. (1966). Determination method of root asphyxiation of some fruit tree rootstocks [in French], *Ann. Amelior. Pl.*, **4**, 367–384.

45. Scotto la Massese, C. (1981). Agronomical involvements of relationships between endoparasitic nematodes and fruit trees [in French], ler Colloque sur les Recherches Fruitières, Bordeaux, pp. 51–59.

46. Stefanov, N., and M. Dimitrova (1979). Effect of rootstocks on the growth and reproductive characteristics of apricot trees during the growth and fruiting periods [in Bulgarian], *Gra. i lozar Nauk*, **16**, 54–62.

47. Taylor, J. B., and N. T. Flentje (1968). Infection, recovery from infection and resistance of apricot trees to *Verticillium albo atrum*, *New Zealand Jour. Bot*, **6**, 477–486.

48. Terblanche, J. H., I. S. de Koch, and J. A. van Zyl (1974). The influence of the physiological properties on the selection of rootstocks for apricot, *Decid. Fruit Grow.*, **24**, 82–86.

49. Vachun Z. (1980). Effects of selected rootstocks from *Armeniaca vulgaris* on the growth and productivity of the apricot cultivar 'Velkopavlovicka' during the first years after planting out. *Acta Univ. Agric.* Brno **28**, 653–664.

50. Vidaud J. (1980). L'abricotier. *C.T.I.F.L* 220 pp.

51. Watt A.T.J. (1952). Rootstocks for apricot *Orchard New Zealand* **30**, 31–32.

52. Weaver G.M. (1966). Breeding winter hardy disease and insect resistant rootstocks for peaches. *Proc. XXII Intern. Hortic. Congress.* Vol. 1.

10

PLUM ROOTSTOCKS

William R. Okie
USDA Southeastern Fruit and Tree Nut Research Laboratory
Byron, Georgia

1. PLUM PRODUCTION

As a group, plums are the most diverse and most widely adapted of the stone fruits, having been domesticated separately in Europe, Asia, and North America. Rehder (78) divides *Prunus* into five subgenera: Prunophora (plums and apricots), Amygdalus (peaches and almonds), Cerasus (sweet and sour cherries), Padus (bird cherries), and Laurocerasus (laurel cherries). Prunophora is divided into three sections: Euprunus (European and Japanese plums), Prunocerasus (American plums), and Armeniaca (apricots).

1.1. Origin In Europe

Prunus domestica, the most important source of fruit cultivars, has apparently been grown in Europe for over 2000 years. Hedrick (38) suggested that *P. domestica* originated in the region of the Caucasus Mountains adjacent to the Caspian Sea as a hybrid between the diploid *P. cerasifera* and the tetraploid *P. spinosa*, both of which occur wild in Europe and western Asia. The triploid hybrid may have doubled its chromosomes, resulting in the hexaploid *P. domestica*. Salesses (86), however, showed that both *P. domestica* and *P. spinosa* appear to carry the *P. cerasifera* genome, suggesting a more complex origin for *P. domestica*.

Cultivars of *P. domestica* widely grown can be classified in the following groups:

a. Reine Claude or Greengage plums—perhaps hybrids with *P. insititia*, discussed below ('Reine Claude', 'Imperial Gage')

b. Prunes—in European usage a group of blue-purple freestone cultivars ('French' prune = 'Prune d'Agen' = 'Prune d' Ente,' 'Imperial,' 'Stanley,' 'Italian' prune = 'Fellenberg'). In American usage, any plum that can be dried with the pit intact.

    c. Lombard plums ('Lombard', 'Victoria')
    d. Yellow Egg plums ('Yellow Egg', 'Golden Drop')

A related hexaploid species, *P. insititia*, grows wild throughout Europe and western Asia. Seeds of this species have been found in ancient ruins. It is mentioned by Greek poets in the sixth century B.C., suggesting it was grown before *P. domestica* (38). The variability seen in *P. insititia* is much less than that seen in *P. domestica*. Important groups of cultivars include

    a. Damson and Bullace plums—oval and round shapes, respectively
    b. Mirabelles—round, yellow or gold, of higher quality than other groups
    c. St. Julien—commonly used for rootstocks

In addition to *P. domestica* and *P. insititia*, other species classified in the Euprunus section are listed in Table 1. More complete descriptions of older European plum cultivars are available (38, 99).

1.2. Origin in Asia

A second distinct group of plums (*P. salicina*, formerly *P. triflora*) originated in China and has been grown there since ancient times. About 200–400 years ago, they were brought to Japan and from there spread around the world as "Japanese plums" (38). As a species, *P. salicina* plums are more vigorous, productive, precocious in bearing, and disease resistant than the *P. domestica* plums. Japanese plums and their hybrids tend to be very firm and are widely grown in California for shipping. Although classified in Prunophora along with the European plums (Table 1), *P. salicina* hybridizes more readily with the American plums discussed below because all are diploids. *Prunus simonii* (Table 1) is also common in these hybrids. Important cultivars (including hybrids) are 'Casselman,' 'El Dorado,' 'Friar,' 'Laroda,' 'Red Beaut,' and 'Santa Rosa.'

1.3. Origin in North America

The third plum domestication and source of cultivars has been from the wide range of species native to North America (Table 2). As with the European plums, there are gradations of one species into another, and many cultivars appear to be a mixture of several species (114). Most important for fruit cultivars are *P. americana*, *P. hortulana*, and *P. munsoniana*. Most of the shipping plums of California are hybrids of *P. salicina* or *P. simonii* with one or more of the American species. These native species provide disease resistance, winter hardiness, and the tough skin desirable for shipping. In much of the United States, European and pure Japanese plums are decimated by cold or disease; hence hybrids are the only plums grown (101).

TABLE 1. Important European and Japanese Plum Species—Euprunus Section of *Prunus* (from 78)

Species	Common Name	Origin	Chromosome No.	Subspecies/ Varieties	Uses
P. blireiana Andre (= *P. cerasifera atropurpurea* × *mume*)			16		Ornamental
P. bokhariensis Schneid.		Kashmir			
P. cerasifera Ehrh.	Cherry plum Myrobalan	Europe, Asia	16 (17, 24, 32, 48)	*atropurpurea, divaricata pissardii*	Stock, fruit, ornamental
P. cerasifera × *munsoniana*	Marianna	Texas			Stock
P. cistena Koehne (= *P. cerasifera atropurpurea* × *pumila*)	Purple leaf sand cherry, purple plum				Ornamental
P. cocomilia Ten.	Italian plum	Italy	16		Fruit
P. consociiflora Schneid.		China			
P. curdica Fenzl & Fritsch		Armenia			
P. domestica L.	Garden plum, prune, European plum	Europe, Asia	48		Fruit, ornamental
P. fruticans Weihe		Europe			
P. gigantea Koehne (= *P. cerasifera* × *amygdalo-persica*)					
P. gymnodonta Koehne		Manchuria			
P. insititia L.	Bullace, damson	Europe, Asia Asia Minor	48	*syriaca, italica*	Fruit, stock
P. monticola K. Koch		Greece			
P. pseudoarmeniaca Heldr. & Sart.					
P. salicina Lindl.	Japanese plum	China	16(32)		Fruit
P. simonii Carr.	Apricot plum	N. China			Fruit
P. spinosa L.	Blackthorn, sloe	Europe, Asia	32 (16, 24, 40, 48)		Ornamental
P. sultana Voss (= *P. simonii* × *salicina*)					Fruit
P. thibetica Franch.		W. China			Ornamental
P. ursina		Asia Minor, Syria			Fruit

TABLE 2. Important American Plum Species—Prunocerasus Section of *Prunus* (from 78)

Species	Common Name	Origin[a]	Chromosome No.	Subspecies/ Varieties	Uses
P. alleghaniensis Porter	Allegheny plum, sloe	NE US	16	*davisii*	Fruit, ornamental
P. americana Marsh.	American plum, goose plum, hog plum	E US	16		Fruit, stock
P. angustifolia Marsh.	Chickasaw plum, sand plum	US		*watsonii, varians*	Fruit
P. dunbarii Rehd. (= *P. maritima* × *americana*)					
P. gracilis Engelm. & Gr.	Oklahoma plum	SC US			Fruit
P. gravesii Small		Connecticut			
P. hortulana Bailey	Wild goose plum, hortulan plum	C US	16	*mineri*	Fruit
P. lanata (Sudw.) Mack. & Bush	Inch plum	C US			Fruit
P. maritima Marsh.	Beach plum, shore plum	E US	16		Fruit
P. mexicana S. Wats.	Big-tree plum, Mexican plum	SC US	16		Fruit, stock

Species	Common name	Distribution[a]		Synonyms	Use
P. munsoniana Wight & Hedr.	Wild goose plum	SC US	16		Fruit, ornamental
P. nigra Ait.	Canada plum	US, Canada	16		Fruit
P. orthosepala Koehne (= *P. angustifolia watsonii* × *americana*)		SC US			Fruit
P. reverchonii Sarg.	Hog plum	S US			Fruit
P. rivularis Scheele	Creek plum	Texas	16		Fruit
P. slavinii E. J. Palm. (= *P. angustifolia varians* × *gracilis*)		Kansas, Oklahoma			
P. subcordata Benth.	Pacific plum, Sierra plum	NW US	16	*kelloggii, oregana*	Fruit
P. texana Dietr.	Texas almond cherry, Texas peachbush	Texas			
P. umbellata Ell.	Flatwoods plum, sloe	SE US	16	*injucunda, mitis, tarda*	Ornamental, fruit
P. venulosa Sarg.		Texas			

[a]E = east, W = west, N = north, S = south, C = central.

Although plums and cherries are classified by Rehder into separate subgenera, the Microcerasus section of Cerasus (cherry) appears to be closer botanically to the plums than to the cherries. The sand cherries are more readily hybridized and grafted or budded to plums than to cherries (83). Several of these species have been important as dwarfing rootstocks for plum, and as hybrids, for developing cold-hardy rootstocks and scions (Table 3).

1.4. Current Production

World production of plums is concentrated in the United States, Germany, and Yugoslavia, with additional production throughout Europe and in the Southern Hemisphere (Table 4; 105). Although figures are not available, presumably substantial quantities are produced in China, the Soviet Union, and other communist nations. In the United States, plums are grown in nearly every state for local use, but commercial production is limited to Michigan, California, and the Pacific Northwest. With the exception of the prunes and shipping plums of California, most of this commercial production is *P. domestica* cultivars, which are canned. In the Pacific Northwest, prunes are also dried and sold fresh. California produces 144,000 tons of dried prunes annually, about 70% of the world production, followed by Yugoslavia and France (76). Most of the shipping plums in California, as well as in the Southern Hemisphere, are *P. salicina* and hybrids (58).

2. ROOTSTOCK USAGE

Reviews have been published in several languages that describe the plum rootstock situation in other areas of the world. For Europe, reviews are available in German (35, 49), Dutch (19), Danish (14), Italian (60), Swedish (30), and English (103, 104). The situations in New Zealand (23) and California (17, 69) have also been described. Tukey (102) described available dwarfing rootstocks for plums. Westwood (112) discusses U.S. rootstock usage and lists seed germination information for common rootstock species.

2.1. Europe

Most of the plums grown in Europe belong to *P. domestica* or *P. insititia*. The most successful stocks for these species are the species themselves, Myrobalan (*P. cerasifera*) and Marianna (*P. cerasifera* × *munsoniana*). In countries such as Yugoslavia and Rumania, where clonal stocks are unavailable or unsuitable, Myrobalan seedlings are widely used (34). In Germany, stocks include seed-propagated stocks such as Myrobalan and St. Julien types Black Damas, Damas d'Orleans, and INRA Hybrids #1 and #2. Clonal stocks available include 'Ackermann', 'Brompton', St. Julien A, GF655-2, Marianna, and Marianna GF8-1 (49, 109). Recommended stocks in the United Kingdom are 'Pixy',

TABLE 3. Microcerasus Species Apparently Related to Plums Based on Graft Compatibility and Fertility of Hybridizations (from 78)

Species	Common Name	Origin[a]	Chromosome No.	Subspecies/ Varieties	Uses
P. besseyi Bailey	Western sand cherry	Canada, N US	16		Fruit, stock
P. cistena Koehne (= *P. cerasifera atropurpurea* × *pumila*)	Purple leaf sand cherry, purple plum				Ornamental
P. glandulosa Thunb.	Dwarf flowering almond, Chinese bush cherry	C & N China, Japan	16	*alba, albiplena, rosa, sinensis*	Ornamental
P. humilis Bge.		N China			
P. incana (Pall.) Batsch	Willow cherry	E Europe W Asia	16		Ornamental
P. jacquemontii Hook.	Flowering almond Jacquemont cherry	NW Himalayas			Ornamental
P. japonica Thunb.	Flowering almond Japanese bush cherry	E Asia	16	*thunbergii, englerii, nakaii, kerii*	Fruit, ornamental
P. microcarpa C.A. Mey.		Asia Minor			
P. prostrata Labill.	Rock cherry	Mediterranean, W Asia			Ornamental
P. pumila L.	Sand cherry	N US	16	*depressa, susquehanae, cuneata*	Fruit, stock
P. tomentosa Thunb.	Downy cherry, Manchu cherry, Nanking cherry	N & W China, Japan, Himalayas	16	*leucocarpa, endotricha*	Fruit, stock, ornamental
P. utahensis Dieck (= *P. besseyi* × *angustifolia watsonii?*)	Utah cherry	Nebraska			Ornamental

[a]E = east, W = west, N = north, C = central.

TABLE 4. Estimated 1982 Production of Plums and Prunes (fresh basis)[a]

Continent and Country or State	Thousands of Metric Tons		
North America	567		
Canada		11.0	
United States		556.0	
California			496.0
Idaho			7.0
Michigan			11.0
Oregon			30.0
Washington			11.5
Europe	3381		
Austria		106.0	
Belgium & Luxembourg		6.0[b]	
Bulgaria		200.0[c]	
Denmark		1.5[b]	
France		191.0	
Germany (West)		673.0	
Greece		10.0	
Hungary		163.0[c]	
Italy		170.0	
Netherlands		7.5[b]	
Norway		12.0	
Rumania		700.0[c]	
Spain		98.0	
Sweden		2.0	
Switzerland		14.5[b]	
United Kingdom		32.0	
Yugoslavia		995.0	
Asia	120[b]		
Turkey		120.0[b]	
Southern Hemisphere	150		
Argentina		98.0	
Australia		20.0	
Chile		15.0	
New Zealand		4.0	
South Africa		13.0	
Total (for specified countries)	4318		

[a]Dried prune production multiplied by 3 to convert to fresh basis (from 105). Figures not available for the Soviet Union or China, both major producers.
[b]1980 figures.
[c]1979 figures (from 34).

Common Plum, St. Julien A, 'Pershore', 'Brompton', and Myrobalan B (27, 109). St. Julien A is the most widely used stock in Britain (109). Over the years some European plums have been propagated from suckers dug up from under bearing trees such as 'Pershore Yellow Egg', 'Cambridge Gage', 'Warwickshire

Drooper', 'Aylesbury Prune', and 'Common Mussel'. Subsequently, some of these have been used as stocks for other cultivars. Much of the early selection of plum stocks was done in England, starting with the work of E. G. Hatton in 1914 (36, 37, 102, 103). At the East Malling Research Station, Kent, England, he sorted through the many seedling stocks then in use and made clonal selections. Several of these selections are still in common use: 'Brompton', 'Pershore', St. Julien A, St. Julien K, Myrobalan B, and Black Damas C. Rootstock selections have also been made at the Institut National de la Recherche Agronomique (INRA) Station at La Grande Ferrade, Bordeaux, France, to suit specific soil types such as those with high pH. These clones have the INRA or GF designations: St. Julien selection GF655-2, Damas selection GF1869, French prune selection GF43, and hybrids #1, #2, GF557, GF677, and Marianna GF8-1 (48).

2.2. North America

Most *P. domestica* prunes and plums grown in the United States are grown on stocks of *P. cerasifera* or its hybrids. In California Myrobalan 29C and Marianna 2624 are most popular, with peach and Myrobalan seedlings used in some areas. Peach, Myrobalan 29C, Marianna 4001, and St. Julien A are used in the Northwest. Japanese plums in California are also grown on Myrobalan 29C and Marianna 2624. Japanese plums are compatible with peach, so peach seedlings are commonly used, in particular, 'Lovell' and 'Nemaguard' (58, 76). In northern areas of the United States, native and hybrid cultivars may be grown on *P. americana* for cold hardiness. In the southeastern United States, peach is the primary rootstock used (101), usually 'Lovell', 'Halford', and 'Nemaguard'.

2.3. Southern Hemisphere

Plums grown in the Southern Hemisphere are mostly Japanese type, grown on peach, Myrobalan, or Marianna seedlings or cuttings. In New Zealand, Myrobalan is preferred for European plums and Marianna for the Japanese. Buck plum is used for both where a large tree is desired. Peach has not performed well as a plum stock (23). In South Africa, Marianna and peach seedlings are the primary stocks (44).

2.4. Stocks for Minor Plum Species

Fleming (24) tested four rootstocks for purple plum (*P. cistena*) in Ontario: apricot, Myrobalan, *P. tomentosa*, and *P. dropmoreana* (hybrid cherry). After five years only *P. tomentosa* had a dwarfing effect, but there was scion overgrowth, poor anchorage, and profuse suckering. Trees on Myrobalan were slightly larger than the others and were the latest to leaf out in the spring. Trees on *P. dropmoreana* leafed out the earliest. In Germany, *P. pumila* is used as a stock for *P. cistena* (55).

Roberts and Westwood in Oregon (81) tested rootstocks for clones of Klamath plum (*P. subcordata*) selected for fruit quality. They compared peach, Myrobalan, Marianna, and *P. americana* seedlings as rootstocks. At 4 years trees on peach were biggest, but at 10 years they were smallest. The reverse was true for *P. americana*. Fruit yields per tree and yield efficiency were higher on peach and *P. americana* than on the other two stocks. *Prunus subcordata* seedlings are undesirable as stocks because of suckering and poor transplant survival. However, there is a great deal of seedling variability and better types might be selected (80).

In New Zealand the beach plum (*P. maritima*) has been successfully budded on Myrobalan rootstock, although after several years the stock began to overgrow the scion (22).

Day (17) reported that Marianna was a satisfactory stock for the following American plum species: *P. orthosepala*, *P. alleghaniensis*, and *P. mexicana*. The first two species made smooth unions, in contrast to rough, warty unions for the third. In all cases, the stock was bigger than the scion after 15 years. When Texas almond (*P. texana*) was used as a stock in the nursery, *P. mexicana*, *P. simonii*, *P. munsoniana*, *P. subcordata*, *P. salicina*, and *P. domestica* (most) overgrew the stock. Less scion growth was seen for *P. americana*, *P. angustifolia*, and *P. nigra* (17). *P. besseyi* can also dwarf *P. americana* (15).

3. ENVIRONMENTAL ADAPTATION

3.1. Cold Hardiness

Cold hardiness of specific rootstocks, and of a single scion on different rootstocks, has not been extensively investigated. Early nursery reports of root injury from cold weather noted that European and American plums scions on peach and Myrobalan roots were all dead. On Marianna there was severe injury, while on *P. americana* rootstock there was only slight cold injury to the roots. Sand cherry (*P. besseyi*) was suggested for extreme cold hardiness (15). For the midwestern United States, St. Julien A is more cold hardy than either Marianna or Myrobalan seedlings (25), although Jacob (49) indicated the St. Julien group in general is not very cold hardy in Germany. Controlled freezing tests of seedlings showed cold hardiness increasing as follows: peach < Myrobalan < *P. davidiana* (wild peach) < Marianna < *P. americana* (115). More recent comparisons of tree and bud hardiness indicate *P. besseyi* and *P. nigra* are the most hardy, whereas peach and apricot are much less hardy. *Prunus salicina*, *P. tomentosa*, and hybrids of *P. besseyi* with peach and apricot are intermediate in hardiness (75). Several rootstock selections with sand cherry parentage have been reported to be very cold hardy, although comparative data are not available. These include SVG11-19 (*P. besseyi* × *ussuriensis*) in western Siberia, Kuppers-1 or 'Micronette' (*P.*

pumila) in Germany, and M-13, No. 55, Evraziya 13-27, Evraziya 15-25, AKU-2-31, Chak-5-62, OP-23-23, OP-15-2, and OD2-3 (no species given) from the Soviet Union (56, 57, 73, 107). From Sweden, 'Krikon fran Juo' is reported to be a cold-hardy *P. insititia* type (30).

3.2. Reactions to Soil and Water Problems

As mentioned earlier, French scientists selected several stocks that were able to grow well on soils with high pH. Damas GF1869, St. Julien Hybrid #1, Marianna GF8-1, and peach–almond hybrids GF677 and GF557 all allow trees to be grown on calcareous soils where other stocks fail (48). Nutrient culture trials indicate that GF655-2 is intolerant of high calcium levels; GF1869, tolerant; and Clark Hill Redleaf, resistant (77). There apparently has been no work to select for tolerance to acid soils and the associated aluminum toxicity that may occur, such as in the southeastern United States.

Two studies indicate that Marianna rootstock can tolerate high salt content in the soil better than peach rootstock (5, 68). In California, where higher boron occasionally occurs in orchard soils, almond has been found to tolerate higher levels than Myrobalan, which in turn is more tolerant than peach (32).

Recent work (13) has compared foliar nutrient levels of the same scions on a range of rootstocks. Compared to peach, Myrobalan and Marianna clones will absorb more calcium, nitrogen, potassium, manganese, and zinc, but less boron and magnesium. In another report Marianna 2624 was a more efficient user of nitrogen than Myrobalan 3-J, based on nutrient tank tests (100).

Rom (82) compared several stocks in the nursery for sensitivity to preemergence herbicides commonly used in orchards. Overall the most susceptible were *P. tomentosa*, Myrobalan 29C, and *P. americana*. Marianna and Myrobalan were intermediate, while peach was the most tolerant.

Little has been reported on drought tolerance of specific rootstocks. Almond is probably the most tolerant, at least partly due to a vertically oriented root system (17). Peach and 'Pixy' plum are relatively susceptible, whereas Marianna 4001 and Myrobalan 2-7 are somewhat tolerant under Oregon conditions (98). Westwood (111) harvested 4.5 kg of fruit per tree on peach roots during a dry year in comparison to 18 kg on St. Julien A and Myrobalan 29C, 22.5 kg on Marianna 2624, and 31 kg on Marianna 4001. *P. maritima* has been found to tolerate both very dry as well as wet soil conditions in New Zealand (22). Section 5 describes specific adaptations to light or heavy soils where this is known. Generally, plum rootstocks do better on heavy soils, while peach and almond do better on light soils.

In contrast to drought tolerance, much research has been done on flood tolerance, mostly under controlled conditions. Table 5 summarizes the relative sensitivity of various rootstocks. Some plums are relatively tolerant of flooding, in contrast to peaches, almonds, and apricots. The ability to tolerate waterlogging may be at least partly due to lower levels of endogenous cyanide-releasing compounds in the roots (84, 97).

TABLE 5. Relative Tolerance of *Prunus* to Waterlogging

Least Tolerant	Intermediate	Most Tolerant	Locality	Reference
Peach Apricot Almond[a]	Myrobalan 3J Myrobalan Myrobalan 29C Marianna 2624		California	69, 84
Peach Apricot	Myrobalan	Marianna	South Africa	97
Apricot[a] Peach P. mume P. tomentosa P. pauciflora	P. salicina P. cerasifera	P. japonica	Japan	64, 65
Apricot P. davidiana (wild peach)[a] GF677 (almond × peach)[a] GF557 (almond × peach)[a] Most peach[a] S2540 peach S-37 peach GF305 peach[a] Nemaguard peach[a] Rancho Resistant peach[a]	St. Julien 68 St. Julien A Cirudo 43 GF31[a] Myrobalan (B, P936, P31-6, P566[a] P938, P34, P855, P1254, P1079) St Julien 57-3 GF655-2[a] Brompton[a] P556-1 (Myro × almond) Most Damas Most St. Julien	GF 8-1 Damas 1869 Marianna S2544-2 peach[a]? St. Julien 770-5 Damas Toulouse Marianna P1251	France	85, 88, 89

[a] These stocks were more sensitive than the rest of the group.

3.3. Reaction to Nematodes

The rootstocks used more commonly in California have been tested for reaction to root knot nematodes, *Meloidogyne incognita* (Kofoid and White) Chit., and *M. javanica* (Treub) Chit. Reactions are listed in Table 6. There appear to be many sources of resistance available in *Prunus*. Recent tests from California indicate most species of plum and Microcerasus are resistant or immune, including *P. pumila*, *P. besseyi*, *P. cistena*, *P. japonica*, *P. tomentosa*; and French lines, P322 × S1058, S2535, S3400, GF53.7, and GF43 (D. Ramming, personal communication). Numerous other peach stocks are resistant or immune. A further discussion of them is found in Chapter 6, on peaches.

Rootstock reactions also have been reported for *Pratylenchus*, the root lesion nematode (59). Apricot, 'Bokhara' peach, and some 'Shalil' peach seedlings were resistant to *P. vulnus* Allen & Jensen, while Marianna 2623, Marianna 2624, and Myrobalan 29C were susceptible. Some 'Shalil' peach seedlings were also resistant to *P. penetrans* Cobb (Filip. & Stek.). Peach GF305 may be tolerant of *P. vulnus* (48).

Mojtahedi and Lownsbery (66) compared buildup of ring nematode *Criconemella xenoplax* (Raski) Luc & Raski on 12 different plum rootstocks: Corotto Marianna, Myrobalan seedlings, Myrobalan 3J and 29C, 'Etter's Best' (*P. subcordata* × *P. domestica*), *P. cerasifera atropurpurea*, peach, peach × almond, and Mariannas F, 2624, 2623, and 4001. Because of the amount of variation between pots, there were few significant differences between rootstocks in nematodes per pot. All appeared to be excellent hosts. There were differences in final root weight and therefore in nematodes per gram of root. Peach and Myrobalan seedlings had the lowest root weight and thus the highest number per gram of root (18, 100, and 13,400, respectively). In contrast, 'Etter's Best' and Marianna 4001 had only 1260 and 960 nematodes per gram of root. Parasitism by *C. xenoplax* has been associated with bacterial canker in California and peach tree short life in the southeastern United States (79). Myrobalan is also a good host of the pin nematode, *Paratylenchus neoamblycephalus* Gavaert (8). Both the pin and ring nematodes feed on small feeder roots as ectoparasites. The resulting reduction in feeder roots may lead to waterlogging.

3.4. Reaction to Diseases and Pests

Bacterial canker, caused by *Pseudomonas syringae* pv. *syringae* van Hall and *P. mors-prunorum* Wormw. (considered synonymous by some), is probably the most serious disease of plums worldwide. This disease kills buds and makes girdling cankers that frequently kill entire trees. Table 7 lists reported reactions to the disease. Use of Myrobalan B (resistant) to form the stem and crotch, as well as roots ("stem-building"), has been effective in reducing tree losses from this disease (90). A susceptible scion such as 'Victoria' apparently can make a high-worked rootstock more susceptible, whereas a more resistant

TABLE 6. Galling Reaction of Plums and Plum Rootstock to Root Knot Nematode (*Meloidogyne* spp.)

None	On Some Plants	On all Plants	Reference[a]
Damas C	Cheresoto (= *P. besseyi* × *americana*)	Santa Rosa	18
		St. Julien 3P	
Brompton	Grand Duke		
Improved Wild Goose (= *P. munsoniana*)	Etter's Best (= *P. subcordata* × *domestica*?)	Wickson	
Italian Prune	French Prune		
P. americana	P. bokhariensis		
P. hortulana			
P. munsoniana	Methley		
P. spinosa × domestica			
Robe de Sargent			
St. Julien E, G			
Warwickshire Drooper			
Most apricot cultivars[b]			
Marianna			
Marianna 2623, 2624		Almond	59
Myrobolan 29C, 29, 29D, 29G			
Apricot[b]			
P. hortulana	Blufre prune		110
Nemaguard peach		Lovell peach	69
GF 557 (almond × peach)			48
GF 8-1	(P2069 × 106)4 (= *P. japonica* × *spinosa*)	GF 677	62
(106 × P2175)6 (= *P. spinosa* × *cerasifera*)		P2032 (= *P.cerasifera*)	

[a]Reference 18 did not name species. References 59 and 69 are for *M. incognita* and *M. javanica*. References 62 and 110 include those in 59 plus *M. arenaria*. Reference 48 is for *M. incognita* and *M. arenaria*.
[b]Occasional seedlings susceptible.

TABLE 7. Reaction of Plums to Bacterial Canker Caused by *Pseudomonas syringae* or *P. mors-prunorum*

Highly Susceptible	Susceptible	Intermediate	Resistant	Locality	Reference
California de Agen	Beauti Kashmir	Black Chamba	Burbank's Elephant Heart	India	1
Early Plum	Beauty Pepsu	Burbank's Great Yellow	Chinese Beauty		
Fleming Delicious	Beloved of Favorite	Denniston's Superb	Damson		
Green	Cloth of Gold	Golden	Early Transparent Gage		
Greengage	Grand Duke	Mariposa	Flemish		
Japanese	Monarch	Rasin	Great Yellow		
Methley	Myrobalan	Starking Delicious	Peach Plum		
Prune	Purple Pershore		Plum Zinnouchi		
Santa Rosa	Red Ace		Soldyum Chinese		
Sweet Early	Satsuma		Tambov		
Warwickshire Drooper	Victoria		Wickson		
Yellow Pershore					
Marianna 2624	Myrobalan Seedling	Peach	Almond	California	69
Apricot	Myrobalan 29C				
	Brompton	Warwickshire Drooper	Belgian Myrobalan	U.K.	90,91
	EC3 (Common plum × Brompton)		EA16 (Pershore × Brussels)		
	Pershore Yellow Egg				
	St. Julien A		Myrobalan B		
	14 St. Julien progeny	6 St. Julien progeny	6 St. Julien progeny		

335

scion like 'Early Laxton' can reduce canker on the rootstock. No effect of rootstock on scion susceptibility was detected in Great Britain (91). In contrast, in Oregon the rootstock dramatically influenced scion susceptibility of low-budded trees (9). Rootstock and percentage of scions infected were: *P. tomentosa*, 87%; 'Michaelmas', 77%; Myrobalan seedling, 71%, 'Common Mussel', 62%; St. Julien A, 61%; Myrobalan 29C, 58%; Marianna 2624, 54%; Marianna 4001, 54%; Myrobalan 2-7, 48%; Marianna 2625, 46%; Damas C, 46%; 'Ackermann's', 43%; Myrobalan B, 43%, Myrobalan 5Q, 41%; Marianna 2623, 39%; and 'Lovell' peach, 31%. These trees were apparently not in fully replicated tests. Experience in California (17) agrees that plums on peach roots are less susceptible than those on Myrobalan or Marianna roots. Own-rooted trees of 'Santa Rosa' plum frequently died from bacterial canker while adjacent trees on peach rootstock survived well. Marianna seedlings and some Myrobalan seedlings seem to be more resistant than plants of the clones Myrobalan 29C or Marianna 2624 when topworked (17). Some clones also seem to be capable of "outgrowing" the infection (113).

Prune brownline virus, caused by tomato ringspot virus (TRSV), has become important in California (42, 43). It causes a band of necrotic tissue at the graft union of prune trees. It is only seen on trees propagated on peach or Myrobalan stocks. Marianna 2624 apparently is immune to the virus, so the disease is never seen with this stock. Although no vector has been shown for this disease, in the eastern United States the dagger nematode *Xiphinema americanum* Cobb can transmit TRSV to peach and other stone fruits to cause *Prunus* stem pitting disease. Marianna 2624 also confers immunity to Stanley constriction and decline (SCAD), which is related to TRSV. Own-rooted 'Stanley' also appears resistant to SCAD in contrast to the common *P. domestica* rootstocks (16).

In Europe the Sharka virus causes serious losses in plums. This disease is often called "plum pox" because it causes necrosis and pitting of the fruit, making it unsalable. Leaf chlorosis or flecking may also be seen. Sharka is aphid vectored. Van Oosten (70) tested a wide range of *Prunus* as hosts. All the Laurocerasus (laurel cherries) and Cerasus (true cherries) entries were nonsusceptible except for the two entries from Microcerasus, *P. tomentosa* and *P. cistena*. (This is another clue that they are closer to plums than cherries.) In contrast, all members of Prunophora and most of Amygdalus, except for almond and *P. tenella*, were susceptible. These susceptible hosts include *P. cerasifera*, 'Trailblazer' plum, *P. spinosa*, 'Common Mussel', *P. mume*, *P. triloba*, and peach. A peach × almond hybrid seemed to transmit Sharka to the rootstock but not carry it itself.

Reactions to several other diseases have been reported from California (69). Resistance to the wet-soil problems of crown and collar rots (*Phythophthora*) can be found in Myrobalan seedlings, Myrobalan 29C, and Marianna 2624. In contrast, peach and apricot seedlings are susceptible, and almond highly susceptible. GF43 and St. Julien Hybrid #2 are also resistant to collar rots (48).

Where replanting occurs, oak root rot can become a problem. This disease is caused by two similar organisms, *Armillaria mellea* (Vahl.) Quel. and *Clitocybe tabescens* (Scop.) ex. Fr. These organisms are slow acting and little has been done in controlled tests to compare rootstocks. Experience indicates peach, apricot, almond, and Myrobalan seedlings are susceptible. Myrobalan 29C and Marianna seedlings are less so. Marianna 2624, Marianna GF8-1, Myrobalan GF31, and Reine Claude GF1380 have moderate resistance, and possibly so do 'Etter's Best' plum and *P. subcordata* (17, 20, 69).

Crown gall, caused by *Agrobacterium tumefaciens* (E.F.Sm. and Town.) Conn., is often a problem in nursery sites. Peach and almond are quite susceptible. Myrobalan 29C and Marianna 2624 are less susceptible, while apricot is moderately resistant (69). Several American plum species and *P. pumila* are also resistant (93). In New Zealand, Buck plum is considered moderately resistant (23). Verticillium wilt, caused by *Verticillium albo-atrum* Reinke & Berth, is worst on peach, apricot, and almond. Plum stocks, especially GF1380, GF8-1, and GF31, appear less affected (48, 69). GF677 is reported to be more tolerant of *Fusicoccum* and *Stereum purpureum* (Pers. ex Fr.) Fr. (silver leaf) infection in contrast to GF557 (48).

Little is known of pest resistance in plum rootstocks. Apricot has moderate resistance to peach tree borer (*Synanthedon* spp.), in contrast to plums, almonds and especially peaches, which are susceptible. On the other hand, peaches have some resistance to pocket gophers and mice, whereas plums and especially apricots and almonds are susceptible (69).

4. PHYSIOLOGICAL ADAPTATIONS

4.1. Rootstock Effects on Tree Size and Production

Much effort has been put into finding dwarfing rootstocks for plum, particularly in Europe. Tukey has described much of the early research (102). Table 8 lists relative sizes produced by various stocks. Some discrepancies between reports are due to variation in location and in definitions of *dwarf* or *semidwarf*. In other cases, the degree of dwarfing depends on the scion. Scions have not been included in Table 8.

The greatest dwarfing generally results from use of other species, particularly from the Microcerasus group. Often the dwarf trees are highly productive based on tree size. However, most of these stocks also have anchorage problems, and some are not uniformly compatible. Specific characteristics such as these are listed in Section 5.

In some cases rootstock affects date of spring growth and cessation of growth in the fall, as well as number of years to initial fruiting (precocity). These factors influence both adaptability and productivity of a specific stock. Where known, such effects are noted in Section 5.

TABLE 8. Effect of Rootstock on Tree Size[a]

Most dwarfing	Semidwarfing		Most Vigorous	Locality	Reference
	Black Damas C			Belgium	19
St. Julien K Pixy (= E340.4.6)	Common Plum St. Julien A Aylesbury Prune St. Julien K E478 Ackermann's	Pershore Common Mussel Damas C Marianna	Brompton Myrobalan B St. Julien J	U.K.	27, 28, 36, 37, 108
P. maritima *P. tomentosa* P 2038 (*P. besseyi* × ?)	*P. injucunda* *P. spinosa* *P. besseyi* *P. americana* *P. hortulana*	P 2037 (= *P. besseyi* × ?) F$_1$(P322 × P871)1 [= Belsiana × (Myrobalan × peach)]	Myrobalan 1254	France	4
	GF43	Damas 1869	Peach × almond GF8-1	France	48
	Kuppers-1 (*P. pumila*)			Germany	55
GF43	St. Julien Hyb #2 Pixy Ackermann's	GF655-2 St. Julien Hyb #1	Myrobalan GF31 Marianna Marianna GF8-1	Germany	49
	Common Mussel Broadleaf Mussel	Ackermann Brompton Wangenheim Common Plum Damas Pershore	Myrobalan B St. Julien C Marianna	Poland	31

	Ackermann Common Plum		Myrobalan B Myrobalan Alba Pfrikon	Sweden	14, 30
	St. Julien A Eruni (= BPr32) Marianna				
VPA (*P. besseyi* × *P. armeniaca*)			SKA (= *P. salicina* × *P. cerasifera*)	USSR	29
P. maritima		Marianna Peach	Buck plum	New Zealand	21, 23
P. besseyi *P. tomentosa*	St. Julien K Ackerman St. Julien A Michaelmas Prune		Brompton B Myrobalan 29C	Canada	46, 47
P. cistena *P. tomentosa*	Pershore Ackermann's Marianna 2624 Marianna 2625 Marunke	Peach Michaelmas Myrobalan 5-Q Marianna 2623 St. Julien A Brompton	Common Mussel Myrobalan 2-7 Black Damas C Marianna 4001 Myrobalan 29C Myrobalan seedling Myrobalan B	Oregon	98, 111, 113
Citation (plum × peach)	Myrobalan 100			California	72
P. spinosa *P. angustifolia* *P. tomentosa* Bitter almond		*P. hortulana* *P. besseyi* *P. pumila* Apricot	Lovell peach St. Julien A CEA-2 Myrobalan	Idaho	40

"Comparisons between groupings are more accurate across than down.

4.2. Compatibility

As noted earlier, plum culture involves many species, more than for other fruit crops except perhaps citrus. As a result, incompatibility between stock and scion is very common. Stocks such as Myrobalan and Marianna and their selections have come to predominate in part because they have a wide range of compatibility (67). Less incompatibility is seen when European plums are put on stocks of the same group of species—*P. domestica*, *P. insititia*, and *P. cerasifera*. Similarly the Japanese and American plum species are more congenial with each other and with peach. Between groups more incompatibilities occur. Argles (2) reviewed the literature on specific incompatibilities up until 1937, and listed extensive tables by rootstock and scion species and cultivar. Day (17) documented results in California for both research trials and grower experience, primarily with prune and Japanese plum scions. There is often disagreement between reports. Some disagreement may stem from mislabeled stocks or scions, or from variability seen in seed-propagated stocks. Recent advances in virology make it clear that viruses can affect bud-take and long-term compatibility. Unless the test uses only "virus-free" material, which has been available only recently, past reports of specific incompatibilities should be viewed with virus problems in mind.

5. DESCRIPTIONS OF ROOTSTOCKS FOR PLUMS

5.1. Major Rootstocks

The most important rootstocks for plums are described in Tables 9–12.

5.2. Minor Rootstocks—European Plums and Hybrids

Many other stocks are less widely used, either having fallen from favor for various reasons or being new and under test. In Holland, 'Yellow Kroosje' (*P. domestica*) is a semidwarfing stock, although apparently not compatible with all plums (53). Another local stock in Holland is 'Tonneboer' (71). Kuppers (56) reported the rediscovery of 'Paspartout', a *P. insititia* stock used 150 years ago in Germany. Reportedly, it was widely compatible and itself bore good-quality fruit. Difficulty in propagation may have caused its near elimination.

St. Julien K was another of Hatton's selections from St. Julien. Although it dwarfs European plums, it is not widely used because its thorniness makes nursery operations difficult (27, 28, 47, 109). 'Aylesbury Prune' and Common Plum are other *P. domestica* stocks that are no longer widely used but once were recommended in Britain as semidwarfing stocks (27). Common Plum, which grows semiwild in England, shows good resistance to silver leaf, but suckers badly (31). It is used as a stock for Gages but is incompatible with some other *P. domestica* plums (99).

St. Julien X was selected in New Zealand from English seed and shows potential as a semidwarfing stock for Japanese plums and peaches. It roots easily from hardwood cuttings and is resistant in the nursery to leaf spot (*Coccomyces* spp.) but is thorny. It is well anchored, compatible with 'Ozark Premier' and 'Italian' prune, but may sucker badly (16; Paul Stark III, personal communication). It is commercially available in the United States.

Numerous selections have been made from Myrobalan seedlings. Myrobalan 'Alba' from southern Germany is no longer recommended because it is not widely compatible and is sensitive to cold injury. Myrobalan Wibault from Belgium is very vigorous and delays fruit ripening. Myrobalan INRA1254 from the French program is a vigorous stock for dry and chalky soils. It is used as a stock for Mirabelles and is suited for mechanical harvesting (49). From the United States, Myrobalan selections Ohio 1 and Ohio 2 show good compatibility with 'Stanley', the major eastern cultivar (33). A similar 'Stanley'-compatible clone, M20-3, has been tested at Michigan State University (10, 11, 25). It also promotes earlier defoliation in the fall.

Selections of Marianna have been made in South Africa (44, 45) for semidwarfing (clones 6-46, 8-6, 9-52) and for high yield efficiency (clone 7-2). All of these propagate readily by cuttings. Other hybrid stocks involving *P. cerasifera* include 'Ska' (*P. salicina* × *cerasifera*) from the Soviet Union, which makes a large productive tree with 'Reine Claude' scion (29). The French selection, F_1(P322×P871)1 [Belsiana Japanese plum × (*P. cerasifera* × *persica*)], produces a dwarf to semidwarf tree that grows faster when young than when old. This early rapid growth results in precocious bearing. It is widely compatible except with 'Reine Claude', but does poorly in wet soils. This hybrid looks promising in France but not Germany (4, 35, 49).

Several hybrids have Microcerasus parentage. The most widely tested may be Kuppers 1 or 'Micronette', a *P. pumila* selection out of a seed lot bought as *P. cistena*. It is compatible with at least 14 plum cultivars as well as peach. The plums were dwarfed 30−40% by this stock and showed earlier dormancy and later budbreak. The stock is very cold hardy and adapted to a wide range of soils. Propagation is by layers or cuttings (55, 56). SVG11-19 (*P. besseyi* × *ussuriensis*) is a high-yielding stock cold hardy enough for western Siberia (73). VPA (*P. besseyi* × *armeniaca*) is a highly productive, dwarfing stock that unfortunately has weak anchorage (29). Two French stocks, P2037 and P2038 (*P. besseyi* × *cerasifera* ?), are semidwarfing and very dwarfing, respectively. P2037 is not very productive, in contrast to high productivity and large fruit for P2038. However, P2038 is weakly anchored and is too weak growing for use in dry soils. P2038 propagates readily from softwood cuttings but has some compatibility problems (4, 35, 49). Finally, from the Soviet Union come a series of very hardy, easily propagated stocks which involve crosses of *P. domestica*, *P. americana*, *P. ussuriensis*, *P. salicina*, *P. cerasifera*, and *P. besseyi*. These are Evraziya 43, 13-27, and 15-25; Aku-2-31, Chak-5-62, OP-23-23, OP-15-2, and OD2-3 (107).

TABLE 9. Major Plum Rootstocks and Their Characteristics—*P. domstica* and *P. insititia* Selections

Rootstock	Ackermann's (Marunke)	Black Damas	Brompton	Common Mussel	Pershore (Yellow Egg)	Prune GF43
Species	*P. domestica*	*P. domestica*	*P. domestica*	*P. domestica*	*P. domestica*	*P. domestica*
Origin	Pre-1930 from Ackermann Nursery, Germany	Seedling from France, selected by East Malling, U.K.	East Malling, U.K. selection	U.K. selection	U.K. selection	1952 selection from French prune by INRA, France
Propagation	Cuttings, root suckers	Rooted cuttings	Hardwood cuttings, layers, root sucker	Layers, root cuttings	Suckers, layers; not easy to root	Cuttings
Where used	Europe	Europe	Europe	U.K.	U.K.	France
Characteristics						
Tree size	Semidwarf	Large (except Belgium)	Large–very large	Medium–large	Semidwarf to large	Vigorous but dwarfing; varies with soil
Productivity	Low–medium	Good	Average	Good	—	—
Anchorage	Good	Good	Very good	Good	Good	—
Suckering	Profuse	Profuse	Slight	Profuse	Slight–profuse	None
Compatibility	European plums	Wide	European plums	Good with European	European plums	European plums
Incompatibility	—	—	Some prunes	—	—	—
Resistance	—	Can outgrow bacterial canker	Cold temperatures	—	—	*Phytophthora* root and collar rot, water-logging
Susceptibility	May be cold tender	—	Bacterial canker, Sharka carrier	Bacterial canker, dry soil	—	Chlorotic leafspot
Sites used			Heavy, damp soils			Wet soils
Notes	Thorny, promotes, earlier ripening, virus-free clones available	May delay ripening	Thought to have spread Sharka in Europe, virus-free available	May delay ripening	—	May delay ripening, virus-free available
References	28, 35, 49, 113	19, 35, 36, 37, 113	35, 37, 47, 49, 113	99, 113	37, 113	35, 48, 49

	Pixy	St. Julien A	St. Julien GF655-2	St. Julien hybrid #1	St. Julien hybrid #2	Damas GF1869
Rootstock	Pixy	St. Julien A	St. Julien GF655-2	St. Julien hybrid #1	St. Julien hybrid #2	Damas GF1869
Species	*P. insititia*	*P. insititia*	*P. insititia*	*P. insititia × domestica*	*P. insititia × domestica*	*P. domestica × spinosa*
Origin	Selection from St. Julien from France by East Malling, U.K.	Selection from East Malling, U.K. from local material	Selection from St. Julien K (East Malling) by INRA, France	Cross of St. Julien d'Orleans × Common Mussel by INRA, France	Cross of St. Julien d'Orleans × Brompton by INRA, France	Selection from Damas de Toulouse by INRA, France
Propagation	Layers, hardwood cuttings	Hardwood cuttings, layers	Cuttings—easy, root suckers	F₁ seed (controlled crosses)	F₁ seed (controlled crosses); also cuttings or root suckers	Layers, cuttings
Where used	U.K.	U.K., Germany	Europe	Europe	—	France
Characteristics						
Tree size	Dwarf	Medium, semidwarf	Medium	Average	Semidwarf	Medium
Productivity	Good, precocious	Average, precocious	High, precocious	Average	Good	—
Anchorage	Good but small root system	—	Adequate	—	—	—
Suckering	Slight	Moderate	Slight	—	Slight–moderate	Profuse
Compatibility	Wide with European plums	Wide–European plums	Wide–European plums	European plums, peaches	European plums, peaches	European plums
Incompatibility	—	—	—	—	—	—
Resistance	Bacterial canker (tolerant)	Low winter temperatures	Bacterial canker, viruses	High pH, viruses	Collar rot	High pH, waterlogging, bacterial canker
Susceptibility	Drought	—	High pH	High pH		—
Sites used	Not on dry soils	—	Heavy soils	—	Good soil, not stony light soils	Wet or alkaline soils, heavy
Notes	Promotes earlier ripening, maybe smaller fruit, earlier bloom, thorny, virus-free available	Absorbs calcium well, virus-free clones available, not good with Stanley	May delay bloom and ripening, promising stock, virus-free available	Not well tested	—	(5n) chromosome number, virus-free available, winter hardy
References	35, 49, 98, 108, 109	13, 25, 35, 36, 47, 113	16, 35, 48	48, 49	35, 48, 49	16, 35, 48, 72

TABLE 10. Major Plum Rootstocks and Their Characteristics—Myrobalan Selections

Rootstock	Myrobalan	Myrobalan B	Myrobalan GF31	Myrobalan 29C	Myrobalan 5-Q	Myrobalan 2-7
Species	*P. cerasifera*	*P. cerasifera*	*P. cerasifera*	*P. cerasifera*	*P. cerasifera*	*P. cerasifera*
Origin	Native to Europe and Asia	Selected at East Malling Res. Sta. U.K.	Seedling selection by INRA, France	Selected by Gregory Bros. Nursery, Brentwood, CA	Seedling selected by Univ. of CA	Seedling selected by Univ. of CA
Propagation	Seed	Layers, cuttings (not easy to root)	Cuttings, easy	Cuttings	—	—
Where used	Worldwide	Europe	—	U.S.	—	—
Characteristics						
Tree size	Large	Large (except South Africa)	Large	Large	Medium–large	Very large
Productivity	Good, slightly later ripening	Good, later ripening	Good, precocious yield	Good, slightly later ripening	Later ripening	—
Anchorage	Good	Good	—	Not very good in early years	Good	—
Suckering	Few	Few	—	Medium	Slight	Good
Compatibility	Wide—European and Japanese plums	Wide, especially European plums	European plums, prunes	Wide—Stanley Shiro, Italian prune	—	Prune

Incompatibility	Poor with Stanley, Green-gages, Reine Claude, President	Gages, Mirabelle, Reine Claude	Reine Claude	—	—	—
Resistance	Drought				—	Drought
Susceptibility	Bacterial canker, oak root rot, cold, root knot nematode, prune brownline	—	—	Prune brownline	—	—
Sites used	Light soils	—	Good soils; dry, stony soils	Widely used	—	—
Notes	Best worldwide stock, but highly variable from seed, selections Ohio 1, Ohio 2, compatible with Stanley	Absorbs calcium well, virus-free available	—	Absorbs calcium and potassium very well, virus-free available	—	High K absorber
References	12, 33, 39, 49, 95, 98, 113	13, 14, 21, 36, 47, 49, 108, 113	48, 49	13, 33, 43, 98	113	98, 113

TABLE 11. Major Plum Rootstocks and Their Characteristics—Marianna and Plum Hybrids

Rootstock	Marianna	Marianna 2624	Marianna 4001	Marianna GF8-1	Buck plum (Morrison's plum stock)	Citation
Species	*P. cerasifera* × *munsoniana*	*P. cerasifera* × *munsoniana*	*P. cerasifera* × *munsoniana*	*P. cerasifera* × *munsoniana*	*P. cerasifera* × peach	(Red Beaut plum ×peach)
Origin	Charles Ely, Smith Pt. TX (1884)	Marianna seedling from Univ. of CA	Marianna selection by Univ. of CA	Selection from Marianna, INRA, France	Australia ?	Floyd Zaiger, Modesto, CA (patented)
Propagation	Seed, cuttings—easy	Cuttings	Cuttings	Cuttings—easy; chip buds better than T buds	Cuttings (sterile)	Cuttings (easy)
Where used	Worldwide	U.S.	U.S.	Europe	New Zealand, Australia	—
Characteristics						
Tree size	Medium–large	Smaller than peach	Very large	Large	Very large	Dwarfing
Productivity	Very good, maybe earlier ripening	Good, slightly earlier fruit	High yield, slightly later	Very good, precocious	Very high	Good
Anchorage	Medium	Not very good in early years	Very good	—	Good	Very good
Suckering	Slight unless deep cultivation or shallow soil	Medium	Medium	None	Few	None

Compatibility	Wide—European, Japanese, US plums	Wide	Wide	Wide—European, Japanese plums	All Japanese, most European	Good
Incompatibility	Damson, President, Reine Claude	—	—	Reine Claude	—	—
Resistance	Rosette virus	Waterlogging, oak root rot, tomato ring spot virus, prune brownline, root knot nematodes, Stanley decline (SCAD)	Drought, bacterial canker (can outgrow it)	Waterlogging, viruses, root knot nematode	Crown gall	—
Susceptibility	—	Bacterial canker (may outgrow infection)	—	Silver leaf	Bacterial spot on scion	—
Sites used	Wide range	Wide—does poorly on heavy soils	—	Heavy or sandy, wet, high pH—wide range	Wide, wet soil	—
Notes	Chromosome number ($n = 24$)	Uses nitrogen efficiently; best of this series of selections, virus-free available, winter hardy	Virus-free available	Chromosome number ($n = 24$)	—	Red leaf, upright growth, bloom later, dormant earlier
References	16, 21, 23, 36, 39, 49, 63, 96	16, 17, 98, 100, 113	98, 113	49, 62, 86	23	72, 116

TABLE 12. Major Plum Rootstocks and Their Characteristics—Other Species Including Almond and Peach

Rootstock	Almond	GF677	GF557	Peach	American plum
Species	P. amygdalus	P. amygdalus × persica	P. amygdalus × persica	P. persica	P. americana
Origin	Native to W. Asia	INRA, France	INRA, France	Native to China	Native to U.S.
Propagation	Seed	Cuttings (tissue culture, softwood under mist)	Cuttings, (tissue culture, softwood under mist), easier than GF677)	Seed mostly	Seed (hard-to-root cuttings)
Where used	—	Europe	Europe	U.S.	Northern U.S.
Characteristics					
Tree size	—	Large	Large	Medium–large	semidwarf
Productivity	—	—	—	—	Low
Anchorage	—	—	—	Good	Weak
Suckering	—	—	—	Few to none	Severe
Compatibility	Formosa, French prune, President, other European and Japanese	Good–Japanese and European	Good–Japanese and European	Most European, Japanese, Stanley	U.S. species, possibly French prune
Incompatibility	CA Blue, Robe de Sargent, Santa Rosa Burbank, Wickson	—	—	Lombard, German Prune, Grand Duke, Stanley, Climax, prunes	European & Japanese, Stanley

Resistance	High boron, drought, perhaps Sharka	Drought, *Fusicoccum* and *Stereum*	Drought, root knot nematode, high pH soil	Root knot nematode (some peaches), bacterial canker	Cold, many diseases
Susceptibility	Crown gall, oak root rot, waterlogging	Waterlogging, root knot nematode	Waterlogging (very sensitive)	Prune brownline, oak root rot, waterlogging, Sharka, collar rot	—
Sites used	High boron	Marginal, high pH	Marginal, high pH	—	—
Notes	Scion will overgrow stock	—	—	Mainly Halford, Lovell, Nemaguard, S-37 cultivars are used, virus-free available	Easy to bud, scion overgrowth
References	17, 32, 41, 60	48	48	21, 41, 46, 60, 106, 113	4, 15, 39
Rootstock Species	Beach plum *P. maritima*	Sand cherry *P. besseyi, P. pumila*	Nanking cherry *P. tomentosa*	Sloe *P. spinosa*	
Origin Propagation	Native to U.S. Seed	Native to U.S. Seed, cuttings (often easy)	Native to China Seed, cuttings	Native to Europe Seed	
Where used	—	—	—	—	

(continued)

349

TABLE 12. (continued)

Characteristics				
Tree size	Very dwarf	Mostly very dwarf	Very dwarf	Very dwarf
Productivity	Low	Mostly good	Good, earlier ripening	Low
Anchorage	Weak	Weak	Weak	—
Suckering	Moderate	Severe	Severe	Severe
Compatibility	French prune, Japanese plums Stanley, Shiro, Grand Duke, P. americana	French prune, Italian prune, prune, Italian prune	Shiro, Grand Duke, French some Japanese	French prune, Italian prune,
Incompatibility	—	Damson, Victoria plums	Victoria, prune	Victoria
Resistance	Drought, waterlogging	Very cold hardy, crown gall	Frost, drought	—
Susceptibility	—	—	Bacterial canker	—
Sites used	Very sandy soil	—	—	—
Notes	widely in compatibility-some clones are under test	Seedlings vary widely	Seedlings vary	—
References	4, 22, 23	4, 6, 7, 15, 40, 46, 47, 71, 74, 93	4, 40, 47, 49, 71, 98, 109, 113	4, 40, 71, 81

The University of California recently patented Hansen 536 and Hansen 2168, both peach × almond hybrids (51). These stocks are compatible with plum, propagate easily from cuttings, and produce large, drought-resistant trees. As with other such stocks, they are resistant to root knot nematode and tolerate high-pH soil.

5.3. Minor Stocks—Other Species

None of the cherry species except those in the Microcerasus section have been at all compatible with plums. However, other species related to apricot, peach, and of course, plum have been tested and found partly compatible. These are listed below. Virtually all were tested as seedlings.

P. angustifolia Marsh. Seedlings of this native U.S. plum are dwarfing to Japanese plum and 'Italian' scions. Trees bear early for Japanese plum and are productive for at least 16 years (17, 39, 40). Suckering may be a problem.

P. armeniaca. L. Apricot is occasionally used as a plum rootstock. The compatibility is generally good with prunes and Japanese plums, but scion overgrowth is a serious problem (17). Trees are well anchored and moderately resistant to peach tree borers, crown gall, and several nematode species. Apricot roots are quite susceptible to crown rot and wet soil (69).

P. bokhariensis Schneid. Day (17) described this species as being similar to Marianna plum in both appearance and performance as a rootstock. He reported 'French' prune doing well as a scion after 10 years. The rootstock was vigorous with a deep root system.

P. cistena N. E. Hansen. The purple-leaf sand cherry is sometimes used in Germany as a rootstock for plums (55), but may be difficult to bud (54). It is compatible with European plums as an interstock (54). In the United States, it is reportedly quite dwarfing and produces only a few suckers (98).

P. dasycarpa Ehrh. The purple or black apricot, thought to be a hybrid of *P. armeniaca × cerasifera*, makes a satisfactory stock for 'French' prune in California (17). It appears to be root knot resistant.

P. davidiana Franch. As a rootstock this wild peach from China is compatible with Japanese and European plums (46, 60). It produces a medium-size, not very productive tree. The roots are tolerant of alkaline soils but susceptible to waterlogging, crown gall, and oak root rot.

P. glandulosa Bean. The dwarf flowering almond is compatible as an interstock for European plums (54).

P. hortulana Bailey. This native U.S. plum produces a highly productive, dwarf tree for 'French' prune (4). Some yellowing of the scion foliage occurs and the anchorage is weak, but there are no suckers.

P. injucunda Small. This plum species, related to *P. umbellata*, produces a well-anchored dwarf tree that is highly productive (4). Although compatible with 'French' Prune, some leaf yellowing was seen.

P. mexicana S. Wats. This U.S. plum species is dwarfing to peach, but compatibility is variable (50). It needs to be tested with plum scions.

P. mira Koehne. This wild peach native to China and northern India made satisfactory unions in the nursery with Japanese plums and 'French' prune. It appeared to be resistant to crown gall, but was highly susceptible to oak root rot (17).

P. mume Sieb. (Sieb. & Zucc.). The Japanese apricot makes a poor stock for plums, both Japanese and European (17). The unions are poor and the trees dwarfish. Smith (94) reported similar results unless an interstock such as *P. armeniaca* var. *mandshurica* was used. *Prunus mume* is reportedly resistant to Armillaria root rot, crown gall, and root knot nematode, and perhaps tolerant of wet soils.

P. prostrata Labill. On this low-growing bush cherry from Asia, bud-take of Japanese plums was nil in one trial (52), but in a second trial, scions appeared healthy after one year (92).

P. subcordata Benth. In California, the Sierra plum made a satisfactory stock for 'Burbank' plum. 'French' prune as a scion was greatly dwarfed, but it produced a heavy fruit load. This stock may be very resistant to *Armillaria* root rot (17).

P. tenella Batsch. The Siberian almond is compatible as an interstock for European plums (54).

P. triloba Lindl. This flowering almond is apparently incompatible with 'Victoria' plum, as bud take is very low (71), but it makes a satisfactory interstock for European plums. As such, it provides dwarfing and early dormancy of the scion (54).

Some species have not been tested at all. For example, in the United States there are six *Prunus* species with pubescent fruit that are native to desert areas (61). The desert peach or wild almond of California, *P. andersonii* Gray, has been successfully budded as a scion on Myrobalan (17). The desert almond, *P. fasciculata* Gray, is native across much of the southwestern states. It has been grafted onto almond, but apparently not other *Prunus* (17). The desert apricot,

P. fremontii Wats., is also native to California. In tests there it showed some compatibility with apricot and was quite dwarfing as a rootstock (17). Havard plum, *P. havardii* Wight, is found only in a small area of southern Texas. It shows some potential as an ornamental. The Texas almond or small-flower peachbush, *P. minutiflora* Engelm., is rarely found in limestone soils in Texas. The Texas peachbush, *P. texana* Dietr., is more common in Texas and readily hybridizes with cultivated plums. An additional related species, found only in Mexico, *P. microphylla* Hemsley, the Mexican almond, is even more obscure. Only *P. texana* has been tested as a rootstock for plums, but these species may possess drought or disease resistance that would be very useful. These species are virtually unknown except to a few stone fruit breeders and botanists.

6. FUTURE PROSPECTS FOR PLUM ROOTSTOCKS

It is obvious from the literature cited in this chapter that breeding of new plum rootstocks, as contrasted to testing, is under way in only a few locations. The primary breeding program is at Bordeaux, France. This program combines a study of the genetic relationships between European plum species with the building of new combinations for use as stocks (87). Work there on interspecific stocks for other stone fruits may also be useful for plum (3). Programs at East Malling, United Kingdom, and in Germany are primarily selection programs. Although there are active programs in the Soviet Union, exchange of information and germplasm between these and "Western" programs is limited.

In Germany there is an effort to provide virus-free material for all the commercially used plum rootstocks. Such material would probably reduce some of the virus-related incompatibility problems. Much of the French GF series is already available virus free. In the future, as virus testing is refined, all commercial stocks will likely be available as virus free. In North America virus-free seed and budwood of commonly used stocks are available from the Post Entry Quarantine Station, Sidney, British Columbia; Foundation Plant Materials Center, Davis, California; and the IR-2 Repository, Prosser, Washington.

Other future trends include increased use of own-rooted trees, which eliminates the compatibility problem. Trees of 'Stanley' on their own roots in New York are doing well after 10 years. They are comparable in size to 'Stanley' on St. Julien A, but smaller than 'Stanley' on Myrobalan seedlings (26). These trees are well anchored and productive with little suckering (16). This practice will be more common as propagation methods are refined and automated. Some stocks are already routinely propagated by tissue culture. This method may become routine for both scion-rooted plants and for propagation of rootstocks. Plums have generally been easier to propagate in tissue culture than peaches or almonds.

In the United States most plum rootstock improvement will result from progress made in developing rootstocks for peaches. It is unfortunate that plums, as a crop, receive only minor research effort, because there is a wealth of rootstock germplasm available for development. Many related species and hybrids have useful characters that could be recombined for specific uses.

INTERNATIONAL CATALOG OF ROOTSTOCK BREEDING PROGRAMS

England
East Malling Research Station
East Malling, Maidstone, Kent
EM 19 6BJ, England
Anthony Webster

Soviet Union
Lisavenko Research Institute of Horticulture in Siberia
Barnaul 20
V.S. Pytov

Vavilov All-Union Research Institute of Plant Growing
Krymsk
G.V. Yeremin, Y.A. Gnezdilov

The Voronezh Institute of Agriculture
A.N. Venyaminov, A. G. Turovtseva

Sweden
The Swedish University of Agricultural Sciences
S 230 Alnarp, Sweden
Victor Trajkovski

United States
Zaiger's Genetics
Modesto, CA 95351
Floyd Zaiger

USDA/ARS
Horticultural Crops Research Laboratory
Fresno, CA 93727
David Ramming

USDA/ARS
SE Fruit and Tree Nut Research Laboratory
Byron, GA 31008
W.R. Okie

New York State Agricultural Experimental Station
Geneva, NY 14456
James Cummins

REFERENCES

1. Agarwala, R. K., and V. C. Sharma (1970). Varietal infectibility of stone fruits with bacterial canker, *Indian J. Agr. Sci.*, **40**, 559–561.

2. Argles, G. K. (1937). *A Review of the Literature on Stock–Scion Incompatibility in Fruit Trees, with Particular Reference to Pome and Stone Fruits*, Imp. Bur. Fruit Prod. Tech. Comm. 9, 115 pp.

3. Bernhard, R., C. Grasselly, and J. Sarger (1976). Possible use of intra- and interspecific hybrids for improving stone fruit rootstocks [in French], Proc. Eucarpia Mtg. Tree Fruit Brdg., Wageningen, Sept. 7–10, 1976, pp. 52–62.

4. Bernhard, R., and Y. Mesnier (1975). Dwarfing rootstock selections for *Prunus domestica*: Preliminary trials [in French], *Acta Hort.*, **48**, 13–19.

5. Bernstein, L., J. W. Brown, and H. E. Hayward (1956). The influence of rootstock on growth and salt accumulation in stone-fruit trees and almonds, *Proc. Amer. Soc. Hort. Sci.*, **68**, 86–95.

6. Brase, K. D. (1953). Western sand cherry—a dwarfing rootstock for prunes, plums, peaches, *Farm Res.*, **19**, 4.

7. Brase, K. D., and R. D. Way (1965). *Rootstocks and Methods Used for Dwarfing Fruit Trees*, N.Y. State Agr. Exp. Sta. Bull. 783, 50 pp.

8. Braun, A. L., H. Mojtahedi, and B. F. Lownsbery (1975). Separate and combined effects of *Pratylenchus neoamblycephalus* and *Criconemoides xenoplax* on Myrobalan plum, *Phytopathology*, **65**, 328–330.

9. Cameron, H. R. (1971). Effect of root or trunk stock on susceptibility of orchard trees to *Pseudomonas syringae*, *Plant Dis. Reptr.*, **55**, 421–423.

10. Carlson, R. F. (1972). Myrobalan selections as rootstocks for plum, *Fruit Var. & Hort. Digest*, **26**, 10–12.

11. Carlson, R. F. (1978). Old and new fruit tree rootstocks, *Compact Fruit Tree*, **11**, 2–6.

12. Carlson, R. F., and J. Hull, Jr. (1975). *Rootstocks for Fruit Trees*, Mich. St. Univ. Ext. Bull. E-851, 2 pp.

13. Chaplin, M. H., M. N. Westwood, and A. N. Roberts (1972). Effects of rootstock on leaf element content of 'Italian' prune (*Prunus domestica* L.), *J. Amer. Soc. Hort. Sci.*, **97**, 641–644.

14. Christensen, J. V. (1965). Rootstocks for plum trees, II [in Danish], *Tidsskr. Planteavl.*, **69**, 201–205 (*Hort. Abstr.*, **36**, 2487 [1966]).

15. Craig, J. (1900). Observations and suggestions on the root killing of fruit trees, *Iowa Agr. Exp. Sta. Bull.*, **44**, 179–213.

16. Cummins, J. N. (1984). Fruit tree rootstocks introduced and soon to be introduced, *Compact Fruit Tree*, **17**, 57–63.

17. Day, L. H. (1953). *Rootstocks for Stone Fruits*, Cal. Agr. Exp. Sta. Bull. 736, 76 pp.

18. Day, L. H., and W. P. Tufts (1944). *Nematode-resistant Rootstocks for Deciduous Fruit Trees*, Univ. Cal. Agr. Exp. Sta. Circ. 359, 16 pp.

19. Deckers, J. C., and J. Keulemans (1980). Plum rootstocks [in Dutch], *Boer en de Tuinder*, **86**(2), 27–28, 31.

20. Duquesne, J., H. Gall, and J. M. Delmas (1977). New observations on the reactions of apricot rootstocks to Armillaria root rot [in French], *Pomol. Fr.*, **19**(6), 95–98.

21. Farmer, A. (1955). Comparison of four rootstocks for Sultan plums in the Auckland District, *N. Z. J. Sci. Tech.*, **A37**, 318–322.

22. Farmer, A. (1964). The beach plum in Auckland, *N. Z. Pl. Gdns.*, **5**, 303–305.

23. Farmer, A. (1970). A guide to plum rootstocks, *N. Z. J. Agr.*, **120**(3), 90–95.

24. Fleming, R. A. (1962). Rootstocks for ornamental trees, *Hort. Exp. Sta. Prod. Lab. Rept. 1962*, pp. 46–49.

25. Frecon, J. (1981). Rootstocks and interstems for the home orchardist, *N. Amer. Pomona*, **14**, 119–122.

26. Frecon, J. L. (1983). Self-rooted deciduous fruit trees, *Hort. News*, **63**, 9, 10–11, 16.

27. Glenn, E. M. (1961). Plum rootstock trials at East Malling, *J. Hort. Sci.*, **36**, 28–39.

28. Glenn, E. M. (1968). Two semi-dwarfing plum rootstock trials, *Rept. E. Malling Res. Sta. for 1967*, pp. 85–87.

29. Gnezdilov, Y. A. (1979). Trials with the clonal rootstocks of stone fruit crops [in Russian], *Trudy Po Prik. Bot., Gen., I Sel.*, **65**, 130–137 (*Hort. Abstr.*, **51**, 8398 [1981]).

30. Goldschmidt-Reischel, E. (1982). *Plum Cultivar and Rootstock Trials 1954–1980* [in Swedish], Sver. Lant. Inst. Trad. Rept. 25, 34 pp. (*Biol. Abstr.*, **78**, 67439 [1984]).

31. Grzyb, Z. S., A. Jackiewicz, and A. Czynczyk (1984). Results of the 18-years evaluation of rootstocks for Italian Prune cultivar, *Fruit Science Reports*, **11**(3), 99–104.

32. Hansen, C. J. (1948). Influence of the rootstock on injury from excess boron in French (Agen) Prune and President plum, *Proc. Am. Soc. Hort. Sci.*, **51**, 239–244.

33. Hartman, F. O. (1969). Plums for Ohio, *Ohio Rept. on Res. & Dev.*, **54**(3), 45–46.

34. Hartmann, W. (1982). Stone-fruit production in Eastern Europe [in German], *Erwerbsobstbau*, **24**, 188–192.

35. Hartmann, W. (1984). Rootstocks for plums and damsons [in German], *Deut. Baum.*, **36**, 245–249 (*Hort. Abstr.*, **54**, 6063 [1984]).

36. Hatton, R. G. (1921). Stocks for the stone fruits, *J. Pomology*, **2**, 209–245.

37. Hatton, R. G, J. Amos, and A. W. Witt (1928). Plum rootstocks, *J. Pomology*, **7**, 63–99.

38. Hedrick, U. P. (1911). *The Plums of New York*, Rept. N.Y. State Agr. Exp. Stn., 616 pp.

39. Hedrick, U. P. (1923). *Stocks for Plums*, N.Y. Agr. Exp. Stn. Bull. 498, 19 pp.

40. Helton, A. W. (1976). Effects of selected rootstocks on growth and productivity of two cultivars of *Prunus domestica*, *Can. J. Plant Sci.*, **56**, 185–191.

41. Heppner, M. J., and R. D. McCallum (1927). *Grafting Affinities with Special Reference to Plums*, Cal. Agr. Exp. Sta. Bull. 438, 20 pp.

42. Hoy, J. W., and S. M. Mircetich (1984). Prune brownline disease: Susceptibility of prune rootstocks and tomato ringspot virus detection, *Phytopathology*, **74**, 272–276.

43. Hoy, J. W., J. S. M. Mircetich, R. S. Bethell, J. E. DeTar, and D. M. Holmberg (1984). The cause and control of prune brownline disease, *Cal. Agr.*, **38**(7), 12–13.

44. Hurter, N., F. J. Calitz, and M. J. Van Tonder (1981). An improved rootstock of much promise for Santa Rosa plum, *Agroplantae*, **13**, 15–16.

45. Hurter, N., M. J. Van Tonder, and F. J. Calitz (1978). New plum rootstocks: Real economic advantages, *Decid. Fruit Grower*, **28**, 249–256.

46. Hutchinson, A. (1965). Plum rootstock trials at Vineland to 1964, *Ontario Hort. Exp. Sta. Rept. 1964*, pp. 25–32.

47. Hutchinson, A. (1976). *Rootstocks for Fruit Trees*, Ontario Min. Agr. and Food Publ. 334, 22 pp.

48. Institut National de la Recherche Agronomique (INRA) (1978). *Station de recherches d'aboriculture fruitiere—La Grande Ferrade*, Bordeaux, France, 22 pp.

49. Jacob, H. (1980). The present situation with regard to the use of plum rootstocks [in German] *Obstbau*, **5**, 469–470 (*Hort. Abstr.*, **51**, 4407 [1981]).

50. Johnston, S. (1938). *Prunus mexicana* and *Prunus hortulana* as rootstocks for peaches, *Mich. Agr. Exp. Sta. Quart. Bull.*, **21**, 17–18.

51. Kester, D. E. and R.N. Asay (1986). 'Hansen 2168' and 'Hansen 536': Two new *Prunus* rootstock clones, *Hortscience* **21**, 331–332.

52. Kishore, D. K., and S. S. Randhawa (1983). Note on the graft compatibility of native wild species. II, *Plum. Sci. Hort.*, **19**, 251–255.

53. Knuth, F. M. (1964). Yellow Kroosje plum as a rootstock for peach and plum, *Fr. Var. & Hort. Digest*, **18**, 44.

54. Kuppers, H. (1981). Results of examinations aiming at detection of weakly growing rootstocks from the section Microcerasus (Rehder) for plum and damson trees [in German], *Mitt. Klosterneuburg*, **31**(2), 71–76.

55. Kuppers, H. (1982). *Prunus pumila*: A dwarf plum rootstock? [in German], *Erwerbsobstbau*, **24**, 92–96 (*Hort. Abstr.*, **52**, 5973 [1982].

56. Kuppers, H. (1983). Two new plum-rootstocks—'Paspartout' and 'Micronette' [in German], *Deut. Baum.*, **35**, 115–117 (*Hort. Abstr.*, **53**, 5752 [1983]).

57. Kur'Yanov, M. A. (1981). Preliminary results of studies on new rootstocks for stone fruits and pears [in Russian], *Sbor. Nauch. Trud., Vses. Nauch. Issl. Inst. Sad. Mich.*, **34**, 75–77 (*Hort. Abstr.*, **54**, 3219 [1984]).

58. LaRue, J. H., and M. H. Gerdts (1976). *Commercial Plum Growing in California*, Univ. Cal. Leaflet 2458, 22 pp.

59. Lownsbery, B. F., E. F. Serr, and C. J. Hansen (1959). Deciduous fruit and nut trees—root-knot nematode on peach and root-lesion nematode on walnut cause serious problems for California orchardists, *Cal. Agr*, **13**(9), 19–20.

60. Manaresi, A. (1951). Fruit tree rootstocks used in the Emilia and Romagna districts [in Italian], *Riv. Fruttic.*, **13**, 157−166 (*Hort. Abstr.*, **21**, 2325 [1951]).

61. Mason, S. C. (1913). The pubescent-fruited species of *Prunus* of the southwestern states, *J. Agr. Res.*, **1**, 147−178.

62. Massese, C. S., C. Grassely, J. C. Minot, and R. Voisin (1984). Differential responses of 23 clones and hybrids of *Prunus* regarding to four *Meloidogyne* species [in French], *Rev. Nematol.* **7**, 265−270 (*Biol. Abstr.*, **79**, 80665 [1985]).

63. McClintock, J. A. (1948). A study of uncongeniality between peaches as scions and the Marianna plum as a stock, *J. Agr. Res.*, **77**, 253−260.

64. Mizutani, F., M. Yamada, A. Sugiura, and T. Tomana (1979). Differential water tolerance among *Prunus* species and growth of peach scions on various kinds of rootstocks as affected by waterlogging [in Japanese] *Studies Inst. Hort. Kyoto Univ.*, **9**, 28−35.

65. Mizutani, F., M. Yamada, and T. Tomana (1982). Differential water tolerance and ethanol accumulation in *Prunus* species under flooded conditions, *J. Japan Soc. Hort. Sci.*, **51**, 29−34.

66. Mojtahedi, H., and B. F. Lownsbery (1975). Pathogenicity of *Criconemoides xenoplax* to prune and plum rootstocks, *J. Nematol.*, **7**, 114−119.

67. Mowry, J. B. (1958). Rootstocks for propagating *Prunus* selections, *Fruit Var. & Hort. Digest*, **13**, 23−25.

68. Nasr, T. A., E. M. El-Azab, and M. Y. El-Shurafa (1977). Effect of salinity and water table on growth and tolerance of plum and peach, *Sci. Hort.*, **7**, 225−235.

69. Norton, R. A., C. J. Hansen, H. J. O'Reilly, and W. H. Hart (1963). *Rootstocks for Plums and Prunes in California*, Cal. Agr. Exp. Sta. Leaflet 158, 8 pp.

70. Oosten, H. S. van (1975). Susceptibility of some woody plant species, mainly *Prunus* spp., to sharka (plum pox) virus, *Neth. J. Plant Path.*, **81**, 199−203.

71. Oosten, H. J. van (1979). Fruit tree rootstocks from the Dutch research viewpoint, *Compact Fruit Tree*, **12**, 11−19.

72. Perry, R. L., and R. F. Carlson (1983). Fruit tree rootstocks—current cultivar performance, *Compact Fruit Tree*, **16**, 99−102.

73. Putov, V. S. (1981). Selection of clonal plum rootstocks in the Altai [in Russian], *Sbornik Nauch. Trud.*, **34**, 71−75 (*Hort. Abstr.*, **54**, 3257 [1984]).

74. Putov, V. S., and I. A. Puchkin (1982). *Cerasus besseyi* (Bailey) Lunell for breeding clone stocks of plum in Western Siberia [in Russian], *Biull. Vses. Inst. Rast.*, **123**, 48−50.

75. Quamme, H. A., R. E. C. Layne, and W. G. Ronald (1982). Relationship of supercooling to cold hardiness and the northern distribution of several cultivated and native *Prunus* species and hybrids, *Can. J. Plant Sci.*, **62**, 137−148.

76. Ramos, D. E., ed. (1981). *Prune Orchard Management*, Univ. Cal. Spec. Publ. 3269, 156 pp.

77. Reeves, D. W., J. H. Edwards, J. M. Thompson, and B. D. Horton (1985). Influence of Ca concentration on micronutrient imbalances in *in vitro* propagated *Prunus* rootstock, *J. Plant Nutr.* **8**, 289−302.

78. Rehder, A. (1954). *Manual of Cultivated Trees and Shrubs Hardy in North America*, Macmillan, New York, 966 pp.

79. Ritchie, D. F., and C. N. Clayton (1981). Peach tree short life—a complex of interacting factors, *Plant Dis.*, **65**, 462–469.

80. Roberts, A. N., and L. A. Hammers (1951). *The Native Pacific Plum in Oregon*, Ore. Agr. Exp. Stn. Bull. 502, 22 pp.

81. Roberts, A. N., and M. N. Westwood (1981). Rootstock studies with peach and *Prunus subcordata* Benth., *Fruit Var. J.*, **35**, 12–20.

82. Rom, R. C. (1972). Susceptibility of *Prunus* rootstocks to herbicide injury, *Fr. Var. and Hort. Digest*, **26**, 54–58.

83. Ronald, W. G. (1982). Interspecific hybridization in *Prunus*, *Rept. Proc. 38th Ann. Mtg., Western Canadian Soc. Hort.*, **38**, 93–100.

84. Rowe, R. N., and P. B. Catlin (1971). Differential sensitivity to waterlogging and cyanogenesis by peach, apricot and plum roots, *J. Am. Soc. Hort. Sci.*, **96**, 305–308.

85. Rowe, R. N., and D. V. Beardsell (1973). Waterlogging of fruit trees, *Hort. Abstr.*, **43**, 534–548.

86. Salesses, G. (1977). Research about the origin of two *Prunus* rootstocks, natural interspecific hybrids: An illustration of a cytological study carried out in order to create new *Prunus* rootstock [in French], *Ann. Amel. Plantes*, **27**, 235–243.

87. Salesses, G. (1981). Plum interspecific hybridization and cytology [in French], Premier Colloque sur les Recherches Fruitières, Bordeaux, 1981, pp. 120–128.

88. Salesses, G., R. Saunier, and A. Bonnet (1970). Waterlogging of fruit trees [in French] *Fr. Min. Agr. Bull Tech. Inf.*, **251**, 403–415.

89. Saunier, R. (1966). A method of determining the resistance to root asphyxia of certain fruit-tree species [in French], *Ann. Amel. Plantes*, **16**, 367–384.

90. Schofield, E. R., and L. F. Clift (1959). Trials of the influence of stem builders on bacterial canker of plum in the West Midlands, *Plant Path.*, **8**, 115–120.

91. Shanmuganathan, N., and J. E. Crosse (1963). Experiments to test the resistance of plum rootstocks to bacterial canker, *Rept. E. Malling Res. Sta. 1962*, pp. 101–104.

92. Singh, R. N., and P. N. Gupta (1971). Rootstock problem in stone fruits and potentialities of wild species of *Prunus* found in India, *Punjab Hort. J.*, **11**, 157–175.

93. Smith, C. O. (1924). The study of resistance to crown gall in *Prunus*, *Phytopathology*, **14**, 120 (Abstr).

94. Smith, C. O. (1928). The Japanese apricot as a rootstock, *Proc. Am. Soc. Hort. Sci.*, **25**, 183–187.

95. Soneo, V., M. Botez, and F. Lupescu (1968). The response of Myrobalan as plum rootstock [in Romanian], *Rev. Hort. Vitic.*, **17**(5), 108–115.

96. Stassen, P. J. C., and C. W. J. Bester (1981). Rootstocks for plums: A provisional evaluation, *Decid. Fruit Gr.*, **31**, 201–205.

97. Stassen, P. J. C., and H. J. Van Zyl (1982). Sensitivity of stone fruit rootstocks to waterlogging, *Decid. Fruit Grower*, **32**, 270–275.

98. Stebbins, R. L. (1981). A review of rootstocks for stone fruits, *Ore. Hort. Soc. Ann. Rept.*, pp. 23–30.

99. Taylor, H. V. (1949). *The Plums of England*, Crosby, Lockwood and Son, London, pp. 13–18.

100. Therios, I. N., S. A. Weinbaum, and R. F. Carlson (1979). Nitrate uptake effectiveness and utilization efficiency of two plum clones, *Physiol. Plant.*, **47**, 73–76.

101. Thompson, J. M. (1981). The plum industry in the southeastern United States, *Fruit Var. J.*, **35**, 53–55.

102. Tukey, H. B. (1964). *Dwarfed Fruit Trees*, Macmillan, New York, 562 pp.

103. Tydeman, H. M. (1957). A description and classification of certain plum rootstocks, *Rept. E. Malling Res. Sta. 1956*, pp. 75–80.

104. Tydeman, H. M. (1962). "Rootstocks: IV. Plums" in *Handbuch der Pflanzenzuchtung*, **6**, 562–572, Paul Parey, Berlin.

105. U.S. Dept. of Agriculture. (1983) *Agricultural Statistics 1983*, U.S. Government Printing Office, Washington, 603 pp.

106. Upshall, W. H. (1962). Fruit tree rootstock investigations in Vineland, Canada, *Proc. 15th Inter. Hort. Cong. 1958*, pp. 18–22.

107. Ven'Yaminov, A. N. (1981). Plum hybrids as clonal rootstocks [in Russian], *Sbor. Nauch. Trud.*, **34**, 67–71 (*Hort. Abstr.*, **54**, 3256 [1984]).

108. Webster, A. D. (1980). Dwarfing rootstocks for plums and cherries, *Acta Hort.*, **114**, 201–207.

109. Webster, A. D. (1980). 'Pixy', a new dwarfing rootstock for plums, *Prunus domestica* L., *J. Hort. Sci.*, **55**, 425–431.

110. Werner, D. J. (1984). Personal communication.

111. Westwood, M. N. (1974). Performance of prune rootstocks, *Ore. Hort. Soc. Ann. Rept.*, **65**, 93–95.

112. Westwood, M. N. (1978). *Temperate Zone Pomology*. Freeman, New York, 428 pp.

113. Westwood, M. N., M. H. Chaplin, and A. N. Roberts (1973). Effects of rootstock on growth, bloom, yield, maturity, and fruit quality of prune (*Prunus domestica* L.), *J. Amer. Soc. Hort. Sci.*, **98**, 352–357.

114. Wight, W. F. (1915). *Native American Species of* Prunus, U.S. Dept. Agr. Bull. 179, 75 pp.

115. Winklepleck, R. L., and J. A. McClintock (1939). The relative cold resistance of some species of *Prunus* used as stocks, *Proc. Am. Soc. Hort. Sci.*, **37**, 324–326.

116. Zaiger, C. F. (1983). Interspecific rootstock tree 4-G-816, U.S. Patent Office, Plant Patent 5,112.

11

CITRUS ROOTSTOCKS

William S. Castle
Citrus Research and Education Center,
University of Florida,
Lake Alfred, Florida

1. INTRODUCTION

Citrus fruits are grown in nearly 50 countries throughout the world and are known for their fine flavor and quality. They are unlike all other tree fruits because of their juice-containing vesicles, which fill each of the segments comprising a fruit. This and other characteristics have made citrus a highly prized and economically important fruit crop, ranking at the top with apples in world production and trade.

The true citrus fruits are generally considered to be of Old World origin, having evolved in a region bounded by southern China, northeastern India, and the East Indian archipelago. The earliest mention of citrus fruits occurs in Chinese literature dated about 2205 B.C. (58). It was in these areas, particularly China, that citrus was first cultivated. Through explorations of the world, citrus fruits slowly dispersed, primarily in a westerly direction.

The citron, *Citrus medica* L., became established in Iran and is thought to be the first fruit to become known to Western civilization (58). It was followed by the lemon, *C. limon* (L.) Burmann f., and sour orange, *C. aurantium* L. The origin of neither fruit is known, but they were spread throughout the Mediterranean area during the periods of the Roman and Arab empires. It was not until the sixteenth century that sweet oranges, *C. sinensis* (L.) Osb., had achieved commercial status in Europe (58). Their pathway of dispersal is also uncertain; however, they apparently were already established in the Mediterranean regions of Europe at the time of Vasco da Gama's historic voyage around the Cape of Good Hope in 1498.

The mandarin, *C. reticulata* Blanco, is a recent introduction to the Western world. This fruit spread easterly first to Japan, perhaps during the twelfth or thirteenth century, reaching Europe and the Americas after 1800 (58). Earlier, Columbus introduced the lemon, sweet orange, and sour orange to the

West Indies and eventually they reached North America. The pummelo, *C. grandis* (L.) Osb., was also brought to the Caribbean area and apparently hybridized naturally, resulting in the grapefruit, *C. paradisi* Macf., which has far surpassed the pummelo in commercial importance (58, 101).

As citrus fruits became distributed with the travels of explorers, domestication of this crop was extended beyond their centers of origin and cultivars were developed. The resulting commercial industries are located in a belt between 20 and 40° latitude on each side of the equator, where climates range from semitropical to subtropical (36). Orange cultivars are dominant among the citrus grown in these regions, and the United States, Brazil, Spain, Japan, and Italy are the leading producers (U.S. Department of Agriculture, Agricultural Statistics, 1983).

The genus *Citrus* is one of 33 genera in the orange subfamily, Aurantioideae, of the Rutaceae (183, 184); however, only 3 genera, *Citrus*, *Poncirus* (trifoliate orange), and *Fortunella* (kumquat), have been significant as sources of rootstocks or scion cultivars. From these 3 genera have come essentially 5 rootstocks upon which the world's commercial citrus industries are based.

2. HISTORY OF ROOTSTOCK USE

The ease of propagation by seed and their convenience for transport no doubt influenced the expansion of citrus from its native habitats to new areas, where it was grown as seedling trees. Seedlings were common in much of the world until the mid-1800s. They remained important in some areas of Central and South America and in southeastern Asia, where pummelos and mandarins are also grown as air layers. Many 'Persian' lime (*C. latifolia* Tanaka) trees are still propagated as air layers or marcots in Florida because of their earlier fruiting as compared with budded trees (38).

The budding or grafting of citrus was apparently practiced in the fifth century, but it was not until Phytophthora foot rot appeared in the Azores in 1842 that the transition of citriculture from seedling to budded trees began (58). As this disease spread or became recognized, interest in rootstocks greatly increased because of the loss experienced among the susceptible sweet orange seedlings. *Phytophthora* was later noted in all the Mediterranean countries, and by about 1935, it had been observed nearly everywhere (58). Seedlings were gradually replaced so that today virtually all trees are propagated by budding onto rootstock seedlings. Old seedling orchards still exist, most likely as relics from a time in local history when growers began the changeover to budded trees.

Phytophthora foot rot spawned a search for resistant rootstocks, and sour orange became dominant; however, difficulty was encountered with this rootstock in South Africa and Australia. Trees declined within several years after planting. As a result, rough lemon, *C. jambhiri* Lush., became popular in both countries.

In 1946, it was reported that this decline of trees on sour orange was presumably caused by a virus because it could be graft-transmitted (133). The disease, tristeza, unfortunately had already affected millions of trees budded to sour orange in South Africa, Australia, South America, and California. As a result, massive rootstock trials were initiated to screen for tristeza-resistant rootstocks. Over 100 field plantings have been established in Brazil and its neighboring countries since the late 1920s (142, 156). Rangpur lime, *C. limonia* Osb., was substituted for sour orange in Brazil and Troyer citrange, *C. sinensis × Poncirus trifoliata* (L.) Raf., is now used in California.

Rootstock use and development were considerably accelerated by the two diseases described above. *Phytophthora* was the stimulus for changing to rootstocks, but this change was then responsible for a new array of virus-related problems; however, there were some benefits. Rootstock research in its infant stages involved only a few rootstocks, primarily rough lemon, sour orange, trifoliate orange, sweet orange, and occasionally grapefruit or Cleopatra mandarin. Because of tristeza, germplasm collections were established and breeding efforts intensified.

3. ROOTSTOCK CHARACTERISTICS

The first comprehensive review of citrus rootstocks was published in 1948 (198) and a second review followed in 1979 (204). From these writings it is clear that rootstocks are important because they affect over 20 citrus tree characteristics, primarily horticultural and pathological, with certain aspects of propagation being common to all rootstocks.

3.1. Nucellar Embryony and Propagation

Citrus is unusual in that many species and cultivars, including the commercial rootstocks, reproduce true to type from seed. The seeds are polyembryonic, the extra embryos arising apomictically by nucellar embryony. Some citrus types are monoembryonic. They are usually eliminated as potential rootstocks because they produce only zygotic seedlings, which are highly variable.

The degree of nucellar embryony varies among rootstocks from 100%, where virtually all plants in a stand of seedlings are of nucellar origin, to less than 50%. When many zygotic seedlings are present among nucellar ones, it is important to remove them. Webber clearly demonstrated the horticultural advantages obtained by roguing out off-type seedlings, which is now a standard nursery practice (196).

Scion cultivars are normally propagated by budding (usually the T bud) onto nucellar (clonal) rootstock seedlings. A finished tree may be grown as quickly as one year from seed by greenhouse, containerized methods (49) or, more typically, in two to three years in a field nursery. The field method is fundamentally the same throughout the world, but there are certain regional differences, especially in the way nursery trees are prepared for delivery to the orchard. In humid climates, trees are lifted and shipped bare-root, while in arid climates, they are lifted with a ball of soil, or soil is added later, and then wrapped, often with burlap, to protect the trees from desiccation.

Citrus is sometimes propagated by cuttings, usually for ornamental purposes, or as marcots (38). Grafting is often used for topworking or when budding fails, especially in intergeneric propagation attempts (25, 27).

3.2. Horticultural Characteristics

The horticultural behavior of a citrus tree is determined by the reciprocal interaction of the tree's genetic components. Citrus rootstocks have pronounced effects on scion vigor and size, fruit yield and size, juice quality, tolerance to cold, drought, flooding, salt, and alkalinity, and leaf nutrient content. Also, rootstocks differ in their adaptability to soils with different textures, root distribution, and mycorrhizal dependency (45, 78, 144).

Scion vigor, yield, and juice quality generally are of greatest interest in rootstock development. Yield is the single most important factor because of its strong relationship to profit. Tree growth and yield interact strongly with soil adaptability, often producing contradictory reports of rootstock performance. Growth and yield data from field trials can result in misleading interpretations if considered separately. Growers have sought large, vigorous, highly productive trees in the past. Today, smaller, high-yielding trees are of interest because of their suitability for close planting, but the search for appropriate rootstocks has not been very rewarding (3, 14, 28, 46, 51, 52). Viruses were a complicating factor until their effects on trees budded to intolerant rootstocks became known. This may explain, for example, why *Severinia buxifolia*, Palestine sweet lime, and Cuban Shaddock were once described as dwarfing rootstocks (14). Furthermore, trees on certain rootstocks considered to be poor performers because of low vigor and small tree size have been overlooked even though they were efficient bearers (yield/unit canopy volume) and thus could be high yielding per unit of land area at close spacings.

3.3. Pathological Characteristics

Profitable citriculture is also highly dependent upon rootstock resistance to pests and diseases. Rootstocks differ in their tolerance to Phytophthora foot and root rot, the burrowing [*Radopholus citrophilus* Huettel (formerly *R. similis*)] and citrus (*Tylenchulus semipenetrans* Cobb) nematodes and several virus and viruslike diseases. A disease of unknown cause, blight, is present in Florida and elsewhere, and seems to be rootstock related (63, 222). Thus pathological problems have often overshadowed horticultural ones.

By 1950 it had been reported that two diseases of citrus, psorosis and tristeza, were presumably caused by viruses (201). Psorosis infection can damage citrus trees regardless of the rootstock. Tristeza and two other virus diseases, exocortis and xyloporosis, are only economically harmful when they occur in trees propagated on intolerant rootstocks. Control of psorosis and xyloporosis is being achieved via budwood certification. The spread of exocortis viroid is also normally prevented by certification programs; however, the viroid can be mechanically transmitted unless pruning tools, etc., are properly sterilized. Control of tristeza is more difficult because the virus is transmitted by several species of aphids, most notably *Toxoptera citricidus* Kirk. (133). Only in isolated areas of California and to some extent in Israel is tristeza being controlled by certification.

With the discovery of citrus virus diseases, the 1950s assumed special significance as a temporal dividing line in rootstock study. Virus-contaminated field trials established before this decade could now be interpreted more correctly; also, a new generation of plantings were initiated in order to reexamine susceptible rootstocks.

Phytophthora remains a threat wherever citrus is grown. No rootstocks are immune. Improved cultural practices, higher budding, and chemicals have all reduced disease incidence, but it is a persistent problem that can result in substantial tree loss, particularly among young trees on susceptible rootstocks (199).

Burrowing nematodes are primarily a localized problem in Florida while citrus nematodes occur more universally. Chemicals are also available to help reduce nematode damage, but biological control through rootstock tolerance to nematode and *Phytophthora* diseases is a more desirable solution.

4. CITRUS ROOTSTOCKS

4.1. Sour Orange, Its Variants, and Related Cultivars

Considerable diversity exists among the sour oranges. These differences are reflected in their grouping as common or bitter sour oranges (also known as Seville or Bigarade), the bittersweet sour oranges, variants of/or fruits resembling sour orange, the myrtle-leafed or Chinotto sour orange, and the bergamots, which are considered to be sour orange hybrids (101, 197). The cultivars in each group except for the common and bittersweet sour oranges are generally small to moderate in stature as seedling trees and sometimes with a spreading habit. Their value is primarily ornamental. Some are cultivated for their oil, which is extracted from the peel and petals for use in perfumes. The Chinotto, *C. myrtifolia* Raf., is thought to be a sour orange mutation. Chinotto-like trunk sprouts occasionally arise from trees budded on common sour orange rootstocks (183). The small leaves and short internodes of Chinotto have suggested a potential usefulness as a dwarfing rootstock, but in California, scions on Chinotto were normal sized (22).

Virtually all the commercial sour orange rootstocks are from the common group, which itself is variable. Probably everywhere sour orange has been used as a rootstock, a local selection has been developed, leading to many named and unnamed cultivars. Variant or off-type seedlings, which maintain their differences through propagation, are easy to find in sour orange populations (196). Generally, differences in the field behavior of the various common sour orange clones have not been large (10, 23, 86, 87, 89, 153, 202); however, there have not been any extensive between- or within-group comparisons (91, 202).

4.1.1. *Common or Bitter Sour Orange (C. aurantium* L.). Sour orange is the premier citrus rootstock, common throughout the world where tristeza does not preclude its use and especially valued as a rootstock for producing fresh market fruit. Trees on sour orange are moderately vigorous and their size is generally considered as the standard for comparison; however, relative size can vary significantly depending on the scion and soil type (1, 11, 23, 62, 73, 75, 124). They produce good crops, but few trees on other stocks yield fruit of equal quality. Fruit from trees on sour orange are universally recognized for their high total soluble solids and acid content and for their ability to store on the tree without excessive deterioration or droppage (1, 11, 115, 122, 198). Also, the ascorbic acid (vitamin C) content tends to be high. Fruit size varies from small to large. Grapefruit cultivars in Florida and Texas produce excellent, high-quality, thin-skinned fruit (115, 202). Naval oranges are well suited for sour orange because their relatively poor juice quality and flavor are improved (11, 74). Mandarin suitability depends on the cultivar (97, 107, 113). Naturally small-fruited ones are not recommended for budding on sour orange in Florida (115, 122) but 'Dancy' is commonly grown on sour orange in Texas and Mexico. Clones of 'Clementine' are very popular on sour orange in Spain and elsewhere.

Sour orange can be grown on sandy to loam or clay soils (1, 73, 117, 123, 146). It is well adapted to heavy, often wet types of soil partly because of its moderate resistance to Phytophthora foot rot (95, 109, 198). There is clonal variation in foot rot tolerance (106). Many selections perform poorly in the severe environment of screening tests but survive well in field trials even when inoculated (111, 112, 199). Sour orange is moderately salt tolerant and grows well on calcareous soils (70, 71, 93, 210). Its root system is not unusually widespreading or fibrous but does tend to be deep, often with multiple taproots (45, 49, 54, 170, 198). These characteristics have been observed in clay and sandy soils. Trees on sour orange, however, do not yield as well on coarse sands as those on other stocks (66, 87, 89, 91).

Sour orange, Cleopatra mandarin, trifoliate orange, and Swingle citrumelo are the commercial rootstocks that induce the maximum scion cold hardiness. Trees on these rootstocks survive better than those on other, less tolerant rootstocks, for example, rough lemon, in chronically cold areas such as Texas and parts of Florida (70, 221).

The incidence of blight in Florida has been low among trees on sour orange (63). Sour orange is susceptible to the burrowing and citrus nematode (4, 108) and mal secco, a fungal disease affecting trees on sour orange in Italy and nearby countries (58, 165).

The advantages of sour orange are, unfortunately, almost totally negated when tristeza virus is present. Trees on sour orange are highly susceptible to this virus (17, 32, 92, 151) and, when infected, can decline rapidly, especially if severe strains are present in combination with efficient aphid vectors. Tristeza has most recently caused this stock to be outlawed from nurseries in Spain and Venezuela. Spread of the disease also threatens the continued use of sour orange in other citrus regions. In Florida, tristeza is widespread, but only occasionally do localized outbreaks occur in which trees decline. Sour orange is still used by many growers (55). It is possible that in Florida mild strains infect trees and protect them from later infection by severe strains. Field observations and research studies suggest that intentional inoculation of nursery trees may be a practical technique to assure the continued use of sour orange. The fact that such an approach is being explored only further illustrates the value placed on sour orange's attributes.

Trees on sour orange are essentially unaffected by exocortis or xyloporosis (118, 153). Tristeza susceptibility eliminates sour orange as a stock for most scions except lemons. Lemon trees on sour orange survive in tristeza areas apparently because of a hypersensitive reaction by the lemon that prevents infection. The virus may be acquired, however, through rootstock sprouts or during propagation in the nursery. Sour orange is often a preferred stock for lemons because it reduces the excessive vigor of this scion; but not all lemon clones are compatible with sour orange. 'Eureka' selections are shorter-lived than 'Lisbon' clones (173, 174).

4.1.2. Bittersweet Sour Orange.

The bittersweet sour oranges were brought to Florida and South America by the Spanish. Their name has its origin in the bittersweet taste of the juice. Their place among citrus rootstocks may only be secured at present by a few nurserymen in Florida who use the bittersweet orange exclusively. Bittersweet has substantially the same characteristics as common sour orange except it may react faster to tristeza and have higher *Phytophthora* tolerance (17, 23, 70, 73, 89, 91, 153, 154). In Texas, trees on bittersweet outyielded those on standard sour orange clones (213), but in Florida the results were not as encouraging (89, 91).

4.1.3. Sour Orange Variants or Hybrids.

C. taiwanica Tan. and Shim. became a rootstock of unusual interest when it was described as tristeza tolerant in California and Brazil (16, 17, 92); later it was reported as susceptible to Florida strains (89). Known as the Nanshodaidai in Japan, certain traits of this probable sour orange hybrid are similar to those of common sour orange (4, 16, 108, 221); however, grapefruit, tangelo, sweet orange, and mandarin yields and juice quality on this stock were low (1, 89, 134, 208, 214, 216, 217). Zygotic seedlings of *C. taiwanica* occur frequently and may contribute to differences in field behavior (145).

Abers Narrow Leaf (mistakenly referred to as Willow Leaf) is a probable variant included in Texas rootstock trials without noteworthy results (213). *C. intermedia* Hort. ex Tan. or Yamamikan, another possible sour orange hybrid of minor importance as a rootstock in Japan, did not perform well in Texas (209). It is also tristeza susceptible (20).

The Australian sour orange, or Smooth Flat Seville, is also known as the Appleby after the Australian grower who discovered the selection and noted its supposed tristeza tolerance. It may be a complex hybrid involving sour orange, pummelo, and sweet orange (9, 101). Its seeds are polyembryonic and yield uni- and bifoliate seedlings which could not be distinguished as rootstocks in a lemon trial (189). Reports are conflicting regarding claims that this stock has superior *Phytophthora* and tristeza tolerance as compared to standard sour orange clones (82, 94, 95, 109, 125, 151, 181), but 'Redblush' grapefruit from trees on Smooth Flat Seville had relatively less postharvest decay (128).

China has provided another probable hybrid, Gou-tou, claimed by the Chinese to be tristeza tolerant. It has recently been introduced into Australia and the United States for evaluation.

4.2. Sweet Orange [*C. sinensis* (L.) Osb.]

There are several groups of sweet oranges including the common round, navel, and blood oranges (101). Many were originally grown as seedlings with growers accepting their long juvenile period and erratic bearing. When budding onto rootstocks became commonplace, these problems were largely eliminated and interest turned to developing both sweet orange scion and rootstock cultivars.

From extensive field testing, various sweet orange rootstocks and many named cultivars have demonstrated rather consistently the same characteristics when evaluated under comparable conditions (1, 13, 23, 33, 61, 129). Some cultivars are unstudied because they produce few or no seeds. After lengthy observation it was noted that many cultivars performed well on sweet orange, but it had two serious limitations: intolerance to drought and high susceptibility to *Phytophthora parasitica* Dast. and *P. citrophthora* (Sm. & Sm.) Leonian (33, 117, 129, 146, 198).

Most commercial scion cultivars are long-lived and grow to a large size on sweet orange, producing crops often equal to or better than those from trees on sour orange (1, 23, 33, 66, 71, 87). Valencia and navel oranges, mandarins, and lemons on sweet orange have yielded well, with good fruit size and juice quality (23, 33, 100, 131, 180, 186). Trees are frequently slow growing with a shallow, dense, fibrous root system and perform best on sandy loam soils (33, 50, 132, 136, 181, 198). In heavier soils, tree growth and yield are acceptable, but *Phytophthora* damage can be heavy. Sandy soils accentuate the drought susceptibility of sweet orange and trees tend to wilt before those on other stocks (61).

Sweet orange cultivars are moderately salt and cold tolerant but are not well suited to calcareous soils (16, 70, 71, 86, 88). They are susceptible to burrowing

---

and citrus nematodes except for several clones, Ridge Pineapple, Algerian navel, and Sanguine Grosse Ronde, that are burrowing-nematode tolerant (80, 81). Another cultivar, Precoce de Valence, once considered *Phytophthora* tolerant, has not survived well in some field studies (163, 215). The susceptibility of virtually all cultivars to *Phytophthora* may explain the failure of sweet orange as a rootstock in replant situations (15). Trees on sweet orange stocks are tolerant of tristeza, exocortis, and xyloporosis (33, 61, 118, 140, 146).

Sweet orange is commercially important as a rootstock in Australia. Its widespread use has been discontinued in California, where it is now a minor stock as in Florida (47, 55). The value of this stock could be increased greatly if a foot rot-tolerant selection was found. Its potential is evident in its continued inclusion in rootstock trials. Sweet orange, like sour orange, has been relatively unaffected by blight (61). New fungicides which control *Phytophthora* and the increased availability of irrigation may foster a renewed interest in sweet orange rootstocks.

4.3. Lemon and Related Species

There are numerous cultivars of the so-called true lemon or the lemons of commerce. The lemon rootstocks are separate species, probably of hybrid origin, and therefore have questionable species status (9).

4.3.1. Rough Lemon (C. jambhiri Lush.). Rough lemon has proven to be an excellent rootstock for warm, humid areas with deep, sandy soils. In such an environment, trees on rough lemon grow rapidly and are long-lived and highly productive, yielding large fruit with a low total soluble solids content (33, 62, 66, 87, 100, 131, 135). In arid or coastal environments and clay soils, trees on rough lemon often decline prematurely (16, 19, 33). They are notably drought tolerant because of an efficient, extensive, and dense root system that may extend up to 15 m laterally and 6 m deep in sands (45, 50, 81). Rough lemon is moderately salt and boron tolerant and is acceptable for use in alkaline soils (16, 38, 70, 74).

Rough lemon is regularly cited as being a vigorous rootstock, which is misleading without considering tree size (62, 66, 129, 132, 135). Rough lemon seedlings are not fast growing; however, budded trees on this stock are very vigorous, precocious, and bear heavily as young trees, but they tend to peak earlier than those on other stocks (16, 66, 87). Trees on less vigorous stocks may eventually grow to an equal size or surpass those on rough lemon (66, 87, 132). The vigor of budded rough lemon trees may account for their success when used for replanting in old orchard soils, even though rough lemon is relatively intolerant to *Phytophthora* and is susceptible to the burrowing and citrus nematodes and woody gall (4, 106, 108, 109, 199). Trees on rough lemon are significantly less cold hardy than those on sour orange; thus they are not well suited for use in areas frequently subjected to cold damage (86, 88, 221).

Depending on the extent of damage, however, trees on rough lemon are capable of rapid recovery because of their vigor (67).

There are many selections of rough lemon with morphologically distinct fruit that differ in color from red to yellow (106). Attempts to find a *Phytophthora*-resistant rough lemon have not been successful, although a few were rated as moderately resistant in seedling screening tests (43, 95, 109, 199). A red rough lemon, which survived a California screening program, later proved to be unsatisfactory in field tests (40, 41, 43, 124). Another selection, Soh Jhalia, brought from India and similarly evaluated, has performed well (41, 43). Milam, a possible rough lemon hybrid, is resistant to the burrowing nematode (80, 81, 148). It has evolved essentially as a special-purpose rootstock in Florida, where commercial trees on this stock behave similarly to those on rough lemon (44); however, Milam grew poorly in a calcareous soil in Texas (210).

Rough lemon, like sweet orange and Cleopatra mandarin, is relatively unaffected by the common rootstock-related virus diseases; thus observations from old orchards and field trials are reasonably valid despite virus contamination. Xyloporosis symptoms have occasionally developed in trees on rough lemon. Exocortis and xyloporosis have independently reduced 'Valencia' tree growth and yield, but only by 10 to 15% (71, 154, 179). Rough lemon is tolerant to tristeza, unlike the hypersensitive true lemons (17). Virus tolerance, unfortunately, is offset by intolerance to blight, a serious problem in Florida that has resulted in rough lemon being virtually eliminated as a rootstock for new trees (55, 63, 122, 222).

Orange, grapefruit, and lemon cultivars can be used with rough lemon (1, 62, 115, 120, 135, 188). In humid climates, where the juice content of fruit is high, oranges produce large quantities of total soluble solids per unit of land. Other stocks are better suited for mandarin scions. Fruit from trees on rough lemon are prone to be coarse, thick skinned, and to dry out or granulate early (16, 97, 107, 198). They do not store well on the tree. These characteristics are not compatible with the naturally short on-tree life of many mandarin cultivars (101). The thick peel is desirable for kumquats and may account for the generally low level of fruit splitting among rough lemon–rooted trees.

4.3.2. Volkamer Lemon (C. volkameriana Ten. & Pasq.). Chapot described Volkamer lemon in 1965 as being known for over 300 years, having originated in Italy, and as being a possible lemon–sour orange hybrid tolerant to mal secco and *Phytophthora* (56). Only recently have any field data been reported. Volkamer lemon has many of the same characteristics as rough lemon (19, 20, 96, 137), although trees on this stock are cold hardier and have often yielded more fruit with slightly higher juice quality than those on rough lemon (1, 19, 96, 124, 142). Volkamer lemon is apparently tolerant to *P. parasitica*, but was badly damaged by inoculation with *P. citrophthora* (40, 41, 43, 192). *C. volkameriana* appears to be as susceptible to blight and woody gall as rough lemon (19, 114).

4.3.3. Alemow (C. macrophylla Wester).
Alemow, often called macrophylla, is a classic example of a rootstock that possesses outstanding traits along with poor ones. Trees on alemow are vigorous, precocious, and fruit heavily as young trees (1, 19, 26, 157, 191, 202, 215). They are among the most tolerant of high levels of soil boron, chloride, and calcium and tend to have high levels of leaf manganese (26, 38, 165, 200, 203). *C. macrophylla* is considered *Phytophthora* resistant. Field observations in California and Florida confirm excellent survival of this rootstock when *P. parasitica* is present, but trees are moderately susceptible to *P. citrophthora* (26, 43, 112, 192). Foot rot response can be expected to vary because alemow may be a hybrid involving the monoembryonic citron and pummelo (9). Seedling populations often appear uniform but contain many zygotic plants, explaining why they were rated from resistant to susceptible in a *P. parasitica* screening test (95).

Despite their advantages, trees on alemow produce fruit with a juice quality lower than that for trees on virtually any other rootstock (1, 26, 215). Scion cultivars other than lemons produce fruit of only marginal quality. Moreover, trees on *C. macrophylla* decline from xyloporosis and tristeza. Alemow was once described as tristeza tolerant (17), which strongly contradicts current knowledge (82, 146). Other limitations to the use of alemow include cold sensitivity, intolerance to nematodes, and blight susceptibility (4, 63, 117, 127, 221).

Alemow is popular in California as a stock for lemons, but its future is uncertain. Lemons are short-lived on this stock because of a rootstock necrosis first detected about 1960 (173). All lemon clones on alemow are considered susceptible (174).

4.4. Limes and Related Species

4.4.1. Palestine Sweet Lime (C. limettioides Tan.).
Palestine sweet lime seems to exhibit both lime- and lemonlike characteristics; thus its synonym, Palestine sweet lemon. A recent taxonomic study suggests that sweet lime may be a hybrid of four species including *C. limon* and *C. aurantifolia* (9). It is valued mainly as a rootstock, although it is grown in some countries for fresh consumption.

Palestine sweet lime is best compared with rough lemon (1, 54, 74–76, 119, 202). Small differences have been reported where these two stocks or clonal selections of sweet lime (Colombian, Indian) have been included in a field trial (74–76, 202, 214). Palestine sweet lime is less salt and boron tolerant (70, 200). Fruit size is medium to large for trees on sweet lime and juice quality is generally poor, but has been slightly better in some instances than that recorded for trees on rough lemon (74–76, 214). Nucellar scions perform best on sweet lime. Old-line cultivars infected with xyloporosis, exocortis, or tristeza will reduce tree performance on this stock even though virus effects may require several years to appear (32, 75, 92, 135, 139). Blight has affected trees on sweet lime in Florida, but their relative susceptibility is unknown (63).

4.5. Mandarins and Mandarin Hybrids

The mandarin (*C. reticulata* Blanco) group is large and classification has been difficult because of their diversity and the confusing term *mandarin orange*. The confusion originates with the word *orange*, which has evolved primarily as a synonym for sweet orange, but is also used generically to describe all orange-colored fruit (68); thus there are sweet oranges, sour oranges, and mandarin oranges. Many mandarins have been of sufficient promise to include in rootstock studies, but few have been adopted commercially (23, 102, 113).

4.5.1. Rangpur Lime (C. limonia Osb.).
Rangpur is probably a mandarin hybrid and not a lime like the true acid limes, *C. aurantifolia* (Christm.) Swingle, used as scions. It has been speculated that the parentage of Rangpur may involve rough lemon or sour orange (9, 101). Rangpur was introduced to North and South America probably from India (101) and has been included in many California, Florida, and South American rootstock studies. It was a common stock by 1930 in Brazil, but trees on this stock were observed to decline after several years. Rangpur was considered an unsatisfactory rootstock (142). This rootstock is now used for over 600,000 ha of citrus in Brazil, an apparent paradox that is the result of exocortis infection in trees on Rangpur, which has since been eliminated by the use of virus-free budwood.

Trees on Rangpur, as those on rough lemon, are vigorous and highly productive, particularly as young trees, and yield medium to large size fruit with low to moderate juice quality (1, 13, 60, 100, 167, 215, 216). They are sensitive to cold and *Phytophthora* (60, 86, 88, 109, 221), have excellent drought tolerance because of a deep, vigorous root system, and are not tolerant to either the burrowing or citrus nematode (4, 127). Declinio, a disease similar to blight, is causing heavy tree losses on Rangpur in Brazil (206).

Rangpur differs from rough lemon in several significant ways. Like rough lemon, Rangpur is tristeza tolerant (92), but it is susceptible to exocortis and xyloporosis (60, 118, 124, 139, 142, 149). Both are excellent stocks for deep, sandy soils in humid climates, where grapefruit and sweet orange trees yield well (60, 89, 116). Rangpur also grows well in loam and clay loam soils (73, 83). Juice quality from Rangpur lime-rooted trees is usually above that for rough lemon and has approached the quality of fruit from trees on sour orange and citranges (202, 215). Trees on Rangpur have grown well in areas where rough lemon is short-lived and are very salt and lime tolerant (70, 194, 200). Apparent contradictions in the characteristics of Rangpur may be the result of clonal differences, suggesting that with this and other rootstocks, superior clones may exist.

4.5.2. Cleopatra Mandarin (C. reshni Hort. ex Tan.).
Cleopatra mandarin is one of the most widely and thoroughly studied rootstocks. Among its attributes are tolerance to tristeza, exocortis, xyloporosis, salt, cold, and calcareous soils (10, 70, 86, 117, 118, 146, 198, 200, 211, 221). Blight incidence has generally been

low in Florida (47, 63). Cleopatra is not nematode resistant (109, 127). There are exceptions to the usual description of Cleopatra as *Phytophthora* tolerant, although this rootstock is not as susceptible as rough lemon (16, 109, 165). Furthermore, Cleopatra is chloride tolerant on sandy loam but susceptible on heavier soils (203, 210).

Cleopatra has been included in many rootstock trials, particularly in areas where sour orange was eventually replaced because of tristeza. Nonetheless, it remains a minor rootstock as exemplified by its history in Florida. Cleopatra was introduced to Florida in 1888 and described in 1924 as a replacement for rough lemon of great promise (178); however, it was not until 1952, when tristeza was discovered, that Cleopatra became of interest as a potential substitute for sour orange. Cleopatra gained popularity, but still accounts for only about 10% of rootstock use in Florida (55).

Cleopatra has two major faults as a rootstock. First, trees budded to Cleopatra are capable of achieving a large size, but they are often slow to bear; second, juice quality is excellent but fruit size is small, particularly with 'Valencia' scions (1, 10, 19, 74–76, 99, 113). Low yield usually characterizes only the early bearing years, as trees on Cleopatra improve as they mature (99, 123, 180). Shy bearing is not the result of a lack of flowering or slight suscepti-bility to drought (140). Cleopatra is deep rooted with extensive lateral root development on sandy soils (45, 78); but it thrives best on heavier soils or those with a clay layer close to the soil surface (73, 78). Excessive fruit drop from splitting may be an important factor contributing to low yield with oranges.

Cleopatra is an excellent rootstock for mandarin and related cultivars and 'Pineapple,' 'Hamlin,' and certain other sweet orange cultivars (72, 97, 113). 'Valencia' (especially nucellar clones) and navel yields on Cleopatra are rela-tively poor in Florida and elsewhere (72, 113, 123, 180). Certain cultivars such as 'Temple,' 'Nova,' 'Murcott,' and 'Orlando,' where fruit size is not inherently small, have been satisfactory on Cleopatra (96, 107, 146, 186).

4.5.3. *Citrus sunki* Hort. ex Tan.

Sunki mandarin is less well known than Cleopatra but may have greater rootstock potential (21, 83, 141). It is com-monly used in China but has not been adequately evaluated elsewhere. It is tristeza and xyloporosis tolerant like Cleopatra but is affected by exocortis, which is unusual for a mandarin (70, 71, 91, 92, 142, 153, 166). It has been reported as *Phytophthora* susceptible but has survived well in rootstock trials in other areas (21). Trees on Sunki are highly salt tolerant, moderately cold hardy, and adaptable to calcareous soils (69, 221). Trees were below standard in size on *C. sunki* in Texas even with a nucellar scion (153, 215). Yield and juice total soluble solids content were equal or superior to those for trees on sour orange (202, 215). In Florida, trees budded to Sunki were large with higher juice quality and yield than trees on Cleopatra (89). Similar favorable comparisons in Brazilian trials encourage further testing of Sunki (72, 142, 156). Variations in field results may occur because zygotic seedlings are com-mon even though the seeds are polyembryonic (84, 190). Hybrids of Sunki with trifoliate orange show promise as dwarfing rootstocks, although they vary in tristeza tolerance (20, 46, 52).

4.5.4. Other Mandarins and Mandarin Hybrids. Trees on *C. amblycarpa* Ochse (Nasnaran) outyielded those on Cleopatra in a California experiment, but showed some of the same weaknesses as Cleopatra and were more difficult to handle in the nursery (4, 16, 19, 95, 127). Trees on Nasnaran have produced fruit with a juice quality comparable to that of fruit from trees on sour orange (21, 188) and are well adapted to clay soils (73, 191). *Citrus kinokuni* Hort. ex Tan. is susceptible to tristeza and does not appear promising (32, 216). Changsha is one of the more promising mandarins tested in Texas. Trees on this stock were smaller than those on sour orange. They yielded well for their size, producing good-quality fruit (208, 210, 216, 217).

Tangelos and tangors, hybrids of mandarin with *C. paradisi* and *C. sinensis*, respectively, and other mandarins have given substandard performance when tried as rootstocks, as compared with their normal use as scions (134, 153, 204, 208, 210, 214, 216). Many are susceptible to xyloporosis (139, 149, 152, 169) and other diseases and pests (4, 95). Some are minor commercial rootstocks such as Willow Leaf in Arizona, Emperor in Australia, and Empress [probably identical to Emperor (132)] in South Africa (132).

4.6. Trifoliate Orange and Its Hybrids

4.6.1. Trifoliate Orange. The genus *Poncirus* consists of a single deciduous species, *Poncirus trifoliata* (L.) Raf., and has been a very significant source of rootstock cultivars.

The trifoliate orange is a common ornamental plant in central and northern China, but its use is most notable in Japan, where it is the primary rootstock for their large satsuma or unshiu plantings. When described by Webber in 1948 (198), trifoliate orange was not widely used outside of Japan. He wrote that this rootstock was highly polyembryonic, with upright, thorny, and slow-growing seedlings. Trees budded to trifoliate orange were noted to be cold hardy, standard sized on heavy soils and dwarfed on light soils, and intolerant of calcareous conditions. They were *Phytophthora* tolerant but variable and yielded small, high-quality fruit. Continued study has not greatly altered this assessment, particularly concerning yield and fruit characteristics. Scion performance in many field studies has been consistent with navel and other oranges and grapefruit, mandarin, and kumquat cultivars (13, 19, 23, 31, 33, 52, 65, 105). Occasionally, large fruit size or low juice quality has been reported for different scion cultivars on this stock (23, 208, 216).

Webber's conclusion that trifoliate orange was an erratic stock prevailed until the role of exocortis viroid and morphological variation among trifoliate seedlings became known. Seedlings can be grouped into small- and large-flowered types (176). The latter types are usually more upright and less branched as nursery plants, and trees budded on these cultivars tend to be slightly larger and more productive, although fruit density is lower and fruit mature later than those on small-flowered cultivars (31). These results are contrary to those of Japan, where trifoliate selections are grouped according to leaf size (185).

The perception of trifoliate orange as a dwarfing rootstock is misleading. Trees on this stock are slow growing and normally do not achieve large size, especially on sandy, unirrigated soils, where they often will show drought stress before trees on other stocks (33, 90, 105, 160, 180, 181, 185). They have shallow root systems consisting of weak lateral root development but abundant fibrous roots (45, 55). On loam or clay soils, they can grow to standard size (182). Saline and calcareous conditions will also reduce tree vigor and cold hardiness (67, 70, 158, 165). One selection of trifoliate orange appears to be a true dwarfing rootstock. Trees on Flying Dragon have been consistently small during field trial in Florida and California. Their performance is otherwise essentially the same as for other trifoliate orange selections (3, 28, 46, 64, 108).

The confusion regarding size and vigor of trees on trifoliate orange is primarily related to the exocortis viroid to which this rootstock is highly susceptible. Xyloporosis and tristeza do not affect scions on trifoliate orange, although the latter virus may cause stem pitting (20, 42, 168). Isolates of the exocortis viroid differ in virulence. Mild forms are being explored as dwarfing agents for use with trifoliate orange and other exocortis-sensitive rootstocks (12, 59).

It is incorrect to describe trees on trifoliate orange as cold hardy without considering climate. The trifoliate tree itself is very cold tolerant, which adds to this misunderstanding. Budded trees have superior winter hardiness when grown in cool climates as in Japan; in other areas where the winters are relatively warm but freezes periodically occur, they may be as damaged as those on more tender rootstocks (86, 221).

Trifoliate orange is an excellent replant rootstock because trees on this stock are resistant to the citrus nematode and *Phytophthora*, but some variation exists (4, 5, 95, 108, 109, 111, 112, 127, 181). None of the trifoliate orange strains tolerate the burrowing nematode (109). Trifoliate is considered as susceptible as rough lemon to blight, and trees on trifoliate decline from a similar disease in Uruguay and Argentina called *marchitamiento repentino* (63, 207).

4.6.2. Carrizo, Troyer, and Other Citranges. *Poncirus* hybridizes easily with *Citrus*. The progeny of trifoliate orange crosses with sweet orange (citranges) and grapefruit (citrumelos) have given rise to a new generation of rootstocks that possess attributes which may effectively substitute for the limitations inherent in *Citrus* stocks such as sour orange. Swingle and Webber produced many trifoliate orange hybrids in Florida after the devastating freeze of 1894–1895. Their work was aimed at creating cold-hardier scions; however, a very significant source of new rootstocks was eventually produced instead.

There are many citranges that are quite variable, as exemplified by the

naming of several cultivars which arose as seedlings from a single fruit pollinated by a single flower (197). Their performance as rootstocks was first
studied in California and differed significantly depending on the specific citrange and scion cultivar. Many, such as the Coleman and Cunningham, were
included in later trials and eventually discarded because of poor performance
and a lack of seeds (11, 198). Carrizo and Troyer were given only brief mention
in a 1948 review of rootstocks, but their seediness and good nursery behavior
were noted (198).

The citrange originally considered most promising was Morton. Sweet orange, grapefruit, 'Orlando' tangelo, and satsuma trees on Morton were very
vigorous and productive (23, 90, 105, 198). Juice quality was excellent. Morton
is particularly well adapted to the Texas soils and climate, where trees on this
stock were comparable to those on sour orange (163, 202, 212, 213, 216).
Morton is exocortis sensitive, as is its parent, trifoliate orange, and showed a
stem pitting response to tristeza in California (17); however, trees on Morton
were not affected by aphid inoculation of the tristeza virus is Brazil, where they
have grown well on loamy soils and been rated highly (92, 140–142).

Morton seeds are highly polyembryonic, but the fruit are usually seedless
(105, 143); thus Morton has not been pursued commercially. Other citranges,
for example, Rustic, Savage, and Uvalde, suffer the same disadvantage (143).
Trees on the latter two stocks have yielded well and withstood freezes in several
trials on sandy and loam soils and may be suitable as rootstocks for close
planting (14, 42, 52, 71, 86, 89, 198, 202). Trees on Rusk citrange also tend to
be small, although in several Florida trials they grew rapidly as young trees (66,
87, 89). Small tree size on Rusk is combined with a limited root system and
drought intolerance (50, 103); however, mandarin, sweet orange, and grapefruit cultivars yield good crops of high-quality fruit for the size of the tree (51,
52, 65, 87, 89, 97, 107). Rusk is also apparently exocortis tolerant, even when
infected with severe strains of the viroid (71, 152). Another very promising
citrange with similar parentage to the Rusk is Benton (125). This citrange is *P.
citrophthora* tolerant and may be compatible, unlike other citranges, with
'Eureka' lemon clones (125, 189).

Carrizo and Troyer have emerged as the only commercially important
citrange rootstocks. They are hybrids of trifoliate orange and the navel sweet
orange. Although each is a named cultivar, they may actually only be different
seedlings from the same pollination, as suggested in a study of their history
(171). Horticultural and pathological comparisons do not provide strong evidence to the contrary (126). When nucellar scions were used, tree growth,
yield, and juice quality have been virtually identical for a broad range of
cultivars (1, 19, 65, 90, 105, 107, 181). In general, trees on both citranges are
among the most vigorous, growing well on a wide range of soils (1, 10, 19, 73,
90, 141, 163, 167, 202). They are prone to zinc and manganese deficiency, as
are trees on trifoliate orange (167, 181, 185). Trees on Carrizo in Florida grow
and fruit unusually well in their early years. Fruit quality with both stocks is
often comparable with Cleopatra and sour orange, although sweet oranges
exhibit fruit creasing on Troyer in South Africa (1, 10, 73, 90).

Carrizo and Troyer inherited the deficiencies of trifoliate orange, that is, poor salt tolerance and sensitivity to exocortis and calcareous soils (38, 71, 74, 75, 146). Neither rootstock is affected by xyloporosis (169). Tristeza inoculations in Brazil had little affect on tree behavior, but sweet orange trees in California have declined in coastal areas (37, 42, 92). The primary difference between these two stocks lies in their nematode tolerance. When several Carrizo sources were tested, seedlings from a California source were rated tolerant to the burrowing nematode (81). No tolerant Troyer sources have been found (109). Both citranges have a variable tolerance to citrus nematode and foot rot depending upon nematode biotype and *Phytophthora* species (5, 108, 109, 112, 192, 199). Carrizo and Troyer are susceptible to blight (63, 222).

Citranges were reported in 1948 to be less cold hardy than might be expected because of the trifoliate parent. The observation that "the citranges are all very cold-resistant when in dormant condition but are easily forced into new growth by spells of warm weather" (197) appears correct for budded trees as well. Trees on Carrizo in particular have been consistently less cold tolerant than those on trifoliate orange, sour orange, or Cleopatra mandarin (65, 71, 220, 221).

4.6.3. Swingle and Other Citrumelos.
Swingle citrumelo is named after Walter Swingle, who with H. J. Webber produced a series of citrumelos in Florida during the early 1900s. The rootstock was not "discovered" until about 40 years later, when intensive testing began in Texas, leading to an official release in 1974 (2, 104). From these and other studies, the many highly desirable attributes of Swingle became known (2). In Texas, 'Redblush' grapefruit, 'Orlando' tangelo, and 'Marrs' orange yield and fruit quality almost always surpassed that of trees on most other stocks, including the standard rootstock, sour orange (202, 215−217). In California, Swingle is compatible with 'Lisbon' lemon and has performed well in grower trials (172). Controlled environment studies and field observations indicate that Swingle is one of the rootstocks that induces as much scion cold hardiness as sour orange (2, 48, 218, 221). Little is known about the Swingle root system. Budded Swingle nursery trees grown in sand did not have an abundance of fibrous roots and the root system had a coarse appearance (54).

Swingle is very tolerant to the citrus nematode and *Phytophthora* (2, 43, 127, 147). Trees infected with tristeza, xyloporosis, and exocortis have grown normally (42, 71, 153); however, virus-infected, old-line trees on Swingle were dwarfed in Florida (91) and in Texas and Honduras, where a bud union crease developed (164). In those specific instances, the virus content of the scion was not completely known. Thus it cannot be assumed that any old-line bud source will result in a crease. The citrange stunt virus may be implicated.

Trees on Swingle have shown severe chlorosis on calcareous, clay soils (191), but have also been observed by the author to be growing satisfactorily in Texas on a loam soil of pH of 8.2. The role of soil texture and pH is unclear (164). Also uncertain is the significance of a yield difference noted in California between trees budded on Swingle or C.P.B. 4475 (40, 41). Swingle is the name reserved for seed source trees propagated by budding from the original Swingle tree, while C.P.B. 4475 trees were propagated by seed. In practice, the name Swingle is used for both seed sources.

Sacaton citrumelo has been evaluated in the United States with inconsistent results. Sacaton seeds are polyembryonic but produce about 40% zygotic seedlings (105). Furthermore, Sacaton has been confused with Yuma citrange. They may identical, as suggested by comparable results obtained from *Phytophthora* screening tests and field trials using seedlings from both sources (40, 41, 43). Sacaton may be tristeza susceptible in some environments (30, 42). Trees are small on this stock and have potential for high-density plantings (52, 72, 91, 105). Lemons have performed well in Arizona and California experiments (1, 100).

4.7. Grapefruit, Pummelo, and Related Cultivars

The importance of grapefruit (*C. paradisi* Macf.) and pummelo [*C. grandis* (L.) Osb.] cultivars as scions does not extend to their use as rootstocks. Grapefruit was commonly included in rootstock trials planted in the 1920s and 1930s because it was widely known and seeds were available (130). Trees on this stock were vigorous with very dense but shallow root systems, which may explain its drought intolerance and low yields on sandy soils (11, 66, 87, 89, 91, 103, 135). Grapefruit rootstock is susceptible to damage by *Phytophthora*, tristeza (but not exocortis or xyloporosis), the burrowing and citrus nematodes, and cold and is salt intolerant (4, 16, 17, 33, 92, 221). As a result, grapefruit has been described as a generally unsatisfactory rootstock (16, 23, 38, 130); nevertheless, it still retains interest as a possible blight-tolerant stock.

The pummelo, or shaddock, is considered a parent, along with sweet orange, of the grapefruit (9). Pummelos are monoembryonic, but seedlings often appear uniform (183, 184, 190). As a rootstock, pummelos exhibit characteristics similar to those of grapefruit (135, 198). A probable pummelo−citron hybrid, the Cuban shaddock, was introduced into the United States in 1923 from Cuba (14). Trees on this stock did not perform well in Texas or California (198, 215). Cuban shaddock was described as a dwarfing rootstock, but subsequent studies have shown it to be intolerant to exocortis and xyloporosis (14, 71).

The Poorman Orange, or New Zealand grapefruit, is also a probable pummelo hybrid, as are *C. natsudaidai* Hay. (Natsumikan), a minor rootstock in Japan, and *C. obovoidea* Tak. (Kinkoji), which is citrus nematode-and tristeza-tolerant (16, 20, 127). Little is known about the latter two stocks. Poorman Orange is tristeza susceptible (32, 92) and apparently tolerant of exocortis and xyloporosis in Israel, where the growth and yield of several old-line scions were excellent (136), in contrast to the poor performance of a nucellar 'Marsh' scion in Florida (89).

4.8. *Citrus,* subgenus *Papeda*

The two subgenera of *Citrus* are distinguished by the presence of an acrid oil in the flesh of the *Papeda.* Most species and hybrids of the *Papeda* group were brought to California for tristeza and horticultural evaluation. Only two have gained acceptance as rootstocks, alemow, discussed earlier, and Yuzu (*C. junos* Sieb. ex Tan. or *C. ichangensis* Swing. × *C. reticulata*). Yuzu is popular in Japan for inarching declining satsuma trees on trifoliate orange and apparently is cold, tristeza, and *Phytophthora* tolerant (18, 26, 185, 221). Young trees did not perform well in Florida, Texas, or California on Yuzu (19, 26, 51, 163, 216). Its ability to absorb iron and manganese efficiently suggests that it may be suitable for calcareous soils (26, 200, 203).

Citrus ichangensis and its probable hybrids, including Yuzu, are noted for their extreme cold hardiness as seedlings (26). Swingle, in 1948, listed the Ichang lemon (Ichang pummelo), *C. latipes* (Swing.) Tan., and *C. webberii* Wester, but not *C. hystrix* DC., as promising rootstocks or useful for breeding (183). Further study in California proved some species to be too tristeza sensitive for use as orange rootstocks, but possibly acceptable for lemons (26).

4.9. Miscellaneous Rootstocks

Citrus tachibana (Mak.) Tan. is mostly of historical interest. It is found throughout southern Japan and is described as a wild satellite species of *C. reticulata* (101, 183). Because of its mandarinlike characteristics and being "strongly resistant to frost or snow" (183), it has been tried as a rootstock, but with little success (158, 163, 185, 202, 214). More interest has been shown in a possible tachibana hybrid, *C. depressa* Hayata (Shekwasha), formerly known as *C. pectinifera*, which is moderately cold and salt tolerant but has a high level of tolerance to lime-induced chlorosis (38, 70, 88). Shekwasha is also tolerant of *Phytophthora* and the rootstock-related viruses (17, 112). Seedlings were reportedly variable (38) despite other data indicating a large number of embryos per seed (190). Various scions budded to Shekwasha were average or better in vigor and yield with generally good fruit size and juice quality (19, 38, 89, 165, 202, 215). These characteristics were not improved by hybridizing Shekwasha with rough lemon, Rangpur, or other cultivars (51, 52, 210, 216, 217). Nevertheless, Shekwasha possesses a number of highly desirable characteristics that may be improved upon by further breeding efforts or simply through higher-density plantings. Also, Shekwasha should be further explored as a rootstock for mandarins (185).

C. pennivesiculata (Lush.) Tan. (formerly *C. moi*), common name, Gajanimma, was included in a 'Valencia' trial with 25 rootstocks. It ranked in the top 10 for yield after 15 years, juice quality was low, and it was tristeza tolerant (19).

Kumquats (*Fortunella* spp.) are poor rootstocks because of slow seedling and budded tree growth and incompatibility with many scions (25, 27, 198). Various hybrids have been tried as rootstocks to no avail, except possibly procimequat, a trigeneric hybrid of *F. hindsii* (Champ.) Swing., of interest as a dwarfing rootstock (51). Calamondin, *C. madurensis* Lour., a probable kumquat hybrid, usually fails as a rootstock because of bud union crease, an incompatibility that occurs with most scions except kumquats (149, 152). Calamondin is more popular as a potted ornamental or as a convenient greenhouse research plant. It is also the primary acid fruit of the Phillippines and commonly used as a rootstock in lowland areas along with a selection known as calamandarin.

5. ROOTSTOCKS AS NURSERY PLANTS

Emphasis is usually placed on orchard performance in rootstock development even though most rootstocks do not achieve commercial status without also being acceptable nursery plants. The nursery behavior of several rootstocks when grown in the field or containers, is given in Table 1.

6. *CITRUS* AND ITS RELATIVES AS ROOTSTOCKS OR INTERSTOCKS

The Citrus relatives have been described as more likely to prove valuable as a germplasm resource for breeding rather than directly as rootstocks. Genera such as *Aeglopsis*, *Afraegle*, *Clausena*, *Murraya*, and *Atalantia* are potential sources of resistance to the citrus nematode (4, 24), and *Citropsis* to the burrowing nematode (24, 79). *Merope* is highly salt tolerant (24) and pummelos survive well when grafted to *Feronia* in wet and *Phytophthora*-infected areas of Thailand. These and other genera, however, have limited

TABLE 1. Nursery Characteristics of Citrus Rootstocks

Rootstock	Remarks
Sour orange	Seed are 70 to 85% nucellar and germinate best at about 26°C. Seedlings are thorny, vigorous, with straight trunks and moderate branching, easy to bud and force, cold tolerant, and very susceptible to scab (*Elsinoe fawcetti* Bitancourt and Jenkins). Bud take may be low when ambient temperature is above 30°C.
Rough lemon	Seed are highly nucellar and germinate well. Seedlings have low vigor but scion buds generally force easily and grow rapidly, especially in warm or hot climates. They are susceptible to scab and *Alternaria citri* Ellis

TABLE 1. *(continued)*

Rootstock	Remarks
	and Pierce, leaf spot; seedlings tend to branch close to the ground.
Volkamer lemon	Similar to rough lemon except more vigorous, with straight trunk and relatively unaffected by *Alternaria*. It may be less herbicide tolerant (53).
Rangpur lime	Seed are highly nucellar and germinate readily. Seedlings are vigorous, easy to handle, susceptible to *Alternaria*, bud readily and force easy.
Alemow	Seed are polyembryonic but seedlings are variable, moderately vigorous, and bud and force easily. Those grown in greenhouses tend to lean readily leading to excessive sprout development.
Sweet orange	Seed are 70 to 90% nucellar but this varies with the cultivar. Seedlings are thorny, bushy, and vigorous, and generally easy to bud and force. Leaf diseases are rare.
Cleopatra mandarin	Seed are highly nucellar and germinate best at temperatures above 30°C. Seedlings are vigorous in hot climates but slow growing in cool areas where plastic tunnels or greenhouses can improve germination. Seedlings are unbranched with short internodes, almost thornless, and cold hardy. They are easy to bud but sometimes difficult to force.
Trifoliate orange	Seeds are highly nucellar and may require chilling for best germination. Seedlings have low to moderate vigor, are very thorny, and small-flowered cultivars are bushy, making them difficult to bud. Seedlings respond to prolonged day length (161) but generally go dormant outdoors in the fall. Flying Dragon seedlings must be carefully rogued because of the large number of off-types.
Carrizo and Troyer citrange	Carrizo seeds can transmit psorosis (35, 39). Seedlings show sensitivity to preemergence herbicides (177). Both citranges are excellent nursery plants with highly nucellar seeds and produce uniform, vigorous unbranched seedlings that are easy to bud and force. Tetraploids occur at a rate of about 1 to 2% (8).
Swingle citrumelo	Seed are highly nucellar with an excellent percent germination. Seedlings are very vigorous, uniform, upright, and easy to bud, but forcing may be erratic, with some buds not breaking. Seedlings become dormant in fall and are sensitive to preemergence herbicides (177).

vegetative compatibility with *Citrus* species. Sometimes a successful graft can be made only to fail in reciprocal attempts (25, 27). Specimens of many of these genera have not been readily available for study and their potential is unknown.

Other genera and species have received more than cursory attention. Citrus can easily be grafted or budded onto *Swinglea glutinosa* (Blanco) Merr. and *Hesperethusa crenulata* (Roxb.) Roem., although the resulting trees are not very cold tolerant (24, 193). *Severinia buxifolia* (Poir.) Tenore showed promise as a rootstock for grapefruit in early California trials. Trees were severely dwarfed, probably due to its tristeza intolerance (14, 24, 25). In Texas, they were not dwarfed and yielded average or better crops of high-quality 'Redblush' grapefruit; however, they were slow to begin bearing (202, 208, 215). 'Orlando' tangelo and 'Marrs' orange trees yielded poorly, although juice quality was excellent (216, 217). *Severinia* seed are monoembryonic and seedlings grow slowly (24, 198). They are citrus nematode and *Phytophthora* resistant and very cold hardy (4, 43, 70, 95, 108).

Severinia is best known for its nutritional characteristics. Trees on this stock have low leaf boron levels because *Severinia* is a boron excluder. They also tend to accumulate manganese (210, 217). *Severinia* has excellent salt tolerance, although it is not well adapted to calcareous soils (24, 71, 210). *Eremocitrus* is a monotypic genus closely related to *Citrus*. Seedlings are variable and have excellent cold, drought, salt, and boron tolerance (24, 25, 162, 219). It hybridizes readily with *Citrus* and is susceptible to the citrus nematode and *Phytophthora* (4, 6, 24, 43, 95, 219). *Microcitrus*, another closely related genus, is resistant to *Phytophthora* and the burrowing nematode and propagates easily onto Citrus (24, 25, 27, 43, 148).

Species of *Citrus* and its relatives have been used without much success as interstocks. Trifoliate orange interstocks had little effect on leaf nutrient content (121). Several *Citrus* species did not influence 'Redblush' grapefruit quality (138) nor did sour orange or rough lemon affect sweet orange quality as interstocks (85). Preliminary experiments have, however, demonstrated success in tree size control. The vigor of lemon trees was reduced with a series of Citrus-relative interstocks (25, 27), Satsuma tree size control was achieved in Japan (110) and 'Valencia' trees were dwarfed in an Australian trial with *Microcitrus* (187).

7. ROOTSTOCK–SCION COMPATIBILITY

Compatibility among *Citrus* species has been judged on the basis of bud union smoothness and the existence of anatomical or morphological abnormalities. Straight, smooth bud unions suggested "thorough congeniality" according to a scheme for rating bud unions proposed by Webber (198). Progressively greater disparity in the size of the stock and scion at their union represented decreasing congeniality. The physiological implications of these differences in bud union morphology are unknown. In practice, they may be meaningless and are not predictive for tree health or longevity. Trifoliate orange and its hybrids commonly overgrow scions without any adverse effects; nonetheless, various cultivars, particularly mandarins, lemons, and kumquats on these and other stocks, occasionally develop bud union creases and decline (34, 195). There are instances of satsuma failure on trifoliate orange in California from what appears to be a mechanical or compression girdling due to severe stock overgrowth. 'Murcott' and 'Nova' trees on Carrizo citrange with normal-appearing bud unions have developed a crease and collapsed as soon as eight years after planting in Florida. 'Eureka' lemon clones grow normally for a few years on Troyer or Carrizo citrange or trifoliate orange, then decline from a sieve-tube necrosis (195). These and similar bud union difficulties may be virus related, but nucellar clones and reciprocally budded trees also decline. Furthermore, a sweet orange or grapefruit interstock eliminated the 'Eureka' decline problem, suggesting a localized incompatibility (29, 82, 174).

8. WORLD ROOTSTOCK TRENDS

The major Citrus rootstock trend worldwide is diversification primarily in response to the continued spread of tristeza into areas where sour orange is the dominant rootstock. Because of its susceptibility to this virus, it can no longer be economically used in certain areas. Sour orange was abandoned in Brazil and California, where it was replaced by other stocks (Table 2). Attempts to use sour orange in South Africa and Australia were unsuccessful because of tristeza decline, although the cause was unknown at the time and was thought to be an incompatibility. Rough lemon became the general-purpose rootstock of South Africa and was common in Australia until *Phytophthora* problems led to substitution by trifoliate orange in some areas of the country. In both countries, sweet orange and mandarins stocks are used and citranges are being evaluated.

Sour orange has its greatest stronghold in the Mediterranean region, where it is the main rootstock in virtually every country of this area and the Middle East. Tristeza has been observed in Egypt and may be present in other countries of this region; however, the virus is currently a limiting factor only in parts of Spain and Israel, where rootstocks changes are occurring (Table 2). Palestine sweet lime is a commercial rootstock limited to use in sandy soils in Cyprus, Egypt, and Israel. Volkamer lemon has promise as a rootstock for lemons, Mexican lime, and as an inarch, along with Swingle citrumelo, for 'Shamouti' trees in Israel (124, 175, 191).

Caribbean countries and Central America are also relatively tristeza free. Recently, the continued use of sour orange in these areas has been threatened by the appearance of tristeza in Venezuela and Colombia (206). Venezuelan growers are changing to Cleopatra and Volkamer lemon, while in Colombia rough lemon is replacing sour orange (206).

TABLE 2. Current Status of Commercial Rootstocks in Ten of the Leading Citrus-Producing Countries[a]

Country	Rootstock History
United States	
Florida	Rough lemon and sour orange have been widely used. About 1960, Carrizo began to replace rough lemon because of blight. Rough lemon is no longer planted. About 10% of nursery trees are propagated on Cleopatra, which is important as a stock for mandarin scions. Swingle is being evaluated for a broad range of scions.
California	Sweet orange and sour orange were common until tristeza was discovered. Troyer has gradually replaced sour orange beginning in 1957. Alemow is used for lemons. Trifoliate orange and rough lemon are minor stocks. Swingle is of interest as a stock for Lisbon lemons.
Texas	Sour orange is used almost exclusively because of its superior cold tolerance.
Arizona	Rough lemon is the major stock, with continued interest in alemow, sour orange, Troyer, and Volkamer lemon.
Brazil	In the main citrus area, São Paulo, 95% of the trees are budded to Rangpur. Previously, Rangpur and Caipira sweet orange were used until exocortis eliminated the former. Sour orange replaced Rangpur but was eventually eliminated by tristeza. The use of nucellar budwood allowed the reestablishment of Rangpur. Cleopatra, sweet orange, Volkamer lemon, trifoliate orange, and rough lemon are minor stocks, although they are important in certain regions outside of São Paulo. A decline is affecting Rangpur, so interest has increased in Cleopatra and Sunki mandarin and sweet orange.
Japan	Trifoliate orange is the primary stock for satsuma mandarins. Yuzu is considerably less important except as an inarch for trees on trifoliate orange in order to maintain their vigor.
Spain	Nearly all citrus was grown on sour orange until tristeza was found in some areas. The Ministry of Agriculture now licenses nurseries and has prohibited the selling of trees on sour orange; however, this stock is still used in private nurseries. Troyer has slowly replaced sour orange since about 1972. About 50% of nursery trees are grown on Troyer, 20% on Carrizo, and 20% on Cleopatra. Sour orange and alemow are popular for lemons.
Italy	Sour orange is dominant for all varieties, but Volkamer lemon has performed well as a stock for lemons.
Mexico	Tristeza is not present. Sour orange is the major stock. Carrizo and Troyer citrange have been introduced, as well as alemow, as a stock for the Mexican lime.
Israel	Palestine sweet lime and sour orange have been the traditional rootstocks. Sweet lime was used for 'Shamouti' sweet orange and later inarched with sour orange because of xyloporosis. Tristeza is now

TABLE 2. *(continued)*

Country	Rootstock History
	present, and sour orange along with sweet lime are being replaced by Volkamer lemon, Troyer, and, to a lesser extent, Cleopatra and trifoliate orange. Swingle citrumelo is under grower trial.
Argentina	Tristeza is widespread and eliminated sour orange during the 1940s. Palestine sweet lime was substituted, but xyloporosis prevented its continued use. Trifoliate orange is common today in eastern Argentina, but trees on this stock are declining and being replaced by trees on rough lemon. Sweet orange, Rangpur, and Cleopatra are minor stocks. In northwestern areas, Rangpur and Cleopatra are the dominant rootstocks. Sour orange and Volkamer lemon are used for lemons.
India	Rough lemon, Rangpur lime, and Palestine sweet lime may have originated in India. Rootstock use is relatively new in India, but each of these has attained some commercial importance. Rough lemon has been the principal rootstock since about 1965. Cleopatra, Kharna Khatta, and trifoliate orange have some regional significance.
China	There are many provinces within China where citrus is grown and there is an equally diverse range of rootstocks. Large-flowered and large-leafed strains of trifoliate orange are used for satsumas and kumquats, and *C. sunki* and other mandarins for satsumas and sweet oranges. Citranges are new introductions. Sour orange is common in certain coastal areas. Yuzu, Rangpur lime, pummelo, and *C. ichangensis* are used to a limited extent.

^aInformation compiled from a survey conducted by the author in August 1984.

Diversification is occurring where tristeza restricts or threatens the use of sour orange because no one rootstock can adequately replace it. The only exception has been Rangpur lime in Brazil; but even within this 600,000-ha industry, disease is once again stimulating an interest in other rootstocks (206).

Carrizo and Troyer citrange and Swingle citrumelo have attracted universal interest as promising rootstocks to replace sour orange. In the United States, Corsica, and Spain, the citranges are commercially prominent, while in other areas these three rootstocks are only experimental.

Trees on the citranges and Swingle citrumelo are more vigorous and productive than those on their common parent, trifoliate orange, probably the second most important rootstock worldwide. In addition to being a major stock for the mandarin industries of Japan and China, trifoliate orange is also used in Uruguay, Australia, Argentina, New Zealand, and several Mediterranean countries because of its adaptation to heavy soils and *Phytophthora* resistance. Trees on citrange and citrumelo rootstocks may prove to be satisfactory under similar conditions; thus their greater yield potential may result in new plantings on these stocks rather than on trifoliate orange.

9. FUTURE ROOTSTOCK REQUIREMENTS

The forces that have tended to provide direction to rootstock research and development are basically unchanged since the husbandry of Citrus began. Factors such as disease, parasites, cold, and other environmental problems are still at work. Improved rootstocks are needed to reduce injury from cold, as illustrated by recent major freezes in Texas and Florida. *Phytophthora* and viruses seem to be present wherever Citrus is grown. The burrowing and citrus nematode are serious pests and resistance to both has not been combined into one rootstock.

Citriculture is a dynamic industry. New diseases appear, such as blight, that are rootstock related. Industry expansion in areas throughout the world has resulted in Citrus plantings being established on soils less suitable for local rootstocks. Urbanization often relegates agriculture to poorer land with less water of poorer quality. These events accentuate existing rootstock deficiencies regarding salinity and alkalinity tolerance (56). Most of these needs represent immediate, long-standing requirements. Longer-range goals include size-controlling rootstocks for higher-density plantings and those that also induce scion precocity (159).

The conventional routes to improved rootstocks have been selection and breeding. Selection has been in progress for many years, while breeding specifically for rootstocks is a relatively new activity that has been confined largely to Florida and California (6, 98). Neither approach has yielded many new rootstocks. Cultivars such as Carrizo citrange and Swingle citrumelo arose coincidentally from a scion breeding program (6).

Breeding efforts, which are somewhat less than what the international industry would justify, are hindered by apomixis, long reproductive cycles (typically 1 to 10 years in length), heterozygosity, and the unavailability of germplasm (6). As a result, the combined effects of these factors diminish the probabilities for producing sexual seedlings with the desired recombinations of genetic traits in a single plant; and the result of hybridization cannot be consistently predicted. Moreover, hybrids of *Citrus* with other genera often do not produce viable seed or the seedlings do not survive (6, 7). Nevertheless, the potential for genetic improvement through hybridization is considerable for *Citrus* as compared with other tree fruit genera. Intergeneric gene exchange is relatively rare among cultivated plants, yet success has been achieved in hybridizing several species of *Citrus* with *Eremocitrus*, *Poncirus*, *Fortunella*, and *Microcitrus* (6, 7). Some of the resulting bi- and trigeneric hybrids have already shown promise for improvements in scion cold hardiness (7). A similar breeding approach may also allow desirable rootstock genetic traits in the *Citrus* relatives to be combined with *Citrus* (22, 25, 28).

Rootstock development is a slow and tedious process, but the need for a new rootstock can arise quickly. Therefore the process must be a continuous one in order to avoid placing in jeopardy the considerable importance of rootstocks to citriculture.

INTERNATIONAL CATALOG OF ROOTSTOCK BREEDING PROGRAMS

France
Station de Recherches
Agronomiques de Corse
Cervione, Corsica
Mr. Lelievre

Israel
Agricultural Research Organization
Volcani Center
P.O. Box 6
Bet Dagan, Israel
Pinhas Spiegel-Roy

United States
Department of Botany and Plant Sciences
University of California
Riverside, CA 92521
Mikeal Roose

USDA/ARS
Horticultural Research Laboratory
2120 Camden Road
Orlando, FL 32803
Donald Hutchison

University of Florida, IFAS
Citrus Research and Education Center
700 Experiment Station Road
Lake Alfred, FL 33850
Fred Gmitter, Jude Grosser

REFERENCES

1. Anonymous (1972). Rootstocks for desert citrus, *Citrograph*, **58**(4), 124–125.
2. Anonymous (1973). Swingle citrumelo: An ultraresistant rootstock, *Citrograph*, **59**(11), 387, 388, 391.

3. Anonymous (1982). Dwarf citrus moves from theory to field, *Citrograph*, **68**(11), 259–261.

4. Baines, R. C., W. P. Bitters, and O. F. Clarke (1960). Susceptibility of some species and varieties of citrus and some other rutaceous plants to the citrus nematode, *Plant Dis. Reptr.*, **44**(4), 281–285.

5. Baines, R. C., J. W. Cameron, and R. K. Soost (1974). Four biotypes of *Tylenchulus semipenetrans* in California identified, and their importance in the development of resistant citrus rootstocks, *J. Nematol.*, **6**(2), 63–66.

6. Barrett, H. C. (1977). Intergeneric hybridization of citrus and other genera in citrus cultivar improvement, *1977 Proc. Int. Soc. Citriculture*, **2**, 586–589.

7. Barrett, H. C. (1981). Breeding cold-hardy citrus scion cultivars, *1981 Proc. Int. Soc. Citriculture*, **1**, 61–66.

8. Barrett, H. C., and D. J. Hutchison (1978). Spontaneous tetraploidy in apomitic seedlings of *Citrus*, *Econ. Bot.*, **32**, 27–45.

9. Barrett, H. C., and A. M. Rhodes (1976). A numerical taxonomic study of affinity relationships in cultivated *Citrus* and its close relatives, *System Bot.*, **1**(2), 105–136.

10. Batchelor, L. D., and W. P. Bitters (1951). Two promising rootstocks for citrus in California, *Calif. Citrograph*, **37**(10), 390–391, 409–410.

11. Batchelor, L. D., and M. B. Rounds (1948). "Choice of rootstocks," in *The Citrus Industry*, Vol. II, L. D. Batchelor and H. J. Webber, Eds., Univ. of Calif. Press, Berkeley, pp. 169–222.

12. Bevington, K. B., and P. E. Bacon (1977). Effect of rootstocks on the response of navel orange trees to dwarfing inoculations, *1977 Proc. Int. Soc. Citriculture*, **2**, 567–570.

13. Bevington, K. B., and J. H. Duncan (1977). The influence of rootstock on the performance of Ellendale tangor, *1977 Proc. Int. Soc. Citriculture*, **1**, 124–126.

14. Bitters, W. P. (1949). Dwarfing citrus rootstocks, *Calif. Citrograph*, **34**(12), 516–517, 539–543.

15. Bitters, W. P. (1957). Rootstocks from S to T and XYZ, *Calif. Citrograph*, **43**(7), 258.

16. Bitters, W. P. (1959). Citrus rootstocks for desert area, *Calif. Citrograph*, **45**(11), 349, 363, 364.

17. Bitters, W. P. (1959). "Rootstocks in relation to control of tristeza," in *Citrus Virus Diseases*, J. M. Wallace, Ed., Univ. of Calif. Press, Berkeley, pp. 203–207.

18. Bitters, W. P. (1963). Citus rootstocks and nursery practices in Japan, *Calif. Citrograph*, **49**(5), 205–210.

19. Bitters, W. P. (1967). Valencia orange rootstock trial at South Coast field station, *Calif. Citrograph*, **53**(5), 163, 172–174.

20. Bitters, W. P. (1972). Reaction of some new citrus hybrids and citrus introductions as rootstocks to inoculation with tristeza virus in Calif., *Proc. Fifth Conf. Int. Org. Citrus Virol.*, pp. 112–120.

21. Bitters, W. P. (1973). "World citrus rootstock situation," in *Proc. First Int. Citrus Short Course*, L. K. Jackson, A. H. Krezdorn, and J. Soule, Eds., Gainesville, FL, pp. 1–12.

22. Bitters, W. P. (1973). "Citrus rootstock improvement," in *Proc. First Int. Citrus Short Course*, L. K. Jackson, A. H. Krezdorn and J. Soule, Eds., Gainesville, FL, pp. 92–96.

23. Bitters, W. P., and L. D. Batchelor (1951). Effect of rootstocks on the size of orange fruits, *Proc. Amer. Soc. Hort. Sci.*, **57**, 133–141.

24. Bitters, W. P., J. A. Brusca, and D. A. Cole (1963). The search for new citrus rootstocks, *Calif. Citrograph*, **49**, 443–448.

25. Bitters, W. P., D. A. Cole, and J. A. Brusca (1969). The citrus relatives as citrus rootstocks, *Proc. First Int. Citrus Symp.*, **1**, 411–415.

26. Bitters, W. P., D. A. Cole, and C. D. McCarty (1972). Citrus rootstocks from the papeda group, *Citrograph*, **58**(12), 419, 420, 438, 439.

27. Bitters, W. P., D. A. Cole, and C. D. McCarty (1977). Citrus relatives are not irrelevant as dwarfing stocks or interstocks for citrus, *1977 Proc. Int. Soc. Citriculture*, **2**, 561–567.

28. Bitters, W. P., D. A. Cole, and C. D. McCarty (1978). Facts about dwarf citrus trees, *Citrograph*, **64**(3), 54–56.

29. Bitters, W. P., D. A. Cole, and C. D. McCarty (1981). Effect of height and length of reciprocal interstock insertion on yield and tree size of 'Valencia' oranges, *1981 Proc. Int. Soc. Citriculture*, **1**, 110–113.

30. Bitters, W. P., C. D. McCarty, and D. A. Cole (1973). An evaluation of some new citrus rootstocks with respect to their reaction to tristeza virus, *1973 Proc. First World Congr. Citriculture*, **2**, 557–563.

31. Bitters, W. P., C. D. McCarty, and D. A. Cole (1973). An evaluation of trifoliate orange selections as rootstocks for Washington navel and Valencia orange, *1973 Proc. First World Congr. Citriculture*, **2**, 127–131.

32. Bitters, W. P., and E. R. Parker (1953). *Quick Decline of Citrus as Influenced by Top-root Relationships*, Calif. Agric. Exp. Sta. Bull. 733.

33. Bowman, F. T. (1956). *Citrus Growing in Australia*, Angus and Robertson, Sydney.

34. Bridges, G. D., and C. O. Youtsey (1968). Further studies of the bud-union abnormality of rough lemon rootstocks with sweet orange scions, *Proc. Fourth Conf. Int. Org. Citrus Virol.*, pp. 236–239.

35. Bridges, G. D., C. O. Youtsey, and R. R. Nixon, Jr. (1965). Observations indicating psorosis transmission by seed of Carrizo citrange, *Citrus Industry*, **46**(12), 5, 6, 14.

36. Burke, J. H. (1967). "The commercial citrus regions of the world," in *The Citrus Industry*, Vol. I, W. Reuther, H. J. Webber, and L. D. Batchelor, Eds., Univ. of Calif. Press, Berkeley, pp. 40–89.

37. Calavan, E. C., R. M. Pratt, B. W. Lee, and J. P. Hill (1967). Tristeza related to decline of orange trees on citrange rootstock, *Calif. Citrograph*, **53**(3), 75, 84–88, 90.

38. Campbell, C. W. (1972). Rootstock effects on tree size and yield of 'Tahiti' lime (*Citrus latifolia* Tanaka), *Proc. Fla. State Hort. Soc.*, **85**, 332–334.

39. Campiglia, H. G., C. M. Silveira, and A. A. Salibe (1976). Psorosis transmission through seeds of trifoliate orange, *Proc. Seventh Conf. Int. Org. Citrus Virol.*, pp. 132–133.

40. Carpenter, J. B., R. M. Burns, and J. R. Furr (1975). *Phytophthora*-tolerant rootstocks for lemons, *Plant Dis. Reptr.*, **59**(1), 54–56.

41. Carpenter, J. B., R. M. Burns, and R. F. Sedlacek (1980). *Phytophthora* resistant rootstocks for lisbon lemons in California, *Citrograph*, **66**(12), 287–288, 291–292.

42. Carpenter, J. B., R. M. Burns, and R. E. Sedlacek (1981). Performance of rootstocks inoculated with virus, *Citrograph*, **67**(5), 101–105.

43. Carpenter, J. B., and J. R. Furr (1962). Evaluation of tolerance to root rot caused by *Phytophthora* parasitica in seedlings of citrus and related genera, *Phytopathology*, **52**(12), 1277–1285.

44. Castle, W. S. (1976). Field performance of several common citrus scions on Milam rootstock, *Proc. Fla. State Hort. Soc.*, **89**, 11–14.

45. Castle, W. S. (1978). Citrus root systems: Their structure, function, growth and relationship to tree performance, *1978 Proc. Int. Soc. Citriculture*, **1**, 62–69.

46. Castle, W. S. (1980). Citrus rootstocks for tree size control and higher density plantings in Florida, *Proc. Fla. State Hort. Soc.*, **93**, 24–27.

47. Castle, W. S. (1982). Commercial citrus rootstocks in the United States, *Fruit Varieties J.*, **36**(3), 74–79.

48. Castle, W. S. (1983). Growth, yield and cold hardiness of seven-year-old 'Bearss' lemon trees on twenty-seven rootstocks, *Proc. Fla. State Hort. Soc.*, **96**, 23–25.

49. Castle, W. S., and J. J. Ferguson (1982). Current status of greenhouse and container production of citrus nursery trees in Florida, *Proc. Fla. State Hort. Soc.*, **95**, 42–46.

50. Castle, W. S., and A. H. Krezdorn (1977). Soil water use and apparent root efficiencies of citrus trees on four rootstocks, *J. Amer. Soc. Hort. Sci.*, **102**(4), 403–406.

51. Castle, W. S., and R. L. Phillips (1977). Potentially dwarfing rootstocks for Florida citrus, *1977 Proc. Int. Citriculture*, **2**, 558–561.

52. Castle, W. S., and R. L. Phillips (1980). Performance of 'Marsh' grapefruit and 'Valencia' orange trees on eighteen rootstocks in a closely spaced planting, *J. Amer. Soc. Hort. Sci.*, **105**(4), 496–499.

53. Castle, W. S., and D. P. H. Tucker (1978). Susceptibility of citrus nursery trees to herbicides as influenced by rootstock and scion cultivar, *HortScience*, **13**(6), 692–693.

54. Castle, W. S., and C. O. Youtsey (1977). Root system characteristics of citrus nursery trees, *Proc. Fla. State Hort. Soc.*, **90**, 39–44.

55. Castle, W. S., and C. O Youtsey (1980). Trends in Florida citrus rootstocks, *Citrus Industry*, **61**(7), 10–14.

56. Chapman, H. D. (1968). "The mineral nutrition of citrus," in *The Citrus Industry*, Vol. II, W. Reuther, L. D. Batchelor, and H. J. Webber, Eds., Univ. of Calif. Press, Berkeley, pp. 127–289.

57. Chapot, H. (1965). Le *Citrus volkameriana* Pasquale, *Al Awamia*, **14**, 29–45.

58. Chapot, H. (1975). "The citrus plant," in *Citrus*, Ernst Hafliger, Ed., Technical Monograph No. 4, Ciba-Geigy Agrochemicals, Basel, Switzerland, pp. 6–13.

59. Cohen, M. (1968). Exocortis virus as a possible factor in producing dwarf citrus trees, *Proc. Fla. State Hort. Soc.*, **81**, 115–119.

60. Cohen, M. (1970). Rangpur lime as a citrus rootstock in Florida, *Proc. Fla. State Hort. Soc.*, **83**, 78–84.

61. Cohen, M. (1972). Sweet orange rootstock in experimental trials on the east coast of Florida, *Proc. Fla. State Hort. Soc.*, **85**, 61–65.

62. Cohen, M., and H. J. Reitz (1963). Rootstocks for Valencia orange and Ruby Red grapefruit: Results of a trial initiated at Fort Pierce in 1950 on two soil types, *Proc. Fla. State Hort. Soc.*, **76**, 29–34.

63. Cohen, M., and H. K. Wutscher (1977). Diagnosis of trees with citrus blight (YTD), *1977 Proc. Int. Soc. Citriculture*, **3**, 884–446.

64. Cole, D., and C. D. McCarty (1981). 'Flying Dragon': A potential dwarfing rootstock, *Citrograph*, **67**(4), 71–72.

65. Constantin, R. J., R. T. Brown, and S. Thibodeaux (1979). Performance of 'Owari' satsuma mandarin and 'Washington' navel orange on trifoliate orange and citrange in Louisiana, *J. Amer. Soc. Hort. Sci.*, **104**(1), 120–123.

66. Cook, J. A., G. E. Horanic, and F. E. Gardner (1952). Citrus rootstock trials. A nine-year progress on seven rootstocks on Lakeland fine sand, *Proc. Fla. State Hort. Soc.*, **65**, 69–77.

67. Cooper, W. C. (1952). Influence of rootstock on injury and recovery of young citrus trees exposed to the freezes of 1950–51 in the Rio Grande Valley, *Proc. Rio Grande Valley Hort. Soc.*, **6**, 16–24.

68. Cooper, W. C. (1982). *In Search of the Golden Apple*, Vantage, New York.

69. Cooper, W. C., and E. O. Olson (1951). Influence of rootstock on chlorosis of young Red Blush grapefruit trees, *Proc. Amer. Soc. Hort. Sci.*, **51**, 125–132.

70. Cooper, W. C., E. O. Olson, N. P. Maxwell, and G. Otey (1956). Review of studies of adaptability of citrus varieties as rootstocks for grapefruit in Texas, *J. Rio Grande Valley Hort. Soc.*, **10**, 6–19.

71. Cooper, W. C., E. O. Olson, N. Maxwell, and A. Shull (1957). Orchard performance of young trees of red grapefruit on various rootstocks in Texas, *Proc. Amer. Soc. Hort. Sci.*, **70**, 213–222.

72. da Cunha Sobrinho, A. P., O. S. Passos, W. S. Soares Filho, and Y. S. Coelho (1981). Behavior of citrus rootstocks under tropical conditions, *1981 Proc. Int. Soc. Citriculture*, **1**, 123–126.

73. delValle, N., O. Herrera, and A. Rios (1981). The influence of rootstocks on the performance of 'Valencia' orange under tropical conditions, *1981 Proc. Int. Soc. Citriculture*, **1**, 134–137.

74. Economides, C. V. (1976). Growth and productivity of 'Washington' navel orange trees on six rootstocks in Cyprus, *Hort. Res.*, **16**, 83–88.

75. Economides, C. V. (1976). Performance of Marsh seedless grapefruit on six rootstocks in Cyprus, *J. Hort. Sci.*, **51**, 393–400.

76. Economides, C. V. (1977). The influence of rootstocks on tree growth, yield and fruit quality of 'Valencia' oranges in Cyprus, *J. Hort. Sci.*, **52**, 29–36.

77. El-Zeftawi, B. M., and I. R. Thornton (1975). Effects of rootstocks and fruit stripping on alternate bearing of Valencia orange trees, *J. Hort. Sci.*, **50**, 219–226.

78. Ford, H. W. (1959). Growth and root distribution of orange trees on two different rootstocks as influenced by depth to subsoil clay, *Proc. Amer. Soc. Hort. Sci.*, **74**, 313–321.

79. Ford, H. W., and W. A. Feder (1960). *Citropsis gilletiana*, a citrus relative resistant to the burrowing nematode in laboratory tests, *Proc. Fla. State Hort. Soc.*, **73**, 60−64.

80. Ford, H. W., and W. A. Feder (1964). *Three Citrus Rootstocks Recommended for Trial in Spreading Decline Areas*, Univ. of Fla. Agric. Exp. Sta. Cir. S-151.

81. Ford, H. W., and W. A. Feder (1969). Development and use of citrus rootstocks resistant to the burrowing nematode, *Radopholus similis*, *Proc. First Int. Citrus Symp.*, **2**, 941−948.

82. Fraser, L. R., and P. Broadbent (1979). *Virus and Related Diseases of Citrus in New South Wales*, Surrey Beatty and Sons, N.S.W., Australia.

83. Fung-Kon-Sang, W. E. (1977). Promising citrus rootstocks and scion cultivars in Suriname, *1977 Proc. Int. Soc. Citriculture*, **2**, 648−650.

84. Furr, J. R., and C. L. Ream (1969). Breeding citrus rootstocks for salt tolerance, *Proc. First Int. Citrus Symp.*, **1**, 373−380.

85. Gardner, F. E. (1968). The failure of rough lemon and sour orange interstocks to influence tree growth, yields and fruit quality of sweet orange varieties, *Proc. Fla. State Hort. Soc.*, **81**, 90−94.

86. Gardner, F. E., and G. E. Horanic (1958). Influence of various rootstocks on the cold resistance of the scion variety, *Proc. Fla. State Hort. Soc.*, **71**, 81−86.

87. Gardner, F. E., and G. E. Horanic (1961). A comparative evaluation of rootstocks for Valencia and Parson Brown oranges on Lakeland fine sand, *Proc. Fla. State Hort. Soc.*, **74**, 123−127.

88. Gardner, F. E., and G. E. Horanic (1963). Cold tolerance and vigor of young citrus trees on various rootstocks, *Proc. Fla. State Hort. Soc.*, **76**, 105−110.

89. Gardner, F. E., and G. E. Horanic (1968). Growth, yield, and fruit quality of Marsh grapefruit on various rootstocks on the Florida east coast—a preliminary report, *Proc. Fla. State Hort. Soc.,* 79: 109−114.

90. Gardner, F. E. and G. E. Horanic. 1967. *Poncirus trifoliata* and some of its hybrids as rootstocks for Valencia sweet orange, *Proc. Fla. State Hort. Soc.*, **80**, 85−88.

91. Gardner, F. E., D. J. Hutchison, G. E. Horanic, and P. C. Hutchins (1967). Growth and productivity of virus-infected Valencia orange trees on twenty-five rootstocks, *Proc. Fla. State Hort. Soc.*, **80**, 89−92.

92. Grant, T. J., S. Moreira, and A. A. Salibe (1961). Citrus variety reaction to tristeza virus in Brazil when used in various rootstock and scion combinations, *Plant Dis. Reptr.*, **45**(6), 416−421.

93. Grieve, A. M., and R. R. Walker (1983). Uptake and distribution of chloride, sodium and potassium ions in salt-treated citrus plants, *Aust. J. Agric. Res.*, **34**, 133−143.

94. Grimm, G. R., and S. M. Garnsey (1968). Foot rot and tristeza tolerance of Smooth Seville orange from two sources, *Proc. Fla. State Hort. Soc.*, **81**, 84−90.

95. Grimm, G. R., and D. J. Hutchison (1977). Evaluation of *Citrus* spp., relatives, and hybrids for resistance to *Phytophthora parasitica* Dastur, *1977 Proc. Int. Soc. Citriculture*, **3**, 863−865.

96. Halsey, D. D. (1974). *C. volkameriana* looks like an alternate in desert citrus, *Citrograph*, **60**(12), 427, 441, 442.

97. Hearn, C. J., and D. J. Hutchison (1977). The performance of 'Robinson' and 'Page' citrus hybrids on 10 rootstocks, *Proc. Fla. State Hort. Soc.*, **90**, 44−47.

98. Hearn, C. J., D. J. Hutchison, and H. C. Barrett (1974). Breeding citrus root-stocks, *HortScience*, **9**(4), 357–358.

99. Hilgeman, R. H. (1975). Yield and tree growth of seven 'Valencia' orange type cultivars on four rootstocks, *HortScience*, **10**(1), 66–68.

100. Hilgeman, R. H., D. R. Rodney, J. A. Dunlap, and T. A. Hales (1966). Root-stock evaluation for lemons on two soil types in Arizona, *Proc. Amer. Soc. Hort. Sci.*, **88**, 280–290.

101. Hodgson, R. W. (1967). "Horticultural varieties of citrus," in *The Citrus Industry*, Vol. I, W. Reuther, H. J. Webber, and L. D. Batchelor, Eds., Univ. of Calif. Press, Berkeley, pp. 431–591.

102. Holzhausen, L. C., W. A. Eshuys, and P. J. Muller (1977). *Citrus reticulata*, and a few other species as rootstocks for the nucellar 'Palmer' navel orange, *1977 Proc. Int. Soc. Citriculture*, **2**, 549–557.

103. Horanic, G. E., and F. E. Gardner (1959). Relative wilting of orange trees on various rootstocks, *Proc. Fla. State Hort. Soc.*, **72**, 77–79.

104. Hutchison, D. J. (1974). Swingle citrumelo—a promising rootstock hybrid, *Proc. Fla. State Hort. Soc.*, **87**, 89–91.

105. Hutchison, D. J. (1977). Influence of rootstock on the performance of 'Valencia' sweet orange, *1977 Proc. Int. Soc. Citriculture*, **2**, 523–525.

106. Hutchison, D. J., and G. R. Grimm (1972). Variation in *Phytophthora* resistance of Florida rough lemon and sour orange clones, *Proc. Fla. State Hort. Soc.*, **85**, 38–39.

107. Hutchison, D. J., and C. J. Hearn (1977). The performance of 'Nova' and 'Orlando' tangelos on 10 rootstocks, *Proc. Fla. State Hort. Soc.*, **90**, 47–49.

108. Hutchison, D. J., and J. H. O'Bannon (1972). Evaluating the reaction of citrus selections to Tylenchulus semipenetrans, *Plant Dis. Reptr.*, **56**(9), 747–751.

109. Hutchison, D. J., J. H. O'Bannon, and G. R. Grimm (1972). Reaction of selected citrus rootstocks to foot rot, burrowing and citrus nematodes, *Proc. Fla. State Hort. Soc.*, **85**, 39–43.

110. Iwasaki, T., M. Nishiura, T. Shichijo, and N. Okudai (1961). Double working of satsuma orange. II. Effects of interstocks on tree growth, fruit quality and yield, *J. Jap. Soc. Hort. Sci.*, **30**, 63–72.

111. Klotz, L. J., W. P. Bitters, R. C. Baines, P. W. Moore, T. A. DeWolfe, and R. H. Small (1971). Field testing for resistance to fungi and citrus nematodes, *Calif. Citrograph*, **57**(11), 395–396, 411–413.

112. Klotz, L. J., W. P. Bitters, T. A. DeWolfe, and M. J. Garber (1967). Orchard tests of citrus rootstocks for resistance to *Phytophthora*, *Calif. Citrograph*, **53**(2), 38, 55.

113. Krezdorn, A. H. (1977). Influence of rootstocks on mandarin cultivars, *1977 Proc. Int. Soc. Citriculture*, **2**, 513–518.

114. Krezdorn, A. H. (1977). Citrus in Cuba: The great leap forward, *Citrograph*, **63**(1), 12–15.

115. Krezdorn, A. H. (1979). Selecting citrus rootstocks for fruit size and quality, *Fla. Grower and Rancher*, **72**(4), 16–17.

116. Krezdorn, A. H. (1979). Rootstocks: How they affect yield, *Fla. Grower and Rancher*, **72**(5), 18–19.

117. Krezdorn, A. H. (1979). Citrus rootstocks: Tolerance of environmental and disease problems, *Fla. Grower and Rancher*, **72**(6), 14, 16.

118. Krezdorn, A. H. (1979). Rootstocks: Their relation to virus disease and blight, *Fla. Grower and Rancher*, **72**(7), 25–26.

119. Krezdorn, A. H., and W. S. Castle (1971). Sweet lime, its performance and potential as a rootstock in Florida, *Proc. Fla. State Hort. Soc.*, **84**, 32–38.

120. Krezdorn, A. H., and W. J. Phillips (1970). The influence of rootstocks on tree growth, fruiting and fruit quality of 'Orlando' tangelos, *Proc. Fla. State Hort. Soc.*, **83**, 110–116.

121. Labanauskas, C. K., and W. P. Bitters (1974). The influence of rootstocks and interstocks on the nutrient concentrations in 'Valencia' orange leaves, *J. Amer. Soc. Hort. Sci.*, **99**(1), 32–33.

122. Lawrence, F. P., and G. D. Bridges (1973). *Rootstocks for Citrus in Florida*, Univ. Fla. Coop. Ext. Ser. Cir. 394.

123. Levy, Y., and K. Mendel (1982). Tree development, yield, and fruit quality of two orange cultivars on three rootstocks in the Negev region of Israel, *J. Amer. Soc. Hort. Sci.*, **107**(6), 1001–1004.

124. Levy, Y., and A. Shaked (1980). Tree development, yield and leaf nutrient levels of old-clone lemon trees on eight rootstocks, *Exp. Agric.*, **13**, 49–55.

125. Long, K., L. Frazer, P. Bacon, and P. Broadbent (1977). The Benton citrange: A promising *Phytophthora*-resistant rootstock for citrus trees, *1977 Proc. Int. Soc. Citriculture*, **2**, 541–544.

126. McCarty, C. D., W. P. Bitters, and D. A. Cole (1973). Comparisons between Troyer and Carrizo citrange, *Citrograph*, **59**(9), 294–310.

127. McCarty, C. D., W. P. Bitters, and S. D. Van Gundy (1979). Susceptibility of 25 citrus rootstocks to the citrus nematode, *HortScience*, **14**(1), 54–55.

128. McDonald, R. E., and H. K. Wutscher (1974). Rootstocks affect postharvest decay of grapefruit, *HortScience*, **9**(5), 455–456.

129. Marloth, R. H. (1949). Sweet orange as a rootstock for citrus, *Farming in S. Africa*, **24**, 216–220, 261–265.

130. Marloth, R. H. (1957). Rootstocks for Jaffa and Joppa oranges, *J. Hort. Sci.*, **32**, 162–171.

131. Marloth, R. H. (1958). Rootstocks for grapefruit, *S. African J. Agric. Sci.*, **1**(1), 43–65.

132. Marloth, R. H., and W. J. Basson (1960). Empress mandarin as a rootstock for citrus, *J. Hort. Sci.*, **35**, 282–292.

133. Marti Fabregat, F. (1975). "The tristeza disease of citrus," in *Citrus*, E. Hafliger, Ed., Technical Monograph 4, Ciba-Geigy Agrochemicals, Basel, Switzerland, pp. 51–54.

134. Maxwell, N. P., and H. K. Wutscher (1976). Yield, fruit size, and chlorosis of grapefruit on 10 rootstocks, *HortScience*, **11**(5), 496–498.

135. Mendel, K. (1956). Rootstock–scion relationships in Shamouti trees on light soil, *Israel J. Agric. Res.*, **6**, 35–60.

136. Mendel, K. (1971). 'Poorman': A promising rootstock for Israeli citrus, *HortScience*, **6**(1), 45, 46.

137. Mendel, K., and A. Shaked (1975). *Performance of Valencia Orange Trees Budded on C. volkameriana*, Progress report I, Volcani Center, Bet Dagan, Israel.

138. Mielke, E. A., and S. Issa (1976). Influence of interstocks on grapefruit quality, *Citrograph*, **62**(3), 73.

139. Moreira, S. (1968). Xyloporosis in Brazil, *Proc. Fourth Conf. Int. Org. Citrus Virol.*, pp. 89–91.

140. Moreira, S., T. J. Grant, A. A. Salibe, and C. Roessing (1965). Tristeza tolerant rootstocks—their behavior after twelve years in orchard, *Proc. Third Conf. Int. Org. Citrus Virol.*, pp. 18–24.

141. Moreira, S., and C. Roessing (1965). Behavior of 77 tristeza tolerant rootstocks with old and nucellar clones of Barao orange scions, *Proc. Third Conf. Int. Org. Citrus Virol.*, pp. 299–301.

142. Moreira, S., and A. A. Salibe (1969). The contribution of research for the progressive changes in citrus rootstocks in South America, *Proc. First Int. Citrus Symp.*, **1**, 351–357.

143. Mortensen, E., and C. R. Riecker (1942). Seed production and seedling yields of some citrus varieties of possible value for rootstock purposes, *Proc. Amer. Soc. Hort. Sci.*, **41**, 145–148.

144. Nemec, S. (1978). Response of six citrus rootstocks to three species of *Glomus*, a mycorrhizal fungus, *Proc. Fla. State Hort. Soc.*, **91**, 10–14.

145. Newcomb, D. A. (1973). "Citrus nursery operations," in *Proc. First Int. Citrus Short Course*, L. K. Jackson, A. H. Krezdorn, and J. Soule, Eds., Gainesville, FL, pp. 152–160.

146. Newcomb, D. A. (1978). Selection of rootstocks for salinity and disease resistance, *1978 Proc. Int. Soc. Citriculture*, **1**, 117–120.

147. O'Bannon, J. H., V. Chew, and A. T. Tomerlin (1977). Comparison of five populations of *Tylenchulus semipenetrans* to *Citrus*, *Poncirus* and their hybrids, *J. Nematol.*, **9**(2), 162–165.

148. O'Bannon, J. H., and H. W. Ford (1977). Resistance in citrus rootstocks to *Radopholus similis* and *Tylenchulus semipenetrans* (nematoda), *1977 Proc. Int. Soc. Citriculture*, **2**, 544–549.

149. Olson, E. O. (1954). Some bark and bud union disorders of mandarin and mandarin-hybrid rootstocks in Texas citrus plantings, *Proc. Amer. Soc. Hort. Sci.*, **63**, 131–136.

150. Olson, E. O. (1958). Bud-union crease, a citrus disorder associated with some Kumquat-hybrid rootstocks and scions, *J. Rio. Grande Valley Hort. Soc.*, **12**, 27–34.

151. Olson, E. O. (1960). Evaluation of rootstocks for Valencia orange trees following inoculation under screenhouse conditions with a severe strain of tristeza virus, *J. Rio Grande Valley Hort. Soc.*, **14**, 29–32.

152. Olson, E. O., W. C. Cooper, and A. V. Shull (1957). Effect of bud-transmitted diseases on size of young Valencia orange trees on various rootstocks, *J. Rio Grande Valley Hort. Soc.*, **11**, 28–33.

153. Olson, E. O., W. C. Cooper, N. Maxwell, and A. V. Shull (1962). Survival, size and yield of xyloporosis- and exocortis-infected old-line red grapefruit trees on 100 rootstocks, *J. Rio Grande Valley Hort. Soc.*, **16**, 44–51.

154. Olson, E. O., and A. V. Shull (1962). Size and yield of 12-year-old Valencia orange trees on various rootstocks in presence or absence of exocortis and xyloporosis viruses, *J. Rio Grande Valley Hort. Soc.*, **16**, 40–43.

155. Olson, E. O., and A. V. Shull (1966). A Meyer lemon rootstock trial: Scion-rooting, tree growth, yield, and tree survival after a severe freeze, *J. Rio Grande Valley Hort. Soc.*, **20**, 102–108.

156. Passos, O. S., and A. P. da Cunha Sobrinho (1981). Citrus rootstocks in Brazil, *1981 Proc. Int. Soc. Citriculture*, **1**, 102–105.

157. Peynado, A., and N. J. Sluis (1979). Chloride and boron tolerance of young 'Ruby Red' grapefruit trees affected by rootstock and irrigation method, *J. Amer. Soc. Hort. Sci.*, **104**(1), 133–136.

158. Peynado, A., and R. Young (1969). Relation of salt tolerance to cold hardiness of 'Redblush' grapefruit and 'Valencia' orange trees on various rootstocks, *Proc. First Int. Citrus Symp.*, **3**, 1793–1802.

159. Phillips, R. L. (1969). Dwarfing rootstocks for citrus, *Proc. First Int. Citrus Symp.*, **1**, 401–406.

160. Phillips, R. L., and W. S. Castle (1977). Evaluation of twelve rootstocks for dwarfing citrus, *J. Amer. Soc. Hort. Sci.*, **102**(5), 526–528.

161. Piringer, A. A., R. J. Downs, and H. A. Borthwick (1961). Effects of photoperiod and kind of supplemental light on the growth of three species of citrus and *Poncirus trifoliata*, *Proc. Amer. Soc. Hort. Sci.*, **77**, 202–210.

162. Ream, C. L., and J. R. Furr (1976). Salt tolerance of some *Citrus* species, relatives and hybrids tested as rootstocks, *J. Amer. Soc. Hort. Soc.*, **101**(3), 265–267.

163. Rouse, R. E., and N. P. Maxwell (1979). Performance of mature nucellar 'Redblush' grapefruit on 22 rootstocks in Texas, *J. Amer. Soc. Hort. Sci.*, **104**(4), 449–451.

164. Rouse, R. E., and H. K. Wutscher (1985). Heavy soil and bud union crease with some clones of red grapefruit limit use of Swingle citrumelo rootstock, *HortScience*, **20**(2), 259–261.

165. Russo, F. (1973). Rootstock experiments for lemons and oranges, *1973 Proc. First World Congr. Citriculture*, **2**, 153–162.

166. Salibe, A. A. (1973). "The tristeza disease," in *Proc. First Int. Citrus Short Course*, L. K. Jackson, A. H. Krezdorn, and J. Soule, Eds., Gainesville, FL, pp. 68–76.

167. Salibe, A. A., and S. Moreira (1973). Performance of eight rootstocks with nucellar Baianinha navel orange scion in a sandy soil, *1973 Proc. First World Congr. Citriculture*, **2**, 149–152.

168. Salibe, A. A., S. Moreira, and O. Rodriguez (1972). Performance of selections of trifoliate orange and trifoliate orange hybrids as rootstocks for citrus in the presence of tristeza virus, *Proc. Fifth Conf. Int. Org. Citrus Virol.*, pp. 124–127.

169. Salibe, A. A., O. Rodriguez, and S. Moreira (1972). Studies on xyloporosis of citrus in Brazil, *Proc. Fifth Conf. Int. Org. Citrus Virol.*, pp. 192–194.

170. Savage, E. M., W. C. Cooper, and R. B. Piper (1946). Root systems of various citrus rootstocks, *Proc. Fla. State Hort. Soc.*, **58**, 44–48.

171. Savage, E. M., and F. E. Gardner (1965). The origin and history of Troyer and Carrizo citranges, *Citrus Industry*, **46**(2), 5–7.

172. Schneider, H. (1982). Lisbon lemons compatible with citrumelo rootstocks, *Citrograph*, **68**(2), 37–38, 48.

173. Schneider, H., R. G. Platt, W. P. Bitters, and R. M. Burns (1977). Diseases and incompatibilities that cause decline in lemons, *Citrograph*, **63**(9), 219–221.

174. Schneider, H., and N. J. Sakovich (1984). Compatible rootstocks for lemon trees, *Citrograph*, **70**(1), 17, 19–21.

175. Shaked, A., A. Cohen, M. Hamo, and D. Hasdai (1981). Recent citrus inarching experiments in Israel, *1981 Proc. Int. Soc. Citriculture*, **1**, 144–145.

176. Shannon, L. M., E. F. Frolich, and S. H. Cameron (1960). Characteristics of *Poncirus trifoliata* selections, *J. Amer. Soc. Hort. Sci.*, **76**, 163–169.

177. Singh, M., and N. R. Achhireddy (1984). Tolerance of citrus rootstocks to preemergence herbicides, *J. Environ. Hort.*, **2**(2), 73–76.

178. Skinner, R. E. (1924). Some observations on citrus rootstocks, *Proc. Fla. State Hort. Soc.*, **37**, 17–24.

179. Smith, P. F., S. M. Garnsey, and T. J. Grant (1973). Performance of nucellar 'Valencia' orange trees on 'rough' lemon stock when inoculated with four viruses, *1973 Proc. First World Congr. Citriculture*, **2**, 589–594.

180. Stafford, L. M. (1972). Influence of rootstocks on navel orange yield and tree growth at Mildura, Victoria, *Aust. J. Exp. Agric. Animal Husb.*, **12**, 203–208.

181. Stannard, M. C. (1973). Citrus rootstocks in Australia, *1973 Proc. First World Congr. Citriculture*, **2**, 191–193.

182. Strauss, G. R. (1961). Trifoliate proves excellent rootstock for Ivanhoe planting, *Calif. Citrograph*, **47**(10), 373–376.

183. Swingle, W. T. (1948). "The botany of citrus and its wild relatives of the orange subfamily," in *The Citrus Industry*, Vol. I, H. J. Webber and L. D. Batchelor, Eds., Univ. of Calif. Press, Berkeley, pp. 129–474.

184. Swingle, W. T, and P. C. Reece (1967). "The botany of citrus," in *The Citrus Industry*, Vol. I, W. Reuther, H. J. Webber, and L. D. Batchelor, Eds., Univ. of Calif. Press, Berkeley, pp. 190–480.

185. Tanaka, Y. (1969). Citrus rootstock problems in Japan, *Proc. First Int. Citrus Symp.*, **1**, 407–410.

186. Thornton, I. R. (1977). Comparison of different mandarin scion–rootstock combinations at Mildura, Victoria, *Aust. J. Exp. Agric. Animal Husb.*, **17**, 329–335.

187. Treeby, M. T., and I. R. Thornton (1983). An evaluation of the interaction between interstocks and rootstocks on the yield and tree size of 'Valencia' orange, *Scientia Horticulturae*, **19**, 229–235.

188. Tribulato, E., G. Cartia, A. Catara, and G. Continella (1980). Performance of a Clementine mandarin with cachexia-xyloporosis on eleven rootstocks, *Proc. Eighth Conf. Int. Org. Citrus Virol.*, pp. 232–238.

189. Turpin, J. W., J. E. Cox, and J. H. Duncan (1978). Rootstock trials for lemons in New South Wales, *1978 Proc. Int. Soc. Citriculture*, **1**, 126–128.

190. Ueno, I., M. Iwamasa, and M. Nishiura (1967). Embryo number of various varieties of citrus and its relatives, *Bull. Hort. Res. Sta.*, Okitsu, Japan, Series B., **7**, 11–21.

191. Valdez-Verduzco, J., and V. M. Medina-Urrutia (1981). Influence of rootstocks

on Mexican lime performance in Colima, Mexico, *1981 Proc. Int. Soc. Citriculture*, **1**, 142–144.

192. Vanderweyen, A. (1980). Influence de la varieté d'gommose a Phytophthora, *Al Awamia*, **60**, 65–80.

193. Venning, F. D. (1957). Trials with Swinglea glutinosa (Blanco) Merr. as a rootstock for citrus, *Proc. Fla. State Hort. Soc.*, **70**, 306–307.

194. Walker, R. R., and T. J. Douglas (1983). Effect of salinity level on uptake and distribution of chloride, sodium and potassium ions in citrus plants, *Aust. J. Agric. Res.*, **34**, 145–153.

195. Weathers, L. G., E. C. Calavan, J. M. Wallace, and D. W. Christiansen (1955). A bud-union and rootstock disorder of Troyer citrange with Eureka lemon tops, *Plant Dis. Reptr.*, **39**(9), 665–669.

196. Webber, H. J. (1932). Variations in citrus seedlings and their relation to rootstock selection, *Hilgardia*, **7**(1), 1–79.

197. Webber, H. J. (1948). "Cultivated varieties of citrus," *The Citrus Industry*, Vol. I, H. J. Webber and L. D. Batchelor, Eds., Univ. of Calif. Press, Berkeley, pp. 425–668.

198. Webber, H. J. (1948). "Rootstocks: Their character and Reactions," in *The Citrus Industry*, Vol. II, H. J. Webber and L. D. Batchelor, Eds., Univ. of Calif. Press, Berkeley, pp. 69–168.

199. Whiteside, J. O. (1973). "*Phytophthora* studies on citrus rootstocks," in *Proc. First Int. Citrus Short Course*, L. K. Jackson, A. H. Krezdorn, and J. Soule, Eds., Gainesville, FL, pp. 15–21.

200. Wutscher, H. K. (1973). "Rootstocks and mineral nutrition of citrus," in *Proc. First Int. Citrus Short Course*, L. K. Jackson, A. H. Krezdorn, and J. Soule, Eds., Gainesville, FL, pp. 97–113.

201. Wutscher, H. K. (1977). Citrus tree virus and viruslike diseases, *HortScience*, **12**(5), 478–484.

202. Wutscher, H. K. (1977). The influence of rootstocks on yield and quality of red grapefruit in Texas, *1977 Proc. Int. Soc. Citriculture*, **2**, 526–529.

203. Wutscher, H. K. (1978). Citrus tree rootstocks as a means of coping with marginal nutrient deficiencies and toxicities (Texas and Florida), *Compact Fruit Tree*, **11**, 28–30.

204. Wutscher, H. K. (1979). "Citrus rootstocks," in *Horticultural Reviews*, J. Janick, Ed., Avi Publish., Westport, CT, pp. 237–269.

205. Wutscher, H. K. (1978). Citrus in Argentina, Uruguay and Brazil, *Citrograph*, **64**(11), 279–281.

206. Wutscher, H. K. (1984). Citrus rootstocks in South America, *Proc. Trop. Reg. Amer. Soc. Hort. Sci.* (in press).

207. Wutscher, H. K., H. G. Campiglia, C. Hardesty, and A. A. Salibe (1977). Similarities between marchitamiento repentino disease in Uruguay and Argentina and blight of citrus in Florida, *Proc. Fla. State Hort. Soc.*, **90**, 81–84.

208. Wutscher, H. K., and D. Dube (1977). Performance of young nucellar grapefruit on 20 rootstocks, *J. Amer. Soc. Hort. Sci.*, **102**(3), 267–270.

209. Wutscher, H. K., N. P. Maxwell, and D. Dube (1977). Performance of young

'Fairchild' and 'Bower' mandarin hybrids on six rootstocks, *Proc. Trop. Reg. Amer. Soc. Hort. Sci.*, **20**, 157–165.

210. Wutscher, H. K., E. O. Olson, A. V. Shull, and S. Peynado (1970). Leaf nutrient levels, chlorosis, and growth of young grapefruit trees on 16 rootstocks grown on calcareous soil, *J. Amer. Soc. Hort. Sci.*, **95**(3), 259–261.

211. Wutscher, H. K., A. Peynado, W. C. Cooper, and H. Hill (1973). Method of irrigation and salt tolerance of citrus rootstocks, *1973 Proc. Int. Soc. Citriculture*, **1**, 299–306.

212. Wutscher, H. K., and A. V. Shull (1970). The performance of old line and young line Valencia orange trees on five tristeza tolerant rootstocks in the Rio Grande Valley, *J. Rio Grande Valley Hort. Soc.*, **24**, 12–17.

213. Wutscher, H. K., and A. V. Shull (1972). Performance of 13 citrus cultivars as rootstocks for grapefruit, *J. Amer. Soc. Hort. Sci.*, **97**(6), 778–781.

214. Wutscher, H. K., and A. V. Shull (1973). The performance of Valencia orange tree on 16 rootstocks in south Texas, *Proc. Trop. Reg. Amer. Soc. Hort. Sci.*, **17**, 66–73.

215. Wutscher, H. K., and A. V. Shull (1975). Yield, fruit quality, growth, and leaf nutrient levels of 14-year-old grapefruit *Citrus paradisi* Macf. trees on 21 rootstocks, *J. Amer. Soc. Hort. Sci.*, **100**(3), 290–294.

216. Wutscher, H. K., and A. V. Shull (1976). Performance of 'Orlando' tangelo on 16 rootstocks, *J. Amer. Soc. Hort. Sci.*, **101**(1), 88–91.

217. Wutscher, H. K., and A. V. Shull (1976). Performance of 'Marrs' early orange on eleven rootstocks in south Texas, *J. Amer. Soc. Hort. Sci.*, **101**(2), 158–161.

218. Yelenosky, G. (1976). Cold hardening young 'Valencia' orange trees on Swingle citrumelo (CPB 4475) and other rootstocks, *Proc. Fla. State Hort. Soc.*, **89**, 9–10.

219. Yelenosky, G., H. Barrett, and R. Young (1978). Cold hardiness of young hybrid trees of *Eremocitrus glauca* (Lindl.) Swing., *HortScience*, **13**(3), 257–258.

220. Yelenosky, G., and R. Young (1977). Cold hardiness of orange and grapefruit trees on different rootstocks during the 1977 freeze, *Proc. Fla. State Hort. Soc.*, **90**, 49–53.

221. Young, R. H. (1977). The effects of rootstocks on citrus cold hardiness, *1977 Proc. Int. Soc. Citriculture*, **2**, 518–522.

222. Young, R. H., L. G. Albrigo, M. Cohen, and W. S. Castle (1982). Rates of blight incidence in trees on Carrizo citrange and other rootstocks, *Proc. Fla. State Hort. Soc.*, **95**, 76–78.

12

PECAN ROOTSTOCKS

J. Dan Hanna
Texas A&M University
College Station, Texas

1. ORIGIN AND HISTORY

The pecan, *Carya illinoensis* (Wang) K. Koch, is the largest-growing, longest-lived, and most valuable nut tree native to North America. Its native habitat is centered in the area from Texas in the southwest to Alabama in the southeast, and extends along stream valleys as far north as southern Illinois. The southern extent of the native habitat extends into scattered areas of Mexico as far as 19°N latitude (15).

The native pecan is typically a large monoecious tree, usually alternate or less than alternate in bearing habit, with fruit produced on the terminals of annual shoots. Native pecan groves are largely confined to bottomlands and floodplains bordering rivers and streams, where they tend to dominate and form distinctive forest communities. They prefer alluvial soils that are deep, fertile, rich in organic matter, well drained but water retentive, with a moving (flowing) and relatively nonfluctuating water table (26). Pecans do not tolerate poorly drained soils.

The native pecan tree has a monolayered canopy, which is susceptible to shade and competition, resulting in death of shaded limbs. It has a strong taproot (Fig. 1), usually penetrating to near the water table (preferably 2.4 to 6 m below the surface) and utilizing the water table as a source of survival during drought periods. The upper fibrous root system (Fig. 2) is often aided by symbiotic mycorrhizae and permeates the upper meter of most soils (2). It seems to be the major source of water and nutrients for growth and production of seed.

The seedling pecan has a pronounced and prolonged juvenile state, with delayed production of flowers and nut set. Romberg (17) detailed the characteristics of the juvenile and mature states of growth, which include differences in leaf form, bud size, leaf and shoot color, leaf and shoot pubescence, and bark texture. The progression of a seedling from the juvenile to the mature state

Figure 1. Taproot system of pecan seedlings approximately one month after germination. Photograph courtesy of Dr. J. B. Storey.

depends primarily on height or distance from the crown and varies from 1.5 to 7.5 m. Age of the tissue has no effect on its juvenility or maturity.

Nut set in the pecan is closely associated with an intermediate range of annual shoot growth. As a terminal fruiting species, it does not lend itself well to size control by pruning. The tree will tolerate light to moderate pruning, but responds to heavy pruning with extremely vigorous shoot growth, with no production on compensatory growth for several years. Thus the pecan is quite difficult to fit into the concepts of high-density plantings which have become popular with other deciduous orchards.

The tree exhibits specific ion toxicity to chlorides. Hanna (7) found that leaf analysis of sibling seedlings had a wide range of chloride accumulation, and that leaf micronutrient content varied widely under controlled nutrient culture conditions. The tree often shows zinc deficiency, especially when grown on high-pH soils (19).

Figure 2. A portion of the feeder root system of a six-year-old pecan that had been exposed by the action of floodwaters.

Domestication of the pecan started in the late 1800s with the utilization of seedling orchards planted in place. Propagation of fruiting cultivars on seedling rootstocks was first accomplished in 1846 (2), and has proceeded to the present, with 1012 named cultivars currently recognized (24). With the advent of recognized cultivar proliferation, pecan cultivation has gradually shifted into nonnative areas. Viable pecan orchards on upland soils and in nonnative areas attest to at least a degree of flexibility in cultural requirements. Spread of pecan culture into the irrigated southwestern United States, Israel, South Africa, Australia, and other areas also demonstrates the flexibility of the pecan to adapt to climatic conditions foreign to its indigenous area.

Most commercially important pecan cultivars require a growing season of 180 to 220 days to mature their nuts; however, some cultivars developed in northern areas of the United States require as little as 170 days. It is generally recognized that the pecan tree thrives where the average growing season temperatures are very high, with best growth and fruiting occurring when the mean monthly temperatures range from 24 to 29.5°C, with little diurnal variation (13). In general, the largest nut sizes are attained where heat unit accumulations from April through October total well over 3315 degree days (10°C base) (27). Little research has been done on bud chilling requirements, but most pecan cultivars appear to require at least 400 hours below 7.2°C. Madden (11) observed that seed from cultivars of southern origin had little or no chilling requirement for germination, but that cultivars of northern origin required six weeks of moist cold stratification (0 to −2°C) for good germination.

The pecan flower is wind pollinated, and is genetically very heterozygous, exhibiting a degree of heterodichogamy which usually precludes self-pollination. Madden (12) found that the staminate parent greatly influenced seedling vigor, and that selfing usually reduced vigor. Nut weight was not correlated with either germination or seedling vigor.

2. ROOTSTOCK USAGE

Pecans are commercially propagated by budding or grafting the chosen scion cultivar onto seedling rootstocks (Fig. 3). Thus the trueness to type of a particular cultivar is maintained for that portion of the composite tree above the bud union. The rootstock of the tree, however, is genetically variable, since the seed from which it grew was the progeny of a heterozygous female parent and an unknown heterozygous pollen parent.

The commercial rootstock of greatest use is the pecan. Attempts to find seedling pecan parentage with superior rootstock characteristics—that is, lack of variability in growth, vigor, yield—have been uniformly unsuccessful in obtaining statistical significance. The author has followed the growth of trees growing on 'Burkett' and 'Riverside' rootstocks in far west Texas for 14 years. While there are obvious differences in sizes of trees (those on 'Riverside' average 12% larger in trunk diameter), the results show no statistically significant differences because of the extreme variability within treatment plots. Nurserymen in different regions of the pecan belt have selected and adopted certain cultivar seedling lines which they felt had more vigor and less variation in growth. Thus the 'Riverside' seedling has been the standby of Texas nurserymen for several decades, while the 'Elliott' and 'Curtis' became quite popular in the southeastern states.

Some degree of selection stems from the standard nursery operation, which results in the discard of seedlings too small for budding after the second year of growth. The sale of budlings by size categories further aids in narrowing the variability of trees to be planted in an orchard. It remains quite obvious, however, that the utilization of a heterozygous seedling stock introduces a great deal of variability into the present-day orchard.

The bitter pecan, *Carya aquatica*, is adapted to heavy bottomland soils with

Figure 3. Pecan budling consisting of three-year-old 'Riverside' rootstock and one-year-old 'Western' scion, ready for planting.

low pH, poor drainage, and subject to flooding. Thousands of bitter pecans have been topworked in their native habitat to pecan, with the resulting trees producing satisfactorily (Fig. 4). The growth rate is slightly retarded and might be satisfactory as a moderately dwarfing rootstock (2).

Several species of hickory have been topworked to standard pecan cultivars. *Carya alba* has been used more commonly, and its union with the pecan reflects marked incompatability, with the pecan overgrowing the hickory rootstock. Hickory seedling growth is very slow (2).

There is general agreement that the development of true-to-type rootstocks for pecans would open a whole new realm of possibilities for the pecan grower. Dwarfing, nematode resistance, greater yields, precocity of production, selective ion absorption, salinity tolerance, and a reduction in variability of orchard trees have been among the most often mentioned attributes (2, 16, 28).

Figure 4. Forty-year-old 'San Saba Imp.' pecan scion (*C. illinoensis*) growing on bitter pecan (*C. aquatica*) rootstock.

The primary requirement for the establishment of true-to-type rootstock clones of pecans is an efficient, inexpensive method of asexual propagation that could be adapted by the commercial nurseryman. The first attempt at pecan clonal rootstock propagation was that of Stoutemeyer (22) in 1938, working with rooting of hardwood stem cuttings, followed by Gossard (6), who obtained rooting of trench and air layers of pecans, using toothpicks treated with IBA. He was also successful in utilizing hardwood cuttings two weeks after bud break and softwood cuttings made in late August (5). Romberg (18) utilized nurse seedlings to sustain cuttings until they became established. Later attempts were made by Sparks and Pokorny (21), Whatley, Thompson, and Jefferson (25), Allan et al. (1), and Taylor and Odom (23). These attempts at clonal propagation were all successful in producing rooting of various cuttings, but all encountered problems with survival and establishment of the clone.

The first researchers to report both good rooting and cutting survival were Brutsch, Allan, and Wolstenholme (4) using juvenile stoolbed cuttings taken early in winter. Subsequent investigations have detailed the importance of bud chilling requirement (14) and available carbohydrates (3, 28). With these advances, it is now possible to expect rooting and survival of up to 50% of adult cuttings and 80% of juvenile cuttings. Propagation from the rooting of stem sections (usually 16 in. long) has the disadvantages of being quite extravagant in the use of prospective clonal selection material, having a slow multiplication rate, and being dependent on the season for availability of material with the proper bud conditioning pretreatments.

Recent successes in bud explant tissue culture from pecan seedlings have renewed interest in clonal rootstock potential for pecans. Smith (20) was able to obtain callus formation *in vitro* in 1977. Knox (10) delineated some of the histological and physiological aspects of *in vitro* cell differentiation in 1980. Wood (29) was able to induce shoot proliferation from nodal explant auxillary buds *in vitro* in 1982, but was unable to induce rooting. Hansen (8) induced up to 10 shoots per node utilizing benzyladenine *in vitro*, and then obtained rooting both *in vitro* (Fig. 5) and by transplanting excised shoots to peat pellets following a 10-day IBA treatment. Shoots in the peat pellets were covered with plastic cups for 15 days, after which the cups were perforated. Two months after the original IBA treatment, plantlets were well rooted and greenhouse acclimated. Some problems remain to be solved before these procedures can be commercially feasible, but at last the promise of a fast, effective, economically affordable method of asexual propagation seems imminent.

3. ROOTSTOCK ADAPTATION

With the prospects of efficient propagation methods improving, there is a surge of renewed interest in potential pecan rootstocks. Two relatively small germplasm collections have been made with rootstocks as a specific objective.

Jones (9) at the USDA Pecan Field Station at Brownwood, Texas, has collected and identified several potentially dwarfing stocks, some of which he had grafted to standard cultivars. Most of these stocks are established at Brownwood, but some are also established at Byron, Georgia.

Hanna (7) has collected, identified, and characterized approximately 60 seedlings, each having late bud break, some degree of chloride ion exclusion, and zinc uptake efficiency. Growth of these seedlings ranges from quite dwarf to standard. Material from both of these sources is available.

In summation, research on the pecan has lagged behind that of other fruits and nuts because of its extreme heterozygosity, its long juvenile period, and the lack of an efficient method of asexual propagation. Once a rapid, economically feasible method of clonal propagation is developed, several benefits should occur. First, pecan research will accelerate. Fewer trees with less replication will be necessary to obtain meaningful statistical results as variability of the

Figure 5. *In vitro* clonal propagation of pecan from a nodal explant. Photograph courtesy of Dr. J. E. Lazarte.

seedling rootstock is eliminated. Second, development of clonal rootstocks can proceed, with identification of rootstock germplasm and evaluation of its effects on scion properties. Finally, both pecan growers and the consuming public should benefit from the practical application of expanding research.

INTERNATIONAL CATALOG OF ROOTSTOCK BREEDING PROGRAMS

United States
Texas A&M University
College Station, TX 77843
J. Dan Hanna

W.R. Poage Pecan Field
Station 701, Woodson Rd.
Brownwood, TX 76801
Tommy Thompson

REFERENCES

1. Allan, P., M. O. Brutsch, J. C. LeRoux, B. N. Wolstenholme, and D. Cormack (1968). Rooting of pecan cuttings, *Fmg. S. Afr.*, **44**(9), 15.

2. Brison, F. R. (1974). *Pecan Culture*, Capitol Printing Co., Austin, TX.

3. Brutsch, M. O. (1971). Rooting and early growth of *Carya illinoensis* (Wang) K. Koch stem cuttings, M.S. thesis, University of Natal, Pietermaritzburg.

4. Brutsch, M. O., P. Allan, and B. N. Wolstenholme (1969). Growth of pecan stem cuttings, *Fmg. S. Afr.*, **44**(10), 15.

5. Gossard, A. C. (1944). The rooting of pecan softwood cuttings under continuous mist, *Proc. Amer. Soc. Hort. Sci.*, **44**, 251.

6. Gossard, A. C. (1941). Rooting pecan stem tissue by layering, *Proc. Amer. Soc. Hort. Sci.*, **38**, 213.

7. Hanna, J. D. (1972). Absorption and accumulation of chloride ions by pecan, *Carya illinoensis* (Koch) seedling rootstocks, Ph.D. dissertation, Texas A&M University.

8. Hansen, K. C. (1982). *In vitro* propagation of *Carya illinoensis* (Wang) K. Koch, M.S. thesis, Texas A&M University.

9. Jones, R. W. (1980). Progress report on dwarfing rootstocks, *Proc. S.E. Pecan Growers Assoc.*, **73**, 179.

10. Knox, C. A. (1980). Histological and physiological aspects of growth responses and differentiation of pecan, *Carya illinoensis* (Wang) K. Koch, tissues *in vitro*, Ph.D. dissertation, Texas A&M University.

11. Madden, G. D., and H. W. Tisdale (1975). Effects of chilling and stratification on nut germination of northern and southern pecan cultivars, *HortSci.*, **10**(3), 259.

12. Madden, G. D. (1974). Breeding for the development of the pecan, *Carya illinoensis* (Wang) K. Koch, seedling rootstocks, Ph.D. dissertation, Texas A&M University.

13. Madden, G. D., F. R. Brison, and J. C. McDaniel (1969). "Pecans," in *Handbook of North American Nut Trees*, R. A. Jaynes, Ed., W. F. Humphrey Press, Geneva, NY.

14. McEachern, G. R. (1973). The influence of propagating techniques, the rest phenomenon, and juvenility on the propagation of pecan, Ph.D. dissertation, Texas A&M University.

15. Onderdonk, G. (1908). Texas Dept. of Agri. Bul. No. 2.

16. Romberg, L. D. (1967). Clonal pecan rootstocks, *Proc. Texas Pecan Growers Assoc.*, **46**, 72.

17. Romberg, L. D. (1944). Some characteristics of the juvenile and bearing pecan tree, *Proc. Amer. Soc. Hort. Sci.*, **44**, 255.

18. Romberg, L. D. (1942). Use of nurse seedlings in propagating the pecan from stem cuttings, *Proc. Amer. Soc. Hort. Sci.*, **40**, 298.

19. Smith, M. W., and J. B. Storey (1979). Zinc concentration of pecan leaflets and yield as influenced by zinc source and adjustments, *Jour. Amer. Soc. Hort. Sci.*, **104**, 474.

20. Smith, M. W. (1977). Shoot meristem and callus tissue culture of pecans, *Carya illinoensis* (Wang) K. Koch, Ph.D. dissertation, Texas A&M University.

21. Sparks, D., and F. A. Pokorny (1966). Investigations into the development of a clonal rootstock of pecans by terminal cuttings, *Proc. S.W. Pecan Growers Assn.*, **59**, 51.

22. Stoutemeyer, V. T. (1938). Rooting hardwood cuttings with acids, *American Nurseryman*, **68**(9), 3.

23. Taylor, G. G., and R. E. Odom (1969). Relationship of carbohydrate and nitrogen content to rooting in pecan stem cuttings as influenced by preconditioning treatments, *Plant Propagator*, **15**(3), 5.

24. Thompson, T. E., and F. Young (1985). *Pecan Cultivars-Past and Present*, The Texas Pecan Growers Association, Inc., College Station, TX.

25. Whatley, B. T., S. O. Thompson, and J. H. Jefferson (1966). Propagation of *Carya illinoensis*, (pecan) from cuttings, *Proc. Int. Plant Prop. Soc.*, **16**, 205.

26. Wolstenholme, B. N. (1979). The ecology of pecan trees, Part I, *Pecan Quart.*, **13**(2), 32.

27. Wolstenholme, B. N. (1979). The ecology of pecan trees, Part II, *Pecan Quart.*, **13**(3), 14.

28. Wolstenholme, B. N., and P. Allan (1975). Progress and problems in pecan clonal propagation by stem cuttings, *Gewasproduksie/Crop Production 4*, 29.

29. Wood, B. W. (1982). *In vitro* proliferation of pecan shoots, *HortSci.*, **17**(6), 890.

13

JUGLANS ROOTSTOCKS

Gale H. McGranahan and Peter B. Catlin
USDA/ARS and University of California
Davis, California

The genus *Juglans* contains approximately 21 species native to North, Central, and South America; Eastern Europe; and Asia (70) (Table 1). *Juglans* is a member of the Juglandaceae, which also includes *Platycarya*, *Engelhardia*, *Alfaroa*, *Oreomunnea*, *Carya* (pecan and hickory), and *Pterocarya* (wingnut). *Carya* and *Pterocarya* are thought to be closely related to *Juglans* (70, 136) and intergeneric graft compatibility with *Pterocarya* (135) and alledged crossability with *Carya* (50, 58, 123) have been reported. Within *Juglans*, the important commercial species are limited to *J. regia* (English or Persian walnut) for edible nuts and to *J. nigra* (eastern black walnut) for timber and, to a lesser extent, for nuts. Other walnut species may have indirect economic importance as rootstocks. The emphasis of this chapter will be on *J. regia* and rootstock for *J. regia*. There is, however, a developing trend towards clonal forestry and an increased need for rootstock for *J. nigra* is anticipated (6, 7, 48).

1. ORIGIN AND HISTORY OF PERSIAN WALNUTS

J. regia is native to an area extending from the Carpathian Mountains south through Eastern Europe and east through Turkey, Iraq, Iran, and the countries beyond to the Himalayan Mountains. Vavilov (132) assigned this species to both the central Asiatic and Near Eastern centers of origin. Komanich (50) added a secondary center in Moldavia, U.S.S.R. *J. regia* was reportedly introduced into China from Tibet in 140 to 150 B.C. and was probably brought into parts of India even earlier (23). References to improved sources occur as early as the first century A.D. (Dioscorides cited in DeCandolle, 23). Although native stands of walnut were present in the mountains of Greece, it was not highly valued there until an improved type was introduced from Persia (23). The Romans also cultivated trees of Persian origin and attributed many healthful and practical qualities to their nuts. Walnut culture presumably spread from

TABLE 1. Species of *Juglans* [a]

Species	Common Name	Range
Section a. *Juglans*		
J. regia L.	English or Persian walnut	Southeastern Europe, Iran to Himalayas, and China
subsp. *turcomanica* Popov		
subsp. *fallax* (Dode) Popov		
Section b. *Rhysocaryon* Dode		
J. australis Griseb.		Argentina
J. boliviana (C. DC.) Dode		Western South America
J. californica S. Wats.	Southern California black walnut	California
J. hindsii (Jeps.) Rehder	Northern California black walnut	California
J. hirsuta Mann.		Northeastern Mexico
J. jamaicensis C. DC.	West Indies black walnut	West Indies
J. major (Torr. ex Sitsgr.) Heller	Arizona black walnut	Southwestern United States, northwestern Mexico
var. *major*		South-central Mexico
var. *glabrata* Mann.		
J. microcarpa Berl.	Texas black walnut	Southwestern United States, northwestern Mexico
var. *microcarpa* (*J. rupestris*)		
var. *stewartii* (Johnston) Mann.		Northern Mexico
J. mollis Engelm. ex. Hemsl.		Central Mexico
J. neotropica Diels		Northwestern South America
J. nigra L.	Eastern black walnut	Eastern United States

Taxon	Common name	Distribution
J. olanchana Standl. & L.O. Williams		
var. *olanchana*		Guatemala
var. *standleyi* Mann.		Southeastern Mexico
J. pyriformis Liebm.		Southeastern Mexico
J. soratensis Mann.		Bolivia
J. steyermarkii Mann.		Guatemala
J. venezuelensis Mann.		Venezuela
Section c. *Cardiocaryon* Dode		
J. ailantifolia Carr.	Japanese walnut	
(*J. Sieboldiana* Maxim.)		
var. *ailantifolia*		Japan
var. *cordiformis* (Makino) Rehd.	Heartnut	Japan
J. cathayensis Dode	Chinese walnut	Eastern China, Taiwan
J. mandshurica Maxim.	Manchurian walnut	Manchuria, northeastern China, Korea
Section d. *Trachycaryon* Dode ex Mann.		
J. cinerea L.	Butternut	Eastern United States

[a] Arranged according to the classification in Manning (70).

Italy throughout central and southern Europe and, with colonization, into the Americas. The common name "English Walnut" is probably a result of its importation into their American colonies by the English.

2. WORLD PRODUCTION

English walnuts are now grown in North, Central, and South America; Europe; Asia; and to a limited extent in North Africa and Oceania. World production is estimated by the Food and Agriculture Organization of the United Nations at between 700,000 and 800,000 metric tons per year (27). The United States (primarily California) leads with 25–30% of the world's production, followed by Turkey, 16%; China, 14%; and the Soviet Union, 6%. Altogether Europe accounts for 32% of the total supply, the majority of which is from France, Italy, Romania, Bulgaria, Greece, and Yugoslavia.

Annual production is highly dependent on weather conditions. Because of extensive rains in 1983, for example, the U.S. crop was down 15% from the record-breaking 1982 harvest of over 200,000 metric tons (131). French production was reduced by over 40% from 1982 by high temperatures and dry weather during July and August of 1983, combined with the effects of storm damage of the previous winter. In spite of such setbacks, however, world walnut production is generally increasing (27, 45).

3. ROOTSTOCK USAGE AND HISTORY

Rootstock usage varies by region and degree of development of the industry. Propagation of improved selections or cultivars on seedling rootstocks became increasingly common early in the twentieth century in the United States and parts of Europe, but in many countries the majority of bearing English walnuts are of seedling origin and recommendations for seedling orchards still exist in current literature (97). Current trends, however, are clearly toward increased rootstock usage to take advantage of active cultivar selection programs (28, 29).

In California, which produces 99% of U.S. walnuts, the value of grafted cultivars became apparent after seed and scionwood introduced from France had been evaluated in the late 1800s. By 1912 "no informed grower would plant a seedling orchard" (121). The first rootstocks used in California were seedlings of *J. regia*, but these were found to be suitable only under ideal conditions (121). After several related species and hybrids had been tested, *J. hindsii*, the northern California black walnut, was recommended for rootstock because it showed tolerance to drought and excess moisture, resistance to crown and root rots caused by *Armillaria mellea* (Vahl ex Fr.) Kummer (oak root fungus), and damage by gophers (121). Smith et al. (121) recognized the superiority of certain hybrids but considered the difficulty of obtaining enough hybrid seed a limiting factor for commercial production. Thus for the first half of the twentieth century, the California walnut industry was based almost exclusively on seedling rootstock of *J. hindsii*.

Because research and field observations continued to confirm the superiority of hybrids (116, 117), particularly 'Paradox,' the first-generation hybrid between *J. hindsii* and *J. regia* described by Burbank in 1893 (46), nurseries began to identify sources of *J. hindsii* that naturally produced high percentages of hybrid seed. 'Paradox' has now become commercially available on a relatively large scale and is the rootstock of choice in most of California. Unavailability (and higher price) still limits the planting of 'Paradox' in some years, however, and its hypersensitive reaction to the cherry leaf roll virus (blackline disease) may decrease its usefulness in the future.

In France, which also has a well-developed walnut industry, the first rootstocks available were seedlings of *J. regia*. Evaluation of other species began in the 1920s, and in 1960 a concerted effort in rootstock evaluation and selection was initiated by the Institut National de la Recherche Agronomique in Bordeaux. Until recently, the preference in rootstock has been for *J. nigra* the eastern black walnut, and selected sources of *J. regia* (33). *J. hindsii* was less favored, partly because it does not adapt well to cooler climates and partly because highly productive trees were not available as a seed source. Now, an estimated 20% of the orchard trees in France are grafted onto *J. nigra* (24). Because the blackline disease also affects *J. regia* on *J. nigra*, renewed efforts in rootstock selection are taking place in France.

In China both seedling and grafted trees are grown. Grafting of selected cultivars was practiced in the southern provinces over 400 years ago. The most common rootstock in China is *J. regia*, but the use of *J. mandshurica* for northern provinces, *Pterocarya* in Shandong and Shanxi provinces, and *J. cathayensis* along the Yangtze River has been reported (17). Current breeding and selection programs in China are focused on cultivar rather than rootstock development because of the greater potential gains by cultivar selection. Rootstock development is expected to be emphasized more when improved cultivars are released.

4. ROOTSTOCK SPECIES

Potential rootstocks for *J. regia* include several species and interspecific hybrids. Until now most research has emphasized the variation between species rather than that within species, and usually only a few sources have been used to characterize an entire species. With increasing attention to rootstock improvement and a good potential for clonal propagation (26), selected sources within a species or hybrids are expected to provide the rootstocks of the future.

Currently all rootstocks for *J. regia* in California are derived from open-pollinated seed, and nursery practices for all rootstock species characterized in this section are similar unless otherwise noted. Seed are usually planted in the nursery bed in the fall or stratified in sawdust bins outside and planted just before or shortly after germination in the spring. Seedlings are usually grafted in the nursery row the following spring and sold bare-root when dormant.

Some growers prefer to purchase ungrafted rootstock at the end of the first growing season and have it grafted after it is established in the orchard. Although this method is more economical initially, unless the rootstock is grafted the year it is planted (which is not recommended), an extra year is spent in establishing the orchard. Serr and Rizzi (117) recommend ungrafted seedling rootstock for replants or interplants in mature orchards.

Planting the seed directly in the orchard is another option, but unpredictable germination, nonuniformity, and the intensive culture required make this method less desirable.

4.1. *J. hindsii*

Open-pollinated seedlings of northern California black walnut are the most common rootstock for *J. regia* in California. *J. hindsii* is characterized by 15–19 narrow leaflets per leaf (Fig. 1) and a nearly globose, thick-shelled, slightly grooved nut, 2.5–3.5 cm in diameter (Fig. 2). Nuts require approximately 12 weeks of stratification. *J. hindsii* can be distinguished from *J. regia* and 'Paradox' by isoenzymatic markers in the aspartate amino/transferase and glucose phosphate isomerase systems (1). The visual distinction between the *J. hindsii* rootstock and the *J. regia* scion on a grafted tree is very clear (Fig. 3).

Figure 1. Leaves of *J. regia* (left), 'Paradox' (center), and *J. hindsii* (right).

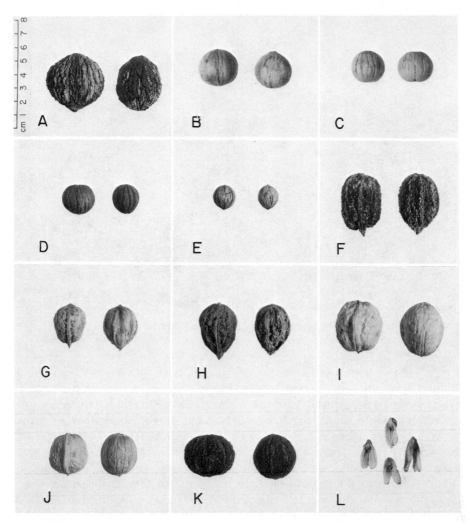

Figure 2. Nuts of *J. nigra* (*A*), *J. hindsii* (*B*), *J. californica* (*C*), *J. major* (*D*), *J. microcarpa* (*E*), *J. cinerea* (*F*), *J. ailantifolia* (*G*), *J. mandshurica* (*H*), *J. regia* (*I*), 'Paradox' (*J. hindsii* × *J. regia*) (*J*), 'Royal' (*J. hindsii* × *J. nigra*) (*K*), and *Pterocarya stenoptera* (*L*).

J. hindsii was first discovered in 1837 by Hinds on the lower Sacramento River. The species was later described by Jepson as *J. californica* var. *hindsii*, because Jepson believed that the differences between *J. californica* and *J. hindsii* could be attributed to differences in climate and soil conditions (121). Now it is generally understood to be a separate species, even though few native populations are known to have predated the European settlement of California (38, 127). In spite of its widespread use as a rootstock and street tree, *J. hindsii* could be considered an endangered species because it hybridizes readily with other black walnut species.

Figure 3. *J. regia* grafted on *J. hindsii*.

To what degree variation within the species exists and whether it is due to natural variation or interspecific hybridization is not yet known. Preliminary studies suggest a significant variation in early height growth due to seed source (McGranahan, unpublished). *J. hindsii* is usually recommended for areas that have deep, fertile soils and are free of *Phytophthora*, root lesion nematode, and blackline disease. This rootstock is intolerant of heavy soils, waterlogging, replant or interplant situations, and soils with either low available phosphorus and/or zinc or high levels of lime (116, 117). Its positive attributes are ready availability, low cost, and low susceptibility to crown gall and *Armillaria*. *J. hindsii* has not been recommended for use outside of California (33, 55).

4.2. *J. regia*

Worldwide, the Persian walnut is probably the most common rootstock for walnut cultivars because of its general availability in walnut-growing regions. *J. regia* is characterized by silvery-gray bark, 5–9 (to 13) leaflets per leaf (Fig. 1), and a thick or thin, wrinkled nutshell (Fig. 2). A stratification period of at least eight weeks is usually required for optimum germination. *J. regia* can be distinguished from *J. hindsii* and 'Paradox' by isoenzymatic markers as described previously (1). Both seed and scionwood of *J. regia* may carry the cherry leaf roll virus, which causes the blackline disease (19, 85, 87).

Although in the past *J. regia* was discarded as a rootstock in California because it compared unfavorably with other species in most studies (117, 121), the epidemic of blackline disease has recently resulted in a reevaluation of this species for rootstock. Although *J. regia* scions grafted on *J. regia* rootstock are not usually subject to girdling by blackline at the graft union, they are sensitive to excess salinity, waterlogging, root knot and root lesion nematodes, *Phytophthora* and *Armillaria* root and crown rots, and crown gall. *J. regia* seedlings also exhibit staggered germination and poor vigor in the seedbed and a lack of uniformity within a seed source (McGranahan, unpublished).

In Oregon, where the blackline disease was first described (108), *J. regia* seedlings from Plant Introduction 18,256 have been recommended since the mid−1950s as a meants to overcome the problem (83, 100). Known as 'Manregian,' this seed source originated in a seed collection obtained from the mountains of western China in 1906 by the Chico Plant Introduction Gardens of Chico, California. One individual of this collection was identified as superior and was maintained and clonally distributed to cooperators throughout the United States (35, 134). Seedlings from this source have been used in several rootstock studies, and in Oregon it has been shown to be superior to other *J. regia* sources in early height growth, but inferior to *J. hindsii* (139, 140). In another Oregon study, *J. regia* grafted on seedling 'Manregian' were reported to be substantially larger trees than those on *J. hindsii* (125; H. B. Lagerstedt, personal communication). In one portion of a rootstock study in California, *J. regia* scions on seedling 'Manregian' roots had a yield efficiency significantly less than *J. regia* scions on *J. hindsii*, *J. microcarpa*, and *J. major*; but in a companion plot with better soil (less salt), the differences were not significant (11). Nurseries in California propagate seedlings from Plant Introduction 18,256 and from the cultivars 'Eureka,' 'Waterloo,' and 'Serr' as possible rootstocks for areas where the cherry leaf roll virus is present. In other countries, specific sources of *J. regia* are used for rootstock, but descriptions are not available in the literature.

No clear consensus on whether *J. regia* is the appropriate rootstock for areas where the blackline disease is present has yet emerged in California because of the poor performance of *J. regia* rootstock, 'Manregian' or other, in comparison with either *J. hindsii* or 'Paradox' in any but ideal soil situations. Unless a source of *J. regia* is identified that can overcome this species weaknesses, *J. hindsii* and 'Paradox' will probably dominate the nursery trade in walnut rootstock for years to come. Because of a great deal of untapped diversity within *J. regia*, including several distinct geographic races (101), however, the outlook for breeding and selection within the species appears promising.

4.3. *J. nigra*

Eastern black walnut is native to the eastern and midwestern United States, south from New Hampshire to Georgia and west to Minnesota and Texas, but selective harvesting of the best trees has destroyed the natural population structure in its native range (7). *J. nigra* was introduced into Europe in the early seventeenth century both for timber and as an ornamental, and it has been planted as far north as southern Scandinavia. A tall tree, it is characterized by 15−23 leaflets per leaf and a large, irregularly ridged nut, 3−4 cm across (Fig. 2). Nuts require 12 to 16 weeks of stratification. Scionwood of *J. nigra* may carry a mycoplasma that causes walnut bunch disease (110) and the fungus *Sirococcus clavigignenti−juglandacearum* Nair et al., which causes butternut canker (128).

Because of its availability, *J. nigra* has been evaluated and, in some cases, recommended as rootstock for *J. regia* in Europe and the Soviet Union (33, 90, 137). *J. nigra* is also frequently used in the eastern United States as a hardy rootstock. The rootstock attributes reported for this species, compared with *J. regia*, are greater tolerance to nematodes, *Armillaria*, *Phytophthora*, and crown gall, earlier fruiting, and decreased tree size (25, 33). In a California field trial, *J. nigra* had the lowest yield efficiency of all the black walnut species tested (*J. hindsii*, *J. microcarpa*, and *J. major*) and is not recommended there (11, 121), although it is offered by at least one nursery. Because of the blackline disease, it is no longer recommended in France (20), but the potential of *J. nigra* hybrids is now under investigation in France (24) and the United States (80). Selected sources of *J. nigra* may be an important rootstock for *J. nigra* cultivars in the future.

4.4. *J. californica*

Southern California black walnut is present in southern California from the Santa Ynez Mountains east to the Santa Ana Mountains (38). *J. californica* differs from *J. hindsii* in having fewer leaflets (11−15), smaller nuts (Fig. 2), a bushier form, earlier foliation in spring, and later defoliation in autumn (121). Approximately 12 weeks of stratification are required.

When there was an active walnut industry in southern California, *J. californica* was considered a satisfactory rootstock and many orchards were grafted on it, although its sensitivity to excess moisture and its poor anchorage were recognized (121). By 1945, growers were being advised against using this species because of its susceptibility to both crown and root rots, and were even cautioned in selecting *J. hindsii* to make sure that the source was not actually *J. californica* (4). Since then, this species has attracted little attention and has been included in few rootstock trials.

4.5. *J. microcarpa* (*J. rupestris*)

Texas black walnut is native to Texas and New Mexico. *J. microcarpa* has 15−23 leaflets per leaf, smaller nuts than other *Juglans* species (1.5 cm across) (Fig. 2), and even in good soil only develops into a small tree. It seems to require a longer period of stratification (up to 27 weeks) than other *Juglans* species (10). Serr and Rizzi (117) suggested that *J. microcarpa* might serve as a dwarfing rootstock, but this has not been confirmed except in a rootstock trial for *J. nigra* cultivars (16). *J. microcarpa* has been used successfully as a rootstock in Texas where the high soil pH limits growth of both *J. regia* and *J. nigra* (119). In a California study, *J. microcarpa* performed well compared with other species under high boron and chloride soil conditions, but under less stressful conditions performed relatively poorly. Yield efficiencies for *J. microcarpa* were similar under both conditions (11). Therefore this species may be useful only under certain soil conditions or in breeding.

4.6. *J. major*

Arizona black walnut is native to New Mexico, Arizona, Colorado, and northern Mexico. *J. major* has 9–13 (to 19) leaflets per leaf and deeply grooved nuts, 2–3 cm across (Fig. 2). Serr and Rizzi (117) suggested that *J. major* might perform well in good soil and Shreve (119) recommended it for high-pH soils. It has performed very well in limited rootstock trials in California, but has shown no attributes that would make it clearly superior to *J. hindsii*. Further evaluations are needed.

4.7. Other Species

The potential of many *Juglans* species as rootstock is still unknown—mostly because insufficient seed has been available for testing. *J. mandshurica* is mentioned in the Chinese literature as a rootstock for plantings in northern latitudes (17). A report from the Soviet Union, however, found it unsatisfactory (90). *J. mandshurica* does cross with *J. regia*; the species *J. hopiensis* may be a hybrid between *J. mandshurica* and *J. regia* (54). In Argentina, *J. australis* is being tested as a rootstock (105). *J. cinerea* has been grafted onto *J. regia* but may not be fully compatible (33, 48). *J. ailantifolia* may have some resistance to *Phytophthora* (76, 78) but also exhibits a lack of compatibility with *J. regia* (33). All Central and South American species are of the black walnut type (67–69) and may provide an important source of rootstock germplasm for the future if they are made available for research.

4.8. Hybrids

Confirmed crossability within *Juglans* is illustrated in Fig. 4 (31). Although many combinations may have potential as rootstock, few have been tested because of the difficulty of obtaining enough hybrid seedlings. Most work has been concentrated on 'Paradox' (*J. hindsii* × *J. regia*), which will be discussed here. 'Royal' (*J. hindsii* × *J. nigra*) and the eastern black and Persian hybrid (*J. nigra* × *J. regia*) have also shown promise (25, 33, 121), but in-depth studies of their performance are not available.

Identification of hybrids is difficult. In California, however, nursery people have become adept at identifying the 'Paradox' present in a row of *J. hindsii* because 'Paradox' appears intermediate between both parents (Fig. 1). Identification of 'Paradox' in a family of *J. regia* is more difficult. In both cases,

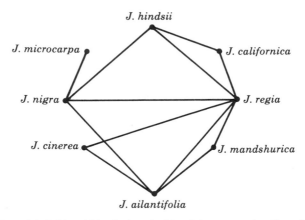

Figure 4. Confirmed hybrids within *Juglans* indicated by connecting lines between species (adapted from Funk [31]).

isoenzymatic markers are needed for positive identification (1). When mature, 'Paradox' is relatively unfruitful, often bearing numerous flowers which abscise early in development. The nuts, when present, are generally intermediate in size between *J. hindsii* and *J. regia* (Fig. 2).

'Paradox' has been shown to be superior to both its parents in many traits including tolerance to both nematodes and *Phytophthora* species. Burbank first noted its more vigorous growth (46) (Fig. 5). Serr and Forde (116)

Figure 5. Relative vigor of seedlings of 'Paradox' (left, 3 trees), *J. hindsii* (center, 2 trees), and *J. regia* (right, 4 trees).

reported that *J. regia* grafted onto 'Paradox' had significantly greater circumference than those grafted on *J. hindsii* when grown together on volcanic hillside soils. Circumference differences were not detected under good soil conditions (deep sandy loam), but replants of 'Paradox' were observed to grow faster than replants of *J. hindsii* on deep soil in old orchards. For these reasons, 'Paradox' is preferred to other rootstocks in California, except where the blackline disease is a major limiting factor.

In addition to a modified hypersensitive reaction to the cherry leaf roll virus, which causes cankers and lethal girdling at and below the graft union, other limitations of 'Paradox' include susceptibility to crown gall, variation in performance, both between and within seed sources, and reduced availability. Selection and micropropagation are expected to overcome both the variation in performance and lack of availability.

4.9. Other Genera

Investigations of the potential of other genera within the Juglandaceae for rootstock has been limited to *Pterocarya stenoptera*. A native of China, wingnut is easily distinguished from *Juglans* by its fruit, which is small (1.5−2 cm long), winged, and borne in racemes 20−30 cm long. The leaves, 20−40 cm long, have a winged rachis and 11−23 leaflets, each 4−10 cm long. Wingnut hybridizes readily within the genus and certain identification is often difficult.

Wingnut was first reported as a potential rootstock for *J. regia* in 1948 (135). Since then it has been shown to have a high level of tolerance to *Phytophthora* (76−78, 84), waterlogging (14), and root lesion nematode (64). It is susceptible to *Armillaria*, intolerant of zinc deficiency, may sucker profusely, and is subject to girdling at the graft union by blackline disease. The major factor limiting its usefulness, however, is a lack of graft compatibility with many *J. regia* cultivars (117). Further work is needed to determine whether it has potential in rootstock breeding.

5. ROOTSTOCK−SCION INTERACTIONS

5.1. Compatibility

Rootstock−scion compatibility research with walnuts has been confounded in many cases by the blackline disease. For years the debilitating and eventually lethal girdle at the graft union of trees with blackline disease was thought to be caused by delayed incompatibility (36, 72, 83). Recently it has been shown that the black line at the graft union is due to a hypersensitive reaction of the rootstock to the virus and is not an indication of incompatibility (85−87).

Successful "takes" have been reported for grafts between *J. regia* and *J. regia, J. cinerea, J. major, J. hindsii, J. microcarpa, J. nigra, J. ailantifolia, J. australis*, several hybrids, and *Pterocarya stenoptera* (11, 48, 105, 117, 135). Overgrowth, swelling at the union, lack of scion vigor, and difficulties in obtaining successful grafts have been evident in many of these combinations and suggest some degree of incompatibility. Whether these possible incompatibilities relate only to the individual combinations or are characteristic of the species in general is unknown. Only in *Pterocarya stenoptera* does it appear that certain *J. regia* cultivars (e.g., 'Eureka,' 'Sorrentino,' and 'Concord') are more compatible than others (117). More research on graft compatibility is needed.

5.2. Size Control

Size control for Persian walnut trees has stimulated little research, although size reduction might be a means to increase yield per hectare and decrease labor costs in close plantings. Several Persian nut cultivars, regardless of rootstock, tend to be smaller trees; the general trend is to use these smaller varieties grafted onto standard vigorous rootstock for close plantings.

A 10–20% reduction in tree size has been reported for trees grafted on *J. nigra* as compared to those on *J. regia* roots (33). In a California trial of seven different rootstocks, by contrast, the size of trees on *J. nigra* was not significantly different from those on *J. regia* roots (11). However, trees on *J. nigra* had the smallest trunk circumference, the lowest yield per tree, and thus the lowest yield efficiency of the seven. That *J. microcarpa* might serve as a dwarfing rootstock has also been suggested, but this has not been confirmed.

J. nigra has been successfully used as a dwarfing interstock for 'Hartley' walnuts on 'Paradox' roots (112, 113). After 13 years, the cross-sectional areas of trunks of trees with *J. nigra* interstocks were about half that of the controls, grafted with *J. hindsii* interstocks. The dwarfed trees also had more catkins and higher yield efficiencies. Increased yield was apparently due to increased pistillate flower production in lateral buds; thus the dwarfing interstock induced a normally terminally bearing cultivar to be a laterally fruitful one. *J. ailantifolia* var. *cordiformis* interstocks also dwarfed the scion, but, unlike the *J. nigra* stem pieces, these interstocks were constricted and some breakage occurred. Interstocks of *J. microcarpa*, *J. ailantifolia*, and 'Paradox' did not reduce scion growth. Whether the observed dwarfing was due to species or source within species is not clear. Dwarfing by interstock has not been recommended for commercial use.

6. ROOTSTOCKS AND THE SOIL ENVIRONMENT

Adaptation of walnut rootstocks and roots to soils can be viewed from two perspectives. One of these is from their nature as the support of a woody perennial plant of potentially great size and longevity. In this sense, walnut root systems must provide the same functions as those of other perennial species: anchorage, the absorption of water and nutrients, sites of hormonal synthesis and conversion of organic constituents, and storage (2, 15). The general form of a walnut root system is shown in Fig. 6. The second perspective, the root or rootstock interaction with the soil environment, is the focus of

Figure 6. General form of a shallow root system of 18-year-old *J. hindsii* in a sandy soil; 90% of roots in surface 1 m, no restrictive strata evident.

this section. A complex of physical, chemical, and biological relationships is important and its impact on root performance can differ markedly with rootstock type.

6.1. Soil Depth

Soils suitable for walnuts are usually described as being deep and of a uniform, medium texture with good internal drainage. Restrictions to deep rooting such as claypans, hardpans, siltstone layers, water tables, and sand lenses can limit the vigor and size of walnut trees (5). For *J. nigra* in the central and eastern United States, a minimum depth of 3 ft to a restrictive or unfavorable soil layer is recommended (31, 59), but a depth of at least 5 ft is considered to be more suitable (34, 95). In California, minimal unrestricted depths of 5−6 ft or more for walnuts, principally *J. hindsii* and 'Paradox' roots, have been recommended (4, 114). That walnut roots can exploit much greater depths has been shown by water extraction at 10−12 ft (133), and *J. hindsii* roots have been observed at depths of 12−13 ft in commercial orchards. That *J. nigra* produced a tap root more than 9 ft deep in three years (95) suggests similar deep-rooting capabilities with this species.

6.2. Soil Texture

Soil texture influences root performance through its effect on aeration, water-holding capacity, resistance to root growth (soil strength), and (in some cases) nutrition. Soils recommended for walnut culture are medium-textured loams, silt loams, or sandy loams (5, 59, 95, 114). Stratified soils or those with abrupt textural changes such as thin sand lenses can restrict root exploration (5). Where walnut orchards have been planted in clay loam or even finer-textured soils in California, shallow root systems and reduced tree size resulted (Fig. 7). With these soils, some compensation for smaller trees can be gained from closer spacing of trees at planting. Reduced vigor and size of walnut trees on fine-textured soils may be largely due to reduced aeration. Soil aeration can be restricted with depth even without increased water content from irrigation or other sources (53, 126). The great sensitivity of walnut roots to low-oxygen stress is discussed later.

Although fine-textured soils may have greater water-holding capacities than coarser soils, the shallow root systems of trees planted on such soils are obviously unable to extract water from below the root zone as the upper strata become depleted (103). Thus the reservoir of available moisture is reduced, increasing the probability of water stress.

Coarse-textured, sandy soils are not desirable for walnuts (5, 59, 95, 114). This appears to be largely due to low water-holding capacities and periodic or long-term water deficits, but the lower mineral-nutrient content of such soils may also be a factor. Tree size has been severely limited where walnuts have been planted on very sandy soils in California. Nevertheless, moderate success can be realized on sandy soils and with shallow root systems with short irrigation intervals (Fig. 6).

Soil strength, the physical resistance to root penetration, can restrict root growth and tree performance. Specific information about soil penetration by walnut roots is lacking, but they may be expected to respond as do other deciduous tree roots (39).

Detailed information about how different rootstock types differ in their response to soil depth and texture does not exist. 'Paradox' hybrid rootstock is generally considered to be superior to either *J. hindsii* or *J. regia* on less than ideal soils (116, 117), but little difference has been observed between 'Paradox' and *J. hindsii*, at least under highly suitable soil conditions. *P. stenoptera* may be more tolerant of fine-textured soils (117).

6.3. Aeration

Juglans species are highly sensitive to inadequate soil aeration (14, 60, 61, 124). Extensive damage in California walnut orchards occurs periodically from either surface flooding or seepage from bodies of water (Fig. 8). Experimental flooding of young seedlings with controlled root temperatures revealed that *J. regia* was most sensitive, *J. hindsii* was next, and 'Paradox' was least sensitive to root zone flooding (14). Considerable seedling variability occurred and differences among rootstock types were relatively small. Nevertheless, less damage or greater survival could be expected with 'Paradox' under orchard conditions, depending on timing and duration of soil saturation.

Figure 7. Ten-year-old Serr/*J. hindsii* with root depth limited to less than 60 cm in a clay soil (upper) and small size (lower). Note 30-cm ruler in both photos.

Figure 8. Soil waterlogging damage to walnuts from seepage from Sacramento River, CA.

P. stenoptera seedlings, although rarely used as rootstocks except in China, were much more tolerant of flooding than *Juglans* seedlings (14). Wingnut seedlings have a weak adaptive capability to form adventitious roots and are able to transport some oxygen internally for short distances (G. B. Vogel and P. B. Catlin, unpublished). This suggests that a potential for adaptive responses to periodic waterlogging may exist among the Juglandaceae. The capacity to develop adventitious roots with aerenchyma (44) would be an important consideration in germplasm evaluation.

A comparison of *J. nigra*, *J. regia* (French source), and *J. ailantifolia* (*J. sieboldiana*) suggested that *J. nigra* might be slightly more tolerant to root zone flooding than the other two species (124). Only three or four plants of each type per treatment were tested, and variation among individuals was substantial. Comparison of *J. nigra* seedlings of four seed parents indicated sensitivity to waterlogging to be the same as for *J. hindsii* (P. B. Catlin, unpublished).

Because all commonly used walnut rootstocks are very sensitive to deficient aeration, this must be carefully considered in irrigation management of new or recent plantings. Growth of small seedlings can be reduced within 24 hours of root zone flooding and stopped within 72 hours (G. B. Vogel & P. B. Catlin, unpublished). Low shoot vigor and severely reduced leaf size in one- to three-year-old trees has been observed in California following wet winters and springs. When soil and air temperatures are high during summer, wilting of leaves and defoliation may occur following irrigation (Fig. 9).

Figure 9. Sudden wilting of leaves of two-year-old (left) and defoliation and regrowth of one-year-old walnut trees (right) following irrigation in August.

A condition occasionally affecting mature walnuts in California and termed "apoplexy" is noted here because it may relate to aeration. Trees exhibit sudden wilting and drying of leaves in summer, typically a few days after surface (usually flood) irrigation (Fig. 10). Death of fine roots may occur, but no pathogens have been implicated (88). This sudden collapse, which can occur within one day, is more common with trees planted in coarser-textured, sandy soils. Essentially no research has been done to determine the cause of this condition, but current speculation is that high soil and air temperatures, depletion of available water to near the wilting point, and inhibited root growth and activity combine with soil saturation to cause sudden wilting. Reduced transpiration has sometimes been shown soon after roots of other tree species were flooded (52). Xylem plugging of walnut trees can occur if inadequate aeration due to flooding of roots occurs (13).

Opinion differs as to whether apoplexy is a greater problem with mature or with young trees. Older trees often recover with formation of new leaves after several weeks or months. Collapse of young trees one to four years of age can be attributed to the cumulative effects of repeated soil saturation during irrigations with insufficient intervening time for root regeneration. Symptoms of apoplexy were especially common with young trees following summer irrigations after the wet winters and springs of 1982 and 1983 in California.

Figure 10. "Apoplexy," or sudden wilting, of mature walnut trees following irrigation in July.

Reduced vigor and much-reduced leaf size before collapse indicated preexisting damage to fine roots. No association of apoplexy with rootstock type has been documented, but if inadequate aeration is a causal component, then trees on 'Paradox' roots would be expected to be somewhat less frequently affected.

6.4. Salinity

Walnuts are considered to be extremely sensitive to salinity in soils or irrigation water (4, 15, 41, 43). In addition to the usual effects of total salts on the osmotic potential of soil water, and thus on water relations, walnuts and most other woody species are subject to specific ion influences (3, 8, 66). Most concern about specific ions focuses on the accumulation of sodium and chloride, but boron is sometimes included as well (3). Accumulation of these ions leads to toxicity symptoms or leaf necrosis. Because of these ion toxicities, the adverse effects of salinity are greater than those expected from osmotic influences alone (66). Although responses have been based upon vegetative growth of young trees, yield reductions of 10% have been predicted at electrical conductivities of soil saturation extracts of 2.3 dS/m (mmho/cm) (3). A further decrease in yield of about 20% was projected for each 1 dS/m increase in conductivity.

Rootstocks for walnut take on special importance for the specific ion effects of salinity, because rootstock types differ in their capacities to absorb and transport salts to the scion and thus influence specific ion toxicity. Whether or not walnut rootstocks differ in their response to the osmotic effects of salinity is not known. The comparisons that follow are based mainly on the accumulation of salts in the scion and on symptoms of leaf toxicity.

Walnuts absorb and transport chloride more readily than sodium (41), as do most other deciduous species (8). Symptoms of leaf toxicity progress from necrosis of leaf tips to that of leaf margins and, in severe cases, to that of entire leaflets. Leaf necrosis usually becomes progressively worse after mid- to late summer and early abscission is common. Leaf symptoms are essentially the same for excessive levels of sodium, chloride, or boron (130). Leaf tissue levels considered to be excessive are 0.3% chloride, 0.1% sodium, and 300 ppm boron on a dry-matter basis (9). Interestingly, the threshold excess level for boron is greater for walnut than for other deciduous fruit-bearing species except fig.

J. regia has been described as being less tolerant to salts than *J. hindsii* and 'Paradox' (4, 117) and 'Paradox' appears to be less tolerant than *J. hindsii* (11). In one orchard, trees on 'Paradox' roots were much more severely affected than those in adjacent rows on *J. hindsii* roots. *J. nigra* has been described as being intolerant of salt (89).

Results from a rootstock plot in California that was irrigated with water high in both chloride and boron allow comparison of several walnut rootstocks (11). All trees had *J. regia* 'Serr' scions. *J. hindsii*, *J. major*, and *J. microcarpa* tended to exclude chloride, with less than toxic threshold levels accumulated in scion leaves. 'Paradox,' 'Royal,' and *J. nigra* absorbed chloride to levels considered to be excessive. Seedlings of one *J. regia* type, 'Manregian' (Plant Introduction 18,256), absorbed still greater amounts of chloride.

All rootstocks in the preceding trials absorbed and transported boron to scion leaves in amounts two to three times those considered excessive. Even though leaf toxicities occurred, trees on *J. microcarpa* and 'Royal' rootstocks accumulated significantly less boron than those on the other rootstocks. Whether this was a real effect due to rootstock type and, if so, whether it has any horticultural significance seems questionable.

Differences in response to salinity may occur among different types of walnut, but not enough solid experimental evidence is available to prove it.

Treatments under reasonably controlled conditions are needed to verify differ-ing capacities to absorb and transport salts. Whether any differences exist in responses to the osmotic component of salinity stress also needs to be deter-mined. Salinity research could be productive for rootstock improvement be-cause responses have been shown to be heritable in plant species (118).

6.5. High-pH and Calcareous Soils

In some locations difficulties are encountered with soils of high pH and the presence of calcium carbonate. So-called lime-induced chlorosis usually in-volves iron deficiency, but can include manganese deficiency as well (12). This problem occurs in small, scattered areas in California (130) and can be accentu-ated by inadequate aeration and/or restricted root growth. Correction of physical problems of soils can sometimes alleviate chlorosis. 'Paradox' is reportedly more tolerant to conditions causing chlorosis than *J. hindsii* (43, 117), and some 'Paradox' selections are considered to be superior in this tolerance (117). Whether the superior performance of 'Paradox' is caused by greater efficiency of iron uptake (81) or by its greater vigor and presumed ability to explore a greater soil volume is not known. The greater tolerance of this rootstock to deficient aeration (14) could also be a factor.

In southwest Texas, seedlings of *J. regia* (Chinese and Hungarian sources) and *J. nigra* rapidly became severely chlorotic when planted in soils of pH 7.5–8.4 (119), requiring correction with chelated iron. In contrast, when *J. regia* or *J. nigra* were grafted onto native *J. microcarpa* (*J. rupestris*) or *J. major*, chlorosis did not occur. Experimental verification of the apparently greater efficiency of iron uptake by these two species and identification of controlling mechanisms would be of value for rootstock improvement.

6.6. Mycorrhizae

Infection of roots by mycorrhizae has been reported for *J. regia* (111), *J. nigra* (51, 96), *J. hindsii* (94, 109), and 'Paradox' (94). Whether dependency upon mycorrhizal association exists with walnut and whether differences in such association occur among rootstock types is unknown. Mycorrhizae may be expected to have greater importance in less fertile soils and where regular nutrient applications are not common (71). Benefit from exogenous inocula-tions might result in nurseries where soils have been fumigated (51, 57). Natural development of mycorrhizal fungi was found in a walnut nursery within one year after fumigation with no supplemental inoculation (94).

Much more comparative research, some under controlled conditions, is needed to clarify the role of mycorrhizae in rootstock response or even in walnut production. In soils where zinc or phosphorous deficiencies may exist (43, 130), the possible mycorrhizal association or lack of it would be of interest.

6.7. Replant Problems

The existence of specific problems when replanting walnuts in soils previously planted with walnuts, such as occur with some other tree species (138), has not been demonstrated. Results from the related practice of interplanting in under-utilized, between-row space in existing orchards show that 'Paradox' seedlings usually perform better than those of *J. hindsii* (117). The greater vigor of 'Paradox' and its generally superior responses to less than ideal soils, *Phytophthora* species, root lesion nematodes, and deficient aeration may explain its superior performance. Insufficient observations have been made on *J. regia* roots under such conditions to assess its response, and no information on *J. nigra* is available.

Slow growth of interplanted trees may, of course, result from competition for water and light with established trees and from management practices favoring the producing trees. Growth of replants or interplants has been enhanced by preplant fumigation or nematicide treatments (117).

Allelopathic effects of walnut on certain other plant species are well known (102), but no data indicates that such possible effects occur with newly planted walnut trees in soils with walnut residues. Disposal of walnut hulls by spreading on the orchard floor, as is sometimes done in California, has been without apparent harm to walnut trees.

6.8. Temperature

Rootstock response to temperature can be viewed from two standpoints: (a) responses of roots to soil temperature and (b) cold hardiness with influences on winter tree survival. Little is known about direct effects of soil temperature on walnut roots or rootstocks. Deciduous tree roots do not experience a rest period as do vegetative buds. In mild climates, low levels of new root growth can be expected throughout the winter months even while shoots undergo rest and dormancy (15). Formation and extension of new roots of *J. hindsii* through-out the winter in California has been observed in a root observation chamber (J. M. Duniway, personal communication). The low-temperature threshold for root growth of walnut may be near 5−7°C, as it is for peach and apple (91). In colder climates with lower soil temperatures, root growth during winter would not be expected.

Optimum soil temperatures for root growth of peach and apple have been reported in the range of 16−24°C (91). In one study with *J. regia*, greatest tree growth occurred with a root temperature of 24°C, but the relatively large increments of temperature measurements in this study make this figure uncer-tain (42). If walnut roots follow patterns shown by peach and apple roots, severe limitations on root growth would occur as temperatures rise above about 27°C, and growth might be expected to stop at about 32°C. Severe limitations of growth of *J. regia* seedlings occurred at root temperatures of

31°C, and plants were killed at root temperatures of 38°C (42). Soil temperatures sufficiently high to limit root growth and viability can readily occur with trees in the first few years after planting, when clean cultivation and minimal leaf canopies can result in temperatures near those of bare soil for the initially shallow root systems. The great potential for damage by summer irrigation, resulting in short periods of deficient aeration under such high temperatures, has been noted previously.

How rootstock species or types within species may affect scion survival in cold climates is not known, but a lack of hardiness of potential rootstocks obviously limits the performance of scions. Discussion here is limited to apparently direct effects of cold on rootstock survival and disregards time of dormancy inception and release in spring. Where direct comparisons can be made, as in California with *J. hindsii* and 'Paradox' rootstocks, time of growth resumption or leafing out appears to depend on the scion cultivar and not on the rootstock.

Among *Juglans* species studied, the most cold hardy is *J. mandshurica*, followed, in decreasing order of hardiness, by *J. cinerea*, *J. nigra*, *J. ailantifolia* var. cordiformis, and *J. regia* (93). Great tolerance to cold probably accounts for the use of *J. mandshurica* stocks in the northern provinces of China. Intraspecific variation in cold tolerance has been reported in *J. regia* and *J. nigra* (31, 32, 56, 93), however, and the Carpathian types of *J. regia* used as rootstocks (40) and other types from this species originating in the cold climates must possess great tolerance. Seedlings of the 'Manregian' type of *J. regia* used in Oregon (56) also have good cold tolerance. Rootstocks used in California—*J. hindsii*, *J. regia*, and 'Paradox'—are thought to be quite cold sensitive; but quantitative comparisons have not been made (29).

6.9. Nutrition

Different rootstock responses to nutrients have been discussed previously for deficiencies of iron and for excessive amounts of sodium, chloride, and boron. For other mineral nutrients, variation among rootstock types in response to fertilizer or to deficiencies has not been reported. Nutrition thus appears to be more affected by the mineral availability in the soil and the amounts needed by the scion than by rootstock characteristics. Reports of nutritional considerations for divergent culture systems in California (9, 43, 92, 130), France (33), and for *J. nigra* (107) have been made recently. Nitrogen is the only nutrient normally applied annually. Other elements are supplied when deficiency symptoms occur or when leaf analyses indicate incipient deficiency. Deficiency symptoms have been described for *J. nigra* (107) and for *J. regia* (43, 92, 130). Color prints of leaves affected by some nutritional problems have been published for *J. regia* (92, 130).

7. ROOTSTOCK–PATHOGEN INTERACTIONS

7.1. Diseases

Diseases have played a major role in rootstock selection because rootstock resistance has been considered a more practical solution to disease problems than chemical control, which is not adequate for many of the rootstock pathogens and because interspecific variation in disease susceptibility has been reported (33, 47, 72, 76, 78, 83, 98, 117, 120). Breeding rootstocks for disease resistance has been given a high priority in the USDA/ARS Walnut Improvement Program at the University of California, Davis (99).

The important root-related diseases confronting walnut growers are root and crown rots caused by *Phytophthora* spp. and *Armillaria mellea*, crown gall caused by *Agrobacterium tumefaciens* (Smith and Town) Conn., and, indirectly, the blackline disease. Most research on walnut root diseases and resulting recommendations have been focused on interspecific, rather than intraspecific, variation. This is partially responsible for the use of different species rather than different sources within the same species for walnut rootstock.

7.1.1. *Phytophthora Species.* *Phytophthora* is considered the most threatening root disease in California orchards today (84). Disease incidence has been increasing due to increased mechanization, multiorchard management, and increased irrigation, which together spread and create a favorable environment for the pathogen. At least eight species of *Phytophthora* have been implicated in the decline and/or death of walnut trees in orchards, and several others are known to be pathogenic on walnuts. The most destructive species in California are *P. citricola* Sawada, *P. cryptogea* Pethyb. and Laff., and *P. cinnamomi* Rands. *P. citricola* as well as *P. cactorum* (Leb. and Cohn) Schroet. and *P. megasperma* Drechsler are widely distributed in California orchards.

Early symptoms of *Phytophthora* infection are poor growth, small, chlorotic leaves, and premature senescence in the scion. These are followed by twig dieback, partial defoliation, and finally death. Rootstock symptoms include cankers at the crown, which may elongate up the stem (crown rot), and death of feeder and small secondary roots (root rot). The presence of cankers may indicate the involvement of *Phytophthora* in tree decline, but root death can arise from several causes. Positive diagnosis requires soil and root sampling and isolation and identification, under laboratory conditions, of the species involved.

Recent studies have shown that at least four *Phytophthora* species (*P. cactorum*, *P. cinnamomi*, *P. citricola*, and *P. cryptogea*) isolated from hosts other than walnut are pathogenic on walnuts (76). This suggests that the disease could spread from adjacent plantings of fruit trees or ornamentals and therefore may be difficult to control without resistant rootstock.

The relative susceptibility of potential rootstock species to infection by

TABLE 2. Response of Walnut and Wingnut to Pathogens[a]

Pathogens	Diseases	J. regia	J. hindsii	Paradox	J. cali-fornica	J. major	J. micro-carpa	J. nigra	J. ailanti-folia	P. steno-ptera
Cherry leaf roll virus	Blackline disease	S	H	H	H	H	H	H	H	H
Phyto phthora[b]	Root and crown rots									
P. citricola	Crown rot	++	++(++)	++(+)	++	++	++	++(-)	++(--)	--(--)
P. cinnamomi	Root rot	++	++(++)	++(++)	.	.	.	++(++)	++(-)	--(--)
P. cryptogea	Root rot	.	++(--)	+(--)	.	.	.	++(--)	++(--)	--
P. cactorum	Crown rot	++	++	-	.	.	.	.	.	.
P. citro-phthora	Root rot	.	++(-)	-(--)	.	.	.	-(--)	--(-)	--(--)
P. megasperma	Root rot	.	-	--	.	.	.	--	.	--
Armillaria mellea (oak root fungus)	Root and crown rot	++	-?	-?	++	.	.	.	.	++
Agrobacterium	Crown gall	++	+,-	++	.	.	.	.	.	.

[a]H (hypersensitive), S (subject to systemic infection), ++ (highly susceptible), + (susceptible), - (resistant or tolerant), -- (highly resistant or highly tolerant)

[b]Responses to *Phytophthora* spp. provided by M. Matheron. Values are derived from greenhouse studies in which seedlings were exposed to periodic flooding. Values in parentheses indicate response without flooding.

436

various *Phytophthora* species has been studied (76–78, 84) (Table 2). Initially *J. regia* was thought to be highly resistant (117, 120, 121), but recently work by Mircetich and Matheron (84) under controlled conditions has shown that *J. regia* and *J. hindsii* are equally susceptible to most *Phytophthora* species studied. 'Paradox,' their hybrid, appears to have more resistance than either parent. This suggests that hybrid vigor and/or complementary genes may be involved in the lower susceptibility observed in hybrids. *J. nigra* and *J. ailantifolia* also exhibit low susceptibility to certain *Phytophthora* species (76), but *P. citricola* has been implicated in death of *J. nigra* in nurseries (37). Wingnut (*Pterocarya stenoptera*) is highly resistant to all species of *Phytophthora* tested (76–78, 84) and may be useful in breeding programs for walnut rootstocks if intergeneric crossability barriers can be overcome.

At present, methods for determining host resistance and/or *Phytophthora* virulence require growing plants in artificially infested soil, usually accompanied by periodic flooding (76, 77, 84). Although the mechanism of resistance is not yet understood, it has been shown that presumably resistant genotypes will develop cankers if mechanically inoculated with the pathogen (76). Many variables—including *Phytophthora* species and isolate, host genotype, temperature, season, and duration of soil flooding—contribute to disease incidence and severity (76). Matheron (76) reported that disease severity increased in *J. hindsii* seedlings growing in either *P. cryptogea*–or *P. citrophthora* (Smith and Smith) Leonian infested soil as the duration of soil flooding was increased from 6 to 48 hours. Similar results were found for 'Paradox' growing in soil infested with *P. citricola*. Whether these results are due to changes in host physiology or inoculum load is unknown. Changes in seasonal susceptibility of *J. hindsii* and 'Paradox' rootstocks exposed to *P. citricola* were also observed with peak susceptibility in June and July.

Because there appears to be a strong interaction between host, pathogen, and environment, disease control is expected to involve both soil water management and improved rootstocks. Of particular importance is avoiding prolonged wetting of the lower trunk and crown and minimizing the duration of soil saturation during irrigations.

Chemical control of *Phytophthora* species affecting walnuts is under investigation (76). Under artificial conditions, metalaxyl and phosethyl Al were found to reduce disease severity in *J. hindsii* and 'Paradox' growing in soil infested with *P. citricola* and *P. cinnamomi*, but these chemicals are not yet recommended for control in walnut orchards and the economics of their use is uncertain.

7.1.2. Armillaria mellea. *Phytophthora* root and crown rots can sometimes be confused with infection caused by *Armillaria mellea* because early symptoms in the scion are similar. Signs of *Armillaria* infection—rhizomorphs, mycelial fans under the bark, and a distinct mushroomlike odor—can be used to distinguish the two diseases (79). Spread of this fungus may occur through root contact between healthy and diseased trees, which creates expanding pockets of infection in otherwise healthy orchards. The fungus can remain viable in infected wood long after the host is dead and thus can be spread by transport of infected wood by man or floodwaters.

Armillaria infection has been recognized as a problem in California walnut orchards since the industry began there and is partially responsible for the California preference of *J. hindsii* over *J. regia* for rootstocks (121). The incidence of *Armillaria* infection is localized, sometimes limited to a few trees, and is not considered a severe threat to the industry in California. In Oregon as well, *Armillaria* is not considered an important problem (83).

Reports from California indicate that *J. hindsii* is resistant to *Armillaria* infection, *J. regia* is susceptible, and 'Paradox' is variable in response, with some clones having more resistance than other (98, 117). Results from surveys in Oregon, however, suggest that *J. regia* has more resistance to *Armillaria* infection than *J. hindsii* (83). Recent observations in California suggest that *Armillaria* can also cause the death of trees on *J. hindsii* rootstock. The contradictory reports in the literature may be due to variability among strains of the pathogen or to intraspecific variation in the host. Additional research is needed to distinguish these factors and develop rootstock recommendations for locations where *Armillaria* might become a problem.

Without an established source of rootstock resistant to the pathogen, management control measures become essential. These include avoidance of planting in areas with known *Armillaria* infection, removal of all tree roots from previous plantings, general sanitation, and controlled irrigation (79). Problems with *Armillaria* in California may increase with replanting or if more virulent strains develop.

7.1.3. Agrobacterium tumefaciens.

Crown gall, caused by the bacterium *Agrobacterium tumefaciens*, is a less serious threat to walnut orchards in California than *Phytophthora*, primarily because effective bio-control measures have been established for new plantings in which crown gall is most damaging. The organism is considered to be ubiquitous in soils, and infection usually occurs through wounds in roots of newly planted trees. Galls, which can vary in size from minute to over 30 cm in diameter, develop at the base of the infected stem or on the roots. Loss of scion vigor and, occasionally, tree death occur as a result of girdling by the undifferentiated gall tissue. Small galls can be removed surgically. Variation in susceptibility to crown gall has been reported (Table 2), with *J. hindsii* considered to be less affected by crown gall than either *J. regia* or 'Paradox' (117). Disease-free nursery stock, good sanitation, avoidance of root wounds, and preplant dips with the antagonistic bacteria *A. radiobacter* have proven to be sufficient control for the present (47), but the development of strains of crown gall resistant to the antagonistic bacteria is already a concern for the future.

7.1.4. Blackline Disease.

The blackline disease, caused by the cherry leaf roll virus, is considered a major threat to walnut production where the predominant rootstock species are *J. hindsii* and 'Paradox,' but the disease is only indirectly a rootstock problem. In diseased trees the rootstock often remains virus free, but a lethal girdle at the graft union causes shoot dieback, decline, and eventual scion death. Northern California black walnut and 'Paradox' (and several other *Juglans* species), cannot be systemically infected by the cherry leaf roll virus, but commercial cultivars of Persian walnut are subject to such systemic infection. Thus the virus, which is pollen borne (74, 75, 104), can enter the Persian scion through the flowers during pollination and travel down the stem until it reaches the black walnut or 'Paradox' rootstock (85, 87). The rootstock resists systemic infection by the virus through a "hypersensitive" reaction in which both the virus and a thin layer of rootstock cells die, causing the black line seen at the graft union. This line may be very narrow, as in *J. hindsii* and some 'Paradox,' or may develop into a canker extending as much as 20 cm below the graft union, as in other 'Paradox' individuals. Profuse suckering of the rootstock may also accompany formation of the black line. The rate of movement of the virus within a tree is now being investigated (S. M. Mircetich, personal communication), and may depend on host genotype and physiology as well as on climatic conditions. Young trees infected in their first few years of flowering theoretically could die within a few years, but large, mature trees could remain productive for many years after infection.

Blackline disease was first observed in Oregon in 1924 (108) and is now known to be present in France, England, Italy, and Hungary, (18, 19, 24, 25, 49, 106). In California it has been gradually spreading from the central coast counties to all the walnut-growing regions. Since trees on their own roots or on *J. regia* rootstock do not exhibit the typical blackline disease symptoms, it is difficult to assess the origin or international range of the disease without controlled studies.

An enzyme-linked immunosorbent assay (ELISA) recently has been developed to detect the presence of the virus (104). Because the virus may be unevenly distributed in mature trees, extensive sampling is required before a Persian walnut tree can be assumed free of the virus. One potential method to reduce spread of the disease to new locations is to graft scionwood (without catkin buds) onto a hypersensitive rootstock. A hypersensitive rootstock should cause rejection of infected scions. In California, where 'Paradox' and *J. hindsii* are the common rootstocks, this precaution is automatic. In countries for which no blackline disease has been reported, such as China, all introduced scionwood should be grafted onto hypersensitive rootstock. Because the virus can also be seed borne in systemically infected hosts (19, 85), precautions with use of seed are also necessary. The ELISA method of detection is more reliable in this case, because the virus is more uniformly distributed in infected seedlings (104).

J. regia, particularly 'Manregian,' has been recommended over other rootstock species to meet the blackline disease problem in Oregon (125). Objections to *J. regia* rootstock for California include its susceptibility to other diseases and soil problems, its lack of vigor, and the possibility that infected trees on *J. regia* rootstock will serve as an inoculum source for many years and thus contribute to spread of blackline disease. There are several reports of blackline disease symptoms in *J. regia*, particularly in seedlings and in mature trees grafted on *J. regia* rootstock (18, 19, 49, 87). Kolber (49) suggests that the cherry leaf roll virus is responsible for leaf and nut symptoms as well as general decline in Persian walnut trees. Whether these effects are due to environment, virus strain, or host genotype is unknown and this uncertainty emphasizes the need to avoid the international spread of virus strains.

Another measure recommended to reduce the impact of blackline disease and still use 'Paradox' and *J. hindsii* rootstock employs high-density plantings to achieve maximum returns before losses from the disease occur (D. E. Ramos, personal communication). To be successful, cultivars selected for such orchards should be precocious, high yielding, and of small stature. The lower cost of *J. hindsii* might make it more economical if *Phytophthora* or other soil problems are not present.

In sparsely infected, mature orchards, removal of infected trees followed by replants on 'Paradox' rootstock is recommended. An alternative is regrafting rootstock suckers. In any case, cultivars selected should not have a female bloom period that coincides with the pollen-shedding period of infected varieties nearby.

Individual selection within *J. regia* both for tolerance to the cherry leaf roll virus and resistance to other diseases may well lead to improved sources of rootstock in the future. Nevertheless, the specter of general decline in *J. regia* grafted on *J. regia* will remain until the effects of the virus on "tolerant" hosts, environmental factors, and variability within virus are better understood.

7.2. Nematodes

Several genera of nematodes parasitize walnut rootstocks in California orchards. The most frequently detected are the root lesion nematode (*Pratylenchus vulnus* Allen and Jenson), ring nematode (*Criconemella xenoplax* Raski), dagger nematode (*Xiphinema americanum* Cobb) and root knot nematode (*Meloidogyne* spp.) (63). The aboveground symptoms of poor growth and dieback resulting from damage to the small feeder roots are similar to the symptoms of other root-related disorders. The root lesion nematode also attacks lateral roots that have undergone secondary enlargement. Black necrotic lesions can exceed several cm^2 in area and extend into the secondary phloem of larger roots. Positive diagnosis of nematode effects can only be accomplished by examination of roots and soils for galls, lesions, and/or presence of nematodes.

P. vulnus is considered the most serious nematode threat to walnuts in California because it is widespread and highly pathogenic. Both the ring and root knot nematodes build up mainly in sandy soils, where walnut is infrequently planted. In addition, black walnut provides resistance to root knot nematode species commonly found in California. The dagger nematode is a virus vector in other tree crop species, but has not been implicated as such in walnuts in California.

The impact of nematodes depends on their concentration in the soil; thus they are most often a problem in replant situations. Control measures include delaying planting of old orchard sites for at least two years, removal of old roots, and preplant fumigation. Mature trees appear to tolerate fairly high populations of nematodes. Adverse effects occur when damage to and death of roots exceed the plant's ability to regenerate them.

Several studies have been undertaken to compare the tolerance of various rootstocks to nematodes, primarily the root lesion nematode (21, 22, 62, 64, 73, 115). No rootstock tested was free of lesions, but differing levels of tolerance were reported. *P. stenoptera* has been shown to be quite tolerant to root lesion nematode (21, 64) even though this tree can be a good host for the nematode (65). Early trials indicated high sensitivity to root lesion nematode and little difference between *J. hindsii* and several sources of *J. regia* seedlings including 'Eureka' (21, 64, 115). In a recent trial employing greater replication and longer exposure than the earlier ones, 'Eureka' seedlings were found to be less affected than *J. hindsii* by root lesion nematode (73). Notable in all reports is a high level of variability within each species. 'Paradox' has been considered to be more tolerant than either of its parent species (64, 117) but recently was reported as equivalent to *J. regia* and more tolerant than *J. hindsii* (73). The 'Paradox' response to root lesion nematode probably depends on the source of seedlings (64, 117).

Limited testing of the following species has shown all to be seriously affected by root lesion nematodes: *J. californica*, *J. neotropica*, *J. ailantifolia*, *J. major*, *J. microcarpa*, and a black walnut type from Mexico (64, 115). *J. nigra* responded poorly (64) or moderately well (115) depending on the study.

J. regia was reported to be possibly more tolerant than *J. hindsii* to the ring nematode, but less tolerant than *J. hindsii* to the root knot nematode (22, 63). Differences were small and considered to be of questionable horticultural importance. In a recent evaluation of tolerance to root knot nematodes, *J. hindsii* and 'Paradox' both performed better than the average *J. regia* source (M. V. McKenry and G. H. McGranahan, unpublished); but again both intraspecific and intrafamily variation was substantial. The results from all these studies suggest that individual selection followed by clonal propagation will be the most successful method for obtaining nematode resistance in walnut rootstock.

8. NEW AND FUTURE ROOTSTOCKS

Four interrelated factors will have a major impact on walnut rootstocks in the future. These are micropropagation, the increasing incidence of blackline disease and *Phytophthora*, and a renewed national emphasis on germplasm collection and maintenance.

The National Clonal Germplasm Repository, Davis, California, is responsible for maintaining a collection of *Juglans*. This collection is expected to provide material (1) for evaluation of the species that previously have not been available, (2) for investigation of intraspecific variation in important traits, and (3) for inter- and intraspecific hybridization. This germplasm collection is not only valuable in research aimed at meeting such current problems as blackline disease and *Phytophthora*, but, more importantly, the collection will be a resource that can be used to meet future problems.

The potential impact of micropropagation is wide ranging. It will allow mass clonal propagation of selected rootstocks or even of own-rooted cultivars. Micropropagated 'Paradox' is expected to be commercially available in California within five years. Micropropagation influences breeding strategies, which can now be aimed at individual selection. Moreover, it increases the potential of hybrids as rootstocks because even those that can only be produced through controlled pollinations can now be screened and released. But if monoculture is practiced, it may also contribute to the genetic vulnerability of walnut orchards.

Technologies related to micropropagation are still under investigation for *Juglans*. These include *in vitro* selection for disease resistance, embryo rescue for wide crosses, and somatic embryogenesis to determine the potential of somaclonal variation (128). Biotechnology is advancing rapidly in woody species, but it is too early to predict when the products of this work will reach the orchard. Certainly the increasing incidence of *Phytophthora* in California and blackline disease worldwide will provide the impetus to use both conventional and novel strategies for rootstock improvement.

Although the industry desires a rootstock with the vigor of 'Paradox,' the tolerance to *Phytophthora*, waterlogging, and nematodes of *Pterocarya*, and the tolerance to blackline disease of *J. regia*, some necessary compromises are to be expected. Thus the walnut rootstocks of the future will still require careful management for optimum performance.

INTERNATIONAL CATALOG OF ROOTSTOCK BREEDING PROGRAMS

Walnuts

France

Institut National de la Recherche Agronomique
B.P. 131-33140
Pont-de-la-Maye, France
Ing. E. Germain

United States
USDA/ARS
Department of Pomology
University of California
Davis, CA 95616
Gale McGranahan

REFERENCES

1. Arulsekar, S., D. E. Parfitt, and G. McGranahan (1985). Isozyme gene markers in *Juglans* species. Inheritance of GPI and AAT in *J. regia* and *J. hindsii*, *J. Hered.*, **76**, 103–106.

2. Atkinson, D. (1980). The distribution and effectiveness of the roots of tree crops, *Hort. Rev.*, **2**, 424–490.

3. Ayers, R. S. (1977). Quality of water for irrigation, *J. Irrig. Drain. Div. Proc. Amer. Soc. Civ. Eng.*, **103**, 135–154.

4. Batchelor, L. D., O. L. Braucher, and E. F. Serr (1945). *Walnut Production in California*, Univ. Calif. Agri. Expt. Sta. Circ. 364.

5. Begg, E. L. (1985). "Identification and evaluation of soils," in *Walnut Orchard Management*, D. E. Ramos, Ed., Div. Agri. Nat. Res. Univ. Calif. Publ. 21410, pp. 20–27.

6. Beineke, W. F. (1981). "New directions in genetic improvement: grafted black walnut plantations," in *Black Walnut for the Future*, USDA For. Serv. Gen. Tech. Rept. NC-74, pp. 64–68.

7. Beineke, W. F. (1983). The genetic improvement of black walnut for timber production, *Plant Breed. Rev.*, **1**, 236–266.

8. Bernstein, L. (1980). *Salt Tolerance of Fruit Crops*, USDA Agri. Infor. Bull. No. 292.

9. Beutel, J., K. Uriu, and O. Lilleland (1978). "Leaf analysis for California deciduous fruits," in *Soil and Plant-tissue Testing in California*, H. M. Reisenauer, Ed., Univ. Calif. Div. Agri. Sci. Bull. 1879, pp. 11–14.

10. Brinkman, K. A. (1974). "*Juglans* L. walnut," in *Seeds of Woody Plants in the United States*, C. S. Schopmeyer, Ed., USDA For. Serv. Handbook 450, U.S. Govt. Print. Off., pp. 454–459.

11. Browne, L. T., L. C. Brown, and D. E. Ramos (1977). Walnut rootstocks compared, *Calif. Agri.*, **31**(7), 14–15.

12. Carlson, R. M. (1985). "Mineral nutrient availability," in *Walnut Orchard Management*, D. E. Ramos, Ed., Div. Agric. Nat. Res., Univ. Calif. Publ. 21410, pp. 110–115.

13. Catlin, P. B., and B. Shaybany (1976). Phenolic compounds as toxins in hypersensitivity of *Juglans* seedlings to waterlogging, *HortScience*, **11**, 306 (Abstr.).

14. Catlin, P. B., G. C. Martin, and E. A. Olsson (1977). Differential sensitivity of *Juglans hindsii*, *J. regia*, Paradox hybrid, and *Pterocarya stenoptera* to waterlogging, *J. Amer. Soc. Hort. Sci.*, **102**, 101–104.

15. Catlin, P. B., and W. R. Schreader (1985). "Root physiology and walnut rootstock characteristics," in *Walnut Orchard Management*, D. E. Ramos, Ed., Div. Agri. Nat. Res., Univ. Calif. Publ. 21410, pp. 75–81.

16. Chase, S. B. (1949). The dwarfing effect of *Juglans rupestris*, *40th Ann. Rept. North. Nut Growers Assoc.*, pp. 158–159.

17. China Trees Committee (1978). *China Main Forest Planting Technology*, Agri. Publ. Co., Beijing, China.

18. Cooper, J. I. (1980). The prevalence of cherry leaf roll virus in *Juglans regia* in the United Kingdom, *Acta. Phytopath. Hung.*, **15**, 139–145.

19. Cooper, J. I., and M. L. Edwards (1980). Cherry leaf roll virus in *Juglans regia* in the United Kingdom, *Forestry*, **53**, 41–50.

20. Crossa-Raynaud, P., and E. Germain (1982). Avenir de la culture des arbres fruitiers à fruits sec dans les pays méditerranéens: Amandier, noyer, noisetier, pistchier, *Fruits*, **37**, 617–626.

21. Day, L. H., and E. F. Serr (1951). Comparative resistance of rootstocks of fruit and nut trees to attack by a root-lesion or meadow nematode, *Proc. Amer. Soc. Hort. Sci.*, **57**, 150–154.

22. Day, L. H., and W. P. Tufts (1944). *Nematode-resistant Rootstocks for Deciduous Fruit Trees*, Univ. Calif. Agric. Exp. Sta. Circ. 359.

23. DeCandolle, A. (1882). *The Origin of Cultivated Plants*, republished 1959, Hafner, New York.

24. Delbos, R., A. Bonnet, and J. Dunez (1984). Le virus de l'enroulement des feuilles du cerisier, largement répandu en France sur noyer, est il à l'origine de l'incompatibilité de greffage du noyer *Juglans regia* sur *Juglans nigra*? [Cherry leaf roll virus: The causal agent of incompatibility of *Juglans regia* grafted on *J. nigra*?], *Agronomie*, **4**, 333–339.

25. Delbos, R., C. Kerlan, J. Dunez, M. Lansac, F. Dosba, and E. Germain (1982). Virus infection of walnuts in France, *Acta. Hort.*, **130**, 123–131.

26. Driver, J. A., and A. H. Kuniyuki (1984). *In vitro* propagation of 'Paradox' walnut rootstock, *HortScience*, **19**, 507–509.

27. Food and Agriculture Organization of the United Nations (1983). *FAO Production Yearbook 36*, FAO, Rome, 320 pp.

28. Forde, H. I. (1975). "Walnuts," in *Advances in Fruit Breeding*, J. Janick and J. N. Moore, Eds., Purdue Univ. Press, West Lafayette, IN, pp. 439–455.

29. Forde, H. I. (1979). "Persian walnut in the western United States," in *Nut Tree Culture in North America*, R. A. Jaynes, Ed., North. Nut Growers Assoc., Hamden, CT., pp. 84–97.

30. Funk, D. T. (1970). *Genetics of Black Walnut*, USDA For. Serv. Res. Pap. WO-10.

31. Funk, D. T. (1979). "Black walnuts for nuts and timber," in *Nut Tree Culture in North America*, R. A. Jaynes, Ed., North. Nut Growers Assoc., Hamden, CT., pp. 51–73.

32. Garavel, L. (1956). A la recherche de noyers résistants au froid, *Rev. Forets Franc*, **8**, 572–575.

33. Germain, E., J. Jalinat, and M. Marchou (1975). "Les porte-greffes du noyer," in

Le Noyer, F. Bergougnoux and P. Grospierre, Eds., Brochure INVUFLEC, Paris, pp. 53–72.

34. Geyer, W. A., R. D. Marquard, and J. F. Barber (1980). Black walnut site quality in relation to soil and topographic characteristics in northeastern Kansas, *J. Soil Water Conser.*, **35**, 135–137.

35. Gibson, M. D. (1967). The Manregian walnut, *58th Ann. Rept. North Nut Growers Assoc.*, pp. 105–109.

36. Glenn, E. M. (1966). Incompatibility in the walnut, *Ann. Rept. E. Malling Res. Sta. for 1965*, pp. 102–103.

37. Green, R. J., Jr., and R. C. Ploetz (1979). "Root rot of black walnut seedlings caused by *Phytophthora citricola*," in *Walnut Insects and Diseases*, USDA For. Serv. Gen. Tech. Rept. NC-52, pp. 5–9.

38. Griffin, J. R., and W. B. Critchfield (1972). *The Distribution of Forest Trees in California*, USDA For. Serv. Res. Pap. PSW-82/1972.

39. Grimes, D. W., P. L. Wiley, and A. B. Carlton (1982). Plum root growth in a variable-strength field soil, *J. Amer. Soc. Hort. Sci.*, **107**, 990–992.

40. Grimo, E. (1979). "Carpathian (Persian) walnuts," in *Nut Tree Culture in North America*, R. A. Jaynes, Ed., North. Nut Growers Assoc., Hamden, CT., pp. 74–83.

41. Haas, A. R. C. (1929). Composition of walnut trees as affected by certain salts, *Bot. Gaz.*, **87**, 364–396.

42. Haas, A. R. C. (1939). Root temperature effects on the growth of walnut and avocado seedlings, *Calif. Avocado Assoc. Yearbook*, pp. 96–102.

43. Hendricks, L. C., R. H. Gripp, and D. E. Ramos (1977). *Walnut Production in California*, Univ. Calif. Div. Agri. Sci. Leaflet 1984.

44. Hook, D. D. (1984). "Adaptations to flooding with fresh water," in *Flooding and Plant Growth*, T. T. Kozlowski, Ed., Academic, New York, pp. 265–295.

45. Horoschak, T. (1972). *The Walnut Industries of the Mediterranean Basin*, USDA For. Agric. Serv. M-245.

46. Howard, W. L. (1945). *Luther Burbank's Plant Contributions*, Calif. Agri. Expt. Sta. Bull. 691.

47. IPM Manual Group (1982). *Integrated Pest Management for Walnuts*, Univ. Calif. Div. Agric. Sci. Publ. 3270.

48. Kaeiser, M., J. H. Jones, and D. T. Funk (1975). Interspecific walnut grafting in the greenhouse, *Plant Propagator*, **20**, 2–7.

49. Kolber, M., M. Nemeth, and P. Szentivanyi (1982). Routine testing of English walnut mother trees and group testing of seeds by ELISA for detection of cherry leaf roll virus infection, *Acta. Hort.*, **130**, 161–171.

50. Komanich, I. G. (1980). *Biology, Cultivation, and Breeding of Walnut*, A. A. Chebotar, Ed., Kishinev, Shtiintsa Publishers (The Botanic Gardens of the Academy of Sciences of the Moldavian SSR).

51. Komanik, P. P., R. C. Schultz, and W. C. Bryan (1982). The influence of vesicular-arbuscular mycorrhizae on the growth and development of eight hardwood tree species, *For. Sci.*, **28**, 531–539.

52. Kozlowski, T. T., and S. G. Pallardy (1984). "Effect of flooding on water,

carbohydrate, and mineral relations," in *Flooding and Plant Growth*, T. T. Kozlowski, Ed., Academic, New York, pp. 165–193.

53. Kramer, P. J. (1983). *Water Relations of Plants*, Academic, New York.

54. Kuang, K., and P. Li (1979). "Juglandaceae," in *Flora Reipublicae Popularis Sinicae*, Vol. 21.

55. Lagerstedt, H. B., and W. W. Roberts (1972). Walnut grafting in Oregon—problems and solutions, *63rd Ann. Rept. North. Nut Growers Assoc.*, pp. 17–23.

56. Lagerstedt,. H. B. (1978). Hardiness in filberts and walnuts, *69th Ann. Rept. North. Nut Growers Assoc.*, pp. 118–121.

57. LaRue, J. H., W. D. McClellan, and W. L. Peacock (1975). Mycorrhizal fungi and peach nursery nutrition, *Calif. Agri.*, **28**(5), 6–7.

58. Lebedinets, L. N. (1968). "Intergenic hybridization in the family Juglandaceae," in *Biol. nauk. v un-takh i ped. in-takh Ukrainy*, pp. 362–363. (Abstr. in D. T. Funk and M. K. Dillow [1977]. *Breeding Nut-producing Species: Juglans and Carya, an Annotated Bibliography*, available from authors.)

59. Losche, C. K. (1973). "Selecting the best available soils," in *Black Walnut as a Crop*, USDA For. Serv. Gen. Tech. Rept. NC-4, pp. 33–35.

60. Losche, C. K., and R. E. Phares (1972). Siltation damage in a black walnut plantation, *J. Soil Water Conser.*, **27**:228–229.

61. Loucks, W. L., and R. A. Keen (1973). *Submersion tolerance of selected seedling trees*, *J. For.*, **71**, 496–497.

62. Lownsbery, B. F. (1956). *Pratylenchus vulnus*, primary cause of the root-lesion disease of walnuts, *Phytopathology*, **46**, 376–379.

63. Lownsbery, B. F. (1985). "Controlling nematodes that parastize roots," in *Walnut Orchard Management*, D. E. Ramos, Ed., Div. Agric. Nat. Res., Univ. Calif. Publ. 21410, pp. 127–129.

64. Lownsbery, B. F., G. C. Martin, H. I. Forde, and E. H. Moody (1974). Comparative tolerance of walnut species, walnut hybrids, and wingnut to the root-lesion nematode, *Pratylenchus vulnus*, *Plant Dis. Rept.*, **58**, 630–633.

65. Lownsbery, B. F., and E. F. Serr (1963). Fruit and nut tree rootstocks as hosts for a root-lesion nematode, *Pratylenchus vulnus*, *Proc. Amer. Soc. Hort. Sci.*, **82**, 250–254.

66. Maas, E. V., and G. J. Hoffman (1977). Crop salt tolerance—Current assessment, *J. Irrig. Drain, Div., Proc. Amer. Soc. Civ. Eng.*, **103**, 115–134.

67. Manning, W. E. (1957). The genus *Juglans* in Mexico and Central America, *J. Arnold Arbor.*, **38**, 121–150.

68. Manning, W. E. (1960). The genus *Juglans* in South America and West Indies, *Brittonia*, **12**, 1–26.

69. Manning, W. E. (1962). Additional notes on *Juglans* and *Carya* in Mexico and Central America, *Bull. Torrey Bot. Cl.*, **89**, 110–113.

70. Manning, W. E. (1978). The classification within the Juglandaceae, *Ann. Mo. Bot. Gdn.*, **65**, 1058–1087.

71. Maronek, D. M., J. W. Hendrix, and J. Kiernan (1981). Mycorrhizal fungi and their importance in horticultural crop production, *Hort. Rev.*, **3**, 172–213.

72. Martin, G. C., and H. I. Forde (1975). Incidence of blackline in *Juglans regia* L. propagated on various rootstock species, *J. Amer. Soc. Hort. Sci.*, **100**, 246–249.

73. Martin, G., B. Lownsbery, and C. Nishijima (1983). Which rootstock for root-lesion nematode, *Sun Diamond Grower*, **3**(6), 30–32.

74. Massalski, P. R., and J. I. Cooper (1982). The association of cherry leaf roll virus with birch and walnut pollen, *Acta Hort.*, **130**, 291–292.

75. Massalski, P. R., and J. I. Cooper (1984). The location of virus-like particles in the male gametophyte of birch, walnut and cherry naturally infected with cherry leaf roll virus and its relevance to vertical transmission of the virus, *Plant Pathol.*, **33**, 255–262.

76. Matheron, M. E. (1984). Factors affecting incidence, severity and control of *Phytophthora* root and crown rot of English walnut trees in California, Ph.D. thesis, Univ. of Calif., Davis.

77. Matheron, M. E., and S. M. Mircetich (1983). Effects of various lengths of soil saturation on severity of crown rot in three English walnut rootstocks caused by *Phytophthora citricola*, *Phytopathology*, **73**, 813 (Abstr.).

78. Matheron, M. E., and S. M. Mircetich (1983). Relative resistance of four different English walnut rootstocks to *Phytophthora citricola*, *Phytopathology*, **73**, 813 (Abstr.).

79. McCain, A. H., and R. D. Raabe (1981). Armillaria *Root Rot: Oak Root Fungus*, Univ. of Calif., Berkeley, Leaflet 2590.

80. McGranahan, G. H., and H. I. Forde (1985). "Genetic improvement," in *Walnut Orchard Management*, D. E. Ramos, Ed., Div. Agric. Nat. Res., Univ. of Calif. Publ. 21410, pp. 8–12.

81. Mengel, K., and E. A. Kirkby (1982). *Principles of Plant Nutrition*, Intl. Potash Inst., Bern, Switzerland.

82. Miller, D. E. (1982). "Physical characteristics of soils as they relate to irrigation and water management," in *Water Management and Irrigation of Tree Fruits*, R. B. Tukey, Ed., Coop. Ext. Wash. State Univ., Pullman, WA, pp. 3–28.

83. Miller, P. W., J. H. Painter, and C. O. Rawlings (1958). *Blackline and Root Rots of Persian Walnut in Oregon*, Oregon State College, Agri. Exp. Sta., Misc. Pap. 55.

84. Mircetich, S. M., and M. E. Matheron (1983). *Phytophthora* root and crown rot of walnut trees, *Phytopathology*, **73**, 1481–1488.

85. Mircetich, S. M., and A. Rowhani (1984). The relationship of cherry leaf roll virus and blackline disease of English walnut trees, *Phytopathology*, **74**, 423–428.

86. Mircetich, S. M., G. A. de Zoeten, and J. A. Lauritis (1980). Etiology and natural spread of blackline disease of English walnut tress, *Acta Phytopath. Hung.*, **15**, 147–151.

87. Mircetich, S. M., R. R. Sanborn, and D. E. Ramos (1980). Natural spread, graft-transmission, and possible etiology of walnut blackline disease, *Phytopathology*, **70**, 962–968.

88. Moller, W. J., and D. E. Ramos (1972). A guide to walnut diseases, *Diamond Walnut News*, **54**(6), 28–29.

89. Monk, R. W., and H. H. Wiebe (1961). Salt tolerance and protoplasmic salt hardiness of various woody and herbaceous ornamental plants, *Plant. Physiol.*, **36**, 478–482.

90. Nedev, N. V. (1973). L'Influence des porte-greffes de noyer commun, noir et mandchou sur les greffons pendant la première et le dèbut de la seconde pèriodes

de developpement. II. Manifestations reproductive des noyers, *Gradinarska i Lozarska Nauka*, **10**, 7–8.

91. Nightingale, C. T. (1935). Effects of temperature on growth, anatomy, and metabolism of apple and peach roots, *Bot. Gaz.*, **96**, 581–639.

92. Olson, W. H., K. Uriu, D. E. Ramos, and J. Pearson (1978). A field guide to walnut nutrition, *Diamond Walnut News*, **60**(3), 39–41.

93. Ozol, A. M. (1949). Winter-hardiness of *Juglans regia* and other species of walnut, *Dokl. Akad. Nauk. SSR*, **66**(4), 725–728. (Abstr. in D. T. Funk, Ed. [1966]. *Annotated Bibiliography of Walnut and Related Species*, U.S. For. Res. Pap. NC-9, N. Central For. Expt. Sta., St. Paul, Minn.)

94. Pelletreau, K. G. (1983). Rooting of leafy walnut cuttings and the role of mycor-rhizae in the survival of transplanted walnut seedlings and rooted cuttings, M.S. thesis, Univ. of Calif., Davis.

95. Ponder, F., Jr. (1981). Some guidelines for selecting black walnut planting sites, *72nd Ann. Rept. North Nut Growers Assoc.*, pp. 112–117.

96. Ponder, F., Jr. (1983). Soil moisture levels and mycorrhizal infection in black walnut seedlings, *Commun. Soil Sci. Plant Anal.*, **14**, 507–511.

97. Popov, S. (1983). Similarity of the fruit of mother trees and their progeny when propagating *Juglans regia* by seed, *Gorskostopanska Novka*, **20**, 12–19.

98. Raabe, R. D. (1979). *Resistance or Susceptibility of Certain Plants to* Armillaria *Root Rot*, Univ. of Calif., Berkeley, Leaflet 2591.

99. Ramos, D. E., G. H. McGranahan, and L. Hendricks (1984). Walnuts, *Fruit Var. J.*, **38**, 112–120.

100. Rawlings, C. O. (1953). Trends with walnut rootstocks, *Ann. Rept. Oregon State Hort. Soc.*, **45**, 216–218.

101. Rehder, A. (1940). *Manual of Cultivated Trees and Shrubs Hardy in North America*, 2nd ed., Macmillan, New York.

102. Rietveld, W. J. (1981). The significance of allelopathy in black walnut cultural systems, *72nd Ann. Rept. North Nut Growers Assoc.*, pp. 117–134.

103. Rizzi, A. D., R. H. Gripp, and N. W. Ross (1967). *Care of a Walnut Orchard*, Univ. of Calif. Agri. Ext. Serv. AXT-238.

104. Rowhani, A., S. M. Mircetich, R. J. Shepherd, and J. D. Cucuzza (1985). Serological detection of cherry leafroll virus in English walnut trees, *Phytopathology*, **75**, 48–52.

105. Sartori, E. (1967). The use of *Juglans australis* in Argentina as rootstock for English walnut, *Plant Propagator*, **13**, 4–7.

106. Savino, B., A. Quacquarelli, D. Gallitelli, P. Piazzola, G. P. Martelli, (1976). Occurrence of two sap transmissible viruses in walnut, *Mitt. biol. Bundesanst. Land Forstwirsch*, **170**, 23–27.

107. Schlesinger, R. C., and D. T. Funk (1977). *Manager's Handbook for Black Walnut*. USDA For. Serv. North. Cent. For. Exp. Stn. Gen. Tech. Rep. NC-38.

108. Schuster, C. E. and P. W. Miller (1933). A disorder of Persian (English) walnuts grafted on black walnut stocks, resulting in girdling, *Phytopathology*, **23**, 408–409.

109. Schuster, C. E., R. E. Stephenson, and W. Evendun (1944). Mycorrhizae of filbert and walnut trees in Oregon orchards, *Bot. Gaz.*, **105**, 388–392.

110. Seliskar, C. E., and C. L. Wilson (1981). "Yellows diseases of trees," in *Myco-plasma Diseases of Trees and Shrubs*, K. Maramorosch and S. P. Raychaudhuri, Eds., Academic, New York, pp. 35–96.

111. Semahanova, N. M., and O. P. Mazur (1968). Mycorrhizae of *Juglans regia* and the conditions for their formation [in Russian, original not seen], *Izv. Akad. Nauk SSR* (ser. biol.), **4**, 517–529.

112. Serr, E. F. (1965). Dwarfing the Persian walnut by use of interstocks, *56th Ann. Rep. North Nut Growers Assoc.*, pp. 106–111.

113. Serr, E. F. (1968). Dwarfing interstocks for Persian walnuts, *Plant Propagator*, **14**, 10–13.

114. Serr, E. F., Jr. (1969). "Persian walnuts in the western states," in *Handbook of North American Nut Trees*, R. A. Jaynes, Ed., North. Nut Growers Assoc., Knoxville, TN, pp. 240–263.

115. Serr, E. F., and L. H. Day (1949). Lesion nematode injury to California fruit and nut trees, and comparative tolerance of various species of the Juglandaceae, *Proc. Amer. Soc. Hort. Sci.*, **53**, 134–140.

116. Serr, E. F, and H. I. Forde (1951). Comparison of size and performance of mature Persian walnut trees on Paradox hybrid and *J. hindsii* seedling rootstocks, *Proc. Amer. Soc. Hort. Sci.*, **57**, 198–202.

117. Serr, E. F., and A. D. Rizzi (1964) *Walnut Rootstocks*, Univ. of Calif. Agric. Ext. Serv. Publ. AXT-120.

118. Shannon, M. C. (1979). In quest of rapid screening techniques for plant salt tolerance, *HortScience*, **14**, 587–589.

119. Shreve, L. W. (1981). Adapting exotic or non-native walnuts and stone fruits to high pH sites by use of native stocks, *72nd Ann. Rept. North Nut Growers Assoc.*, pp. 20–21.

120. Smith, C. O., and J. T. Barrett (1931). Crown rot of *Juglans* in California, *J. Agri. Res.*, **43**, 885–904.

121. Smith, R. E., C. O. Smith, and H. J. Ramsey (1912). *Walnut Culture in California. Walnut Blight*, Univ. Calif. Publ. Bull. 231, Berkeley.

122. Smith, R. E. (1941). *Diseases of Fruits and Nuts*, Calif. Coll. Ext. Circ. 120.

123. Smith, R. L., and H. Allinger (1969). Evaluation of Persian walnuts introduced from Russia, Hungary, and Korea, *60th Ann. Rept. North Nut Growers Assoc.*, pp. 61–65.

124. Solignat, G. (1974). Note sur le comportement à l'asphyxie radiculaire de porte-greffes du noyer *J. regia*, *Pomol. Franc.*, **16**, 75–77.

125. Stebbins, R. L. (1983). A solution to blackline in walnuts, *Nut Grower*, **1** (Jul./Aug.), 10–11, 21.

126. Stolzy, L. H., and H. Fluhler (1978). "Measurement and prediction of anaerobiosis in soils," in *Nitrogen in the Environment*, Vol. 1. *Nitrogen Behavior in Field Soil*, D. R. Nielsen and J. G. MacDonald, Ed., pp. 363–426.

127. Thomsen, H. H. (1963). *Juglans hindsii*, the central California black walnut, native or introduced? *Madroño*, **17**, 1–32.

128. Tisserat, N., and J. E. Kuntz (1984). Butternut canker: Development on individual trees and increase within a plantation, *Plant Dis.*, **68**, 613–616.

129. Tulecke, W., and G. H. McGranahan (1985). Somatic embryogenesis and plantlet regeneration from cotyledons of walnut, *Juglans regia* L., *Plant Sci*, 40:57–63.

130. Uriu, K. (1985). "Mineral nutrient diagnosis: Using soil/plant analyses and symptomology," in *Walnut Orchard Management*, D. E. Ramos, Ed., Div. Agric. Nat. Res., Univ. Calif. Publ. 21410, pp. 115–121.

131. U.S. Department of Agriculture (1983). Foreign Agricultural Circular F-Hort. 11-83, Foreign Agri. Serv., Washington DC.

132. Vavilov, N. E. (1941). The origin, variation, immunity, and breeding of cultivated plants, *Chron. Bot.*, 13 (1949/50), Waltham, MA.

133. Veihmeyer, F. J., and A. H. Hendrickson (1938). Soil moisture as an indication of root distribution in deciduous orchards, *Plant Physiol.*, **13**, 169–177.

134. Webster, B. M. (1970). Manregian walnuts, *61st Ann. Rept. North Nut Growers Assoc.*, pp. 36–40.

135. Whitehouse, W. E., and L. E. Joley (1948). Notes on growth of Persian walnut propagated on rootstocks of the Chinese wingnut *Pterocarya stenoptera*, *Proc. Amer. Soc. Hort. Sci.*, **52**, 103–106.

136. Wing, S. L., and L. J. Hickey (1984). The *Platycarya* perplex and the evolution of the Juglandaceae, *Amer. J. Bot.*, **71**, 388–411.

137. Witt, A. W. (1938). Walnuts, a survey of the investigations on the propagation and testing of walnuts at the East Malling Research Station, *Ann. Rept. for 1937*, pp. 259–266.

138. Yadava, U. L., and S. L. Doud (1980). The short life and replant problems of deciduous fruit trees, *Hort. Rev.*, **2**, 1–116.

139. Zielinski, Q. B. (1956). A three-year comparison of walnut rootstocks, *Nut Growers Soc. Ore. and Wash.*, **42**, 215–216.

140. Zielinski, Q. B. (1957). Comparison of walnut rootstock varieties in Oregon, *Fruit Var. Hort. Dig.*, **12**, 10–11.

14

VITIS ROOTSTOCKS

Gordon S. Howell
Michigan State University
East Lansing, Michigan

1. INTRODUCTION

The most widely planted species of the genus *Vitis* is *V. vinifera* L. Cultivars of *V. vinifera* are found in commercial plantings on all continents, and in nearly every instance the vines are not cultured on their own roots, even though cultivars of the species root easily. Grapevines of *V. vinifera* are nearly all grafted because they are very susceptible to a soil-borne insect—the grape phylloxera.

In the following narrative I shall attempt to provide a brief history of the expansion of grape culture worldwide, the discovery of phylloxera and its devastating effect on *V. vinifera* culture, the evolution of rootstocks to solve phylloxera and other soil problems, and, finally, a brief explanation of the grafting techniques used. In addition, an attempt will be made to raise questions relevant to the assessment of rootstock performance and subsequent rootstock improvement. The narrative will not include detailed ampelographic descriptions of current rootstocks. Readers interested in such information are referred to Perold (24), Galet (9) and Pongracz (27). The narrative will also avoid a tedious listing of the specific response of a specific rootstock on a specific soil in some specific location as reported in the literature. Rather, the emphasis will be to present the current status of our general understanding and to make suggestions about research approaches likely to yield broad general understanding of principles in the future.

2. ORIGIN AND HISTORY OF *VITIS* CULTURE

2.1. Old World Grape Culture

For centuries, grapevines were grown on their own roots. Evidence of grape gathering and grape culture goes back to prehistory and *vinifera* grapes likely evolved and were selected from the wild in the region of Asia Minor. Early human cultures in the area, including the Babylonian, Hebrew, and Egyptian civilizations, all include writings about grape culture and wine production. Grapes and wine also figure extensively in their artworks.

These civilizations were superseded by the Greeks, who were displaced by the Romans. Both these civilizations elevated wine production to the status of a gift of specific pagan gods: Dionysus of the Greeks and Bacchus of the Romans. The spread of grape culture throughout Europe was a by-product of the Roman legions conquering the major landmass of Europe and carrying vines and planting them in the conquered territories. Important areas planted by Romans, in addition to Italy, include the current countries of France, Germany, Spain, Portugal, Switzerland, Austria, Hungary, Bulgaria, and Romania.

The fall of the Roman Empire ushered in the Dark Ages, when grape culture, wine production, and other aspects of educated society were maintained by the monasteries and abbeys of the Church. The Church vineyards became part of very large clerical landholdings that remained until they were greatly reduced by Napoleon.

2.2. New World Grape Culture

With the discovery of the Americas and the trade routes around the Cape of Good Hope to the Orient and Australia, there was a gradual trickle of people from Europe that at first served the purposes of specific European national empires. This trickle gradually became a stream of adventurers and settlers who took, along with a religious commission to save the immortal souls of the pagan heathens and a national commission to expand the empire, grapevines so that wine, a staple in the diet of all Europeans, would be available in the new lands.

The transplantation of European *vinifera* vines had success on all continents, with the singular exception of the early attempts along the eastern portion of North America. In western North America, Spanish missionaries spread viticulture based on *vinifera*, and vines grew with considerable success.

The experience in eastern North America was viewed with surprise by the early settlers since native grapes grew in profusion in the same sites where transplanted *V. vinifera* vines struggled and finally succumbed. It was this native viticultural abundance that had caused Leif Erikson and Viking explorers to name the newly discovered North American continent "Vinland" (Prasta in Vine, 35).

2.3. *Vitis* Culture Worldwide

At present, grape culture exists on every continent and current plantings are estimated at over 10 million ha, making grapes the most widely grown fruit crop in the world.

Grape culture primarily exists between a range of mean annual temperate zone isotherms (10–20°C). Changes in altitude can make grape culture possible in tropical regions, and planting on equator-facing slopes can extend the range to temperature latitude extremes (32).

3. ROOTSTOCKS OF *VITIS* AND THEIR USAGE

3.1. Phylloxera in North America and Its Spread to Europe

The problem of *V. vinifera* mortality in the North American colonies along the Atlantic Ocean was a frustration to the colonists, but the major disaster occurred when similar forms of vine mortality were observed in French vineyards. The cause of this mortality was not understood until Planchon (26) discovered the grape phylloxera (*Dactylosphaera vitifolii* Shimer) on the roots of an infected grapevine in 1868 (27).

Phylloxera were introduced into France in the early 1860s (14) and were discovered in California at about the same time (33). The mode of entry into continental Europe is not known. Some viticulturists speculate that the large number of species native to the North American continent (the largest number of species on any continent) was a fertile opportunity for taxonomists and other grape enthusiasts. These people may have carried the phylloxera on the roots of their specimens back to Europe. Perold (24) suggests that the search for vines resistant to oidium (powdery mildew, *Uncinula necator*) led to the importation of American vines on which lived the grape phylloxera.

3.2. How Phylloxera Damage Grapevines

The type of damage done by the grape phylloxera is associated with the biology of the insect. Phylloxera are aphids in the Order Homoptera. Like many other aphids, they can reproduce both sexually and parthenogenetically (23). Depending on the form of the insect, both leaves and roots may be attacked. An example of the leaf gall stage is shown in Figure 1; this is the most common form observed in eastern U.S. viticulture (23).

Phylloxera attack on roots is the more serious concern. The insect overwinters as either a hibernating nymph or as an egg (19). The egg hatches in the spring as a nymph that attacks the leaves. After several generations on the leaves, some insects drop to the ground, burrow into the soil, and, upon reaching the roots, being feeding, which causes gall forms called "nodosities" and "tuberosities" on the roots (19). This feeding causes damage by depletion of the root nutrients directly and by physical and physiological damage to the root tissues.

Toward the end of the season, a winged variant will emerge from the soil and migrate to a new site for infestation. These migrants produce a generation of true sexual forms. Upon mating, the female lays an egg on the bark of the vine (19). Phylloxera require a soil with sufficient clay content that cracks upon

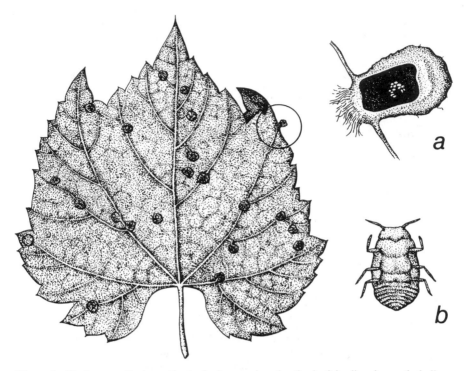

Figure 1. Phylloxera attack on the leaf of grapevines by the leaf feeding form of phylloxera. A cross-section of the leaf gall (a) shows eggs which have been laid by the phylloxera form hatching from the winter egg (produced sexually). The eggs (a) are produced parthenogenetically (asexually) and the hatched form is seen in (b). Drawings by M. Cameron.

drying (33). This provides an easy means of movement by the insect and facilitates its attack upon the root system.

Figure 2 depicts phylloxera attack on vine roots. About a month after initial attack by the insect on the root, the tissues about the nodosity begin to rot and the insect moves to a fresh tissue. This decay, plus the injection of concentrated growth promoter by the insect, results in stunted roots and a destruction of fine feeder roots (33). Root decline precedes vine decline and death.

3.3. Impact of Phylloxera on European Viticulture

Regardless of the mode of entry into continental Europe, the result was devastating. Little (19) says that nearly one-third of all French vineyards were destroyed before control measures were found. Johnson (14) suggests an even worse situation—that within 20 years of infestation the phylloxera had killed "virtually every vine in France." In any event, the impact was striking and dramatic; France has never regained the vineyard area under cultivation during the prephylloxera period.

   Cultural approaches to reduce phylloxera damage were attempted. One approach was to flood vineyards and "drown" phylloxera. This had very limited success. Vineyard site topography and heavy soils were the chief factors limiting the success of this approach.

   The European phylloxera infestation led to a worldwide infestation as European nurseries provided vines for the expansion of non-European produc-

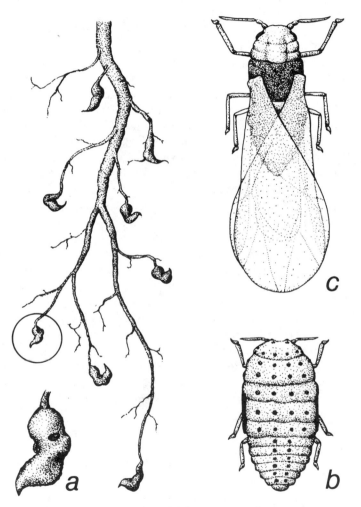

Figure 2. The grape phylloxera. Nodosities (a) form on grapevine roots in response to feeding by the root form (b). To complete the life cycle an acid in insect dispersal, a winged form is produced (c) which moves to the soil surface and flies to a new host vine. Eggs laid by the winged form produce a sexual form which mates. The male dies immediately and the female lays one winter egg and the cycle begins anew. Both (b) and (c) are much enlarged, (b) being less than one mmm and (c) being about two mm in length. Drawings by M. Cameron.

tion. Thus the major purpose of rootstocks for grapevines is to prevent vine mortality resulting from phylloxera attack.

4. ORIGIN OF PHYLLOXERA-RESISTANT ROOTSTOCKS

American species of grapes had evolved over millenia in juxtaposition with the insect, and vine resistance resulted. This provided a means to solve the problem. According to Pongracz (27), Laliman first suggested the grafting of the very susceptible *V. vinifera* cultivars to *Vitis* species native to North America, and by 1878 cuttings from America arrived in France and grafting experiments began. T. V. Munson (22) also encouraged the practice, suggesting that "for northern regions like France, and other temperate climates where *vinifera* grapes endure, for sandy soils, the following: (a) *V. vulpina* (*V. riparia*), (b) *V. rupestris*, (c) *V. Longii*, named in order of preference; and for moderately limy *V. rupestris* and *V. Doaniana*. For very limy soil *V. Champini*, where ground does not freeze over 18 inches deep. All of these just named do well in sandy soils in the regions designated." Key was phylloxera resistance, as stated by Perold (24): "As the European vines are grafted on American stocks to prevent them being destroyed by phylloxera, the resistance of such a stock against phylloxera is one of its essentials."

As suggested by Munson's list, the initial efforts involved grafting *vinifera* vines to pure American grape species selected from wild ecotypes. Initially, clones of *V. riparia* and *V. rupestris* were the major selections. Early attempts to propagate rootstocks from seed were unacceptable. As Pongracz (27) rightly points out, "Each species embraces many different varieties, these varieties differing from one another as much as, for example, *V. vinifera* var. Cabarnet Sauvignon from the seedless *V. vinifera* var. Sultanina."

After much trial and error, certain selections of *V. riparia* and *V. rupestris* were made and two, 'Riparia Gloire' and 'Rupestris St. George' (= duLot) began to be accepted. Currently, however, both are declining in usage in France (9). Galet (9) states that these two species and *V. Berlandieri* are the three most phylloxera-resistant species and that is why the most desirable rootstock cultivars are crosses between these three species.

4.1. Other Soil Problems

4.1.1. Soil pH. The first American species used in Europe were primarily of *V. riparia* and *V. rupestris*. It is from these that the cultivars 'Riparia Gloire' and 'Rupestris St. George' (duLot) were selected. As observed in Table 1, 'Riparia Gloire' possesses very low tolerance to lime (6%) while 'Rupestris St. George' is only slightly better (15%). This posed serious problems in those French growing areas that have soils with a high lime content. Such areas included Graves, Cognac, and Champagne districts (14).

The effect of high lime on nontolerant species is to cause lime chlorosis (24).

TABLE 1. Important Grape Rootstocks, Their Species Background and Their Resistance to Biotic and Abiotic Soil-Borne Stresses[a]

Rootstock	Breeder or Selector	Species	Phylloxera Resistance	Nematode Resistance	Drought Resistance	Lime Resistance (%)	Salt Resistance (g/L)
Riparia Gloire	Portalis	V. riparia	5	2	1	6	0.7
Rupestris St. George	Sijas	V. rupestris	4	2	2	15	—
420 A	Millardet	riparia × berlandieri	4	2	2	20	—
5 BB	Teleki, Kaber	riparia × berlandieri	4	3	1	20	—
SO 4	Teleki	riparia × berlandieri	4	4	1	17	—
5 C	A. Teleki	riparia × berlandieri	4	4	1	17	—
161-49 C	Couderc	riparia × berlandieri	4			25	—
110 R	Richter	rupestris × berlandieri	4	2	4	17	—
99 R	Richter	rupestris × berlandieri	4	3	2	17	—
140 Ru	Ruggeri	rupestris × berlandieri	4	3	4	20	—
1103 P	Paulsen	rupestris × berlandieri	4	2	3	17	0.6
3309 C	Couderc	riparia × rupestris	4	1	1	11	0.4
3306 C	Couderc	riparia × rupestris	4	1	1	11	0.4
101-14	Millardet	riparia × rupestris	4	2	1	9	—
44-53 M	Malegue	riparia × rupestris × cordifolia	4	4	2	10	—
1616 C	Couderc	riparia × Solonis	3	1	1	11	0.8
1202 C	Couderc	rupestris × vinifera	2	1	2	13	0.8
AXR #1	Ganzin	rupestris × vinifera	2	1	2	13	0.8
41 B	Millardet	Berlandieri × vinifera	4	1	3	40	Very sensitive
333 EM	Foex	Berlandieri × vinifera	2	1	2	40	Very sensitive
1613 C	Couderc	Solonis × Othello[b]	2	4	2	low	—

(Continued)

457

TABLE 1. (*Continued*)

Rootstock	Breeder or Selector	Species	Phylloxera Resistance	Nematode Resistance	Drought Resistance	Lime Resistance (%)	Salt Resistance (g/L)
Dogridge	Munson	*V. champini*	2	4	2	?	—
Salt Creek	unknown	*V. champini*	2	4	2	?	—
Harmony	Weinberger and Harmon	1613c × Dogridge	2	4	2	?	—
Freedom	Cain	1613c × Dogridge	2	4	2	?	—

[a]Resistance scale: 5 = very resistant and 1 = very susceptible. After Galet (9); Kasamatis and Lider (15); Perold (24); Pongracz (27); and Winkler et al (34).
[b]Othello = *Labrusca × Riparia × Vinifera*
(—) no data available
(?) unknown

In this case, the high pH reduces the availability of iron to the vine, resulting in the symptoms of iron deficiency. Thus, when phylloxera killed vines in the important viticultural districts just mentioned, the grafting of the *V. vinifera* cultivars to selections of either *V. riparia* or *V. rupestris* resulted in another problem to be resolved, iron chlorosis.

Early efforts, recognizing that *V. vinifera* was lime tolerant, were made by crossing *riparia* or *rupestris* with *vinifera*. Examples of such crosses are 1202 C and AXR #1. The results were not superior in phylloxera resistance (Table 1). Either the *vinifera* predominated and phylloxera resistance was inadequate or the American species predominated and the lime tolerance was inadequate.

This was rectified by the use of *V. berlandieri*, which was collected by Berlandier in western Texas in about 1883 (22). Munson (22) states the virtues of this species: "As a stock to succeed in very dry calcareous soils, resist phylloxera, and live to a great age, we probably have no better species in the United States." This praise might cause one to wonder why pure species stocks of *berlandieri* were never popular. It is explained by Munson: "it has one generally serious fault—the difficulty of propagation from cuttings." For this reason, most of the rootstocks with *berlandieri* as a part of the parentage are crossings with *riparia*, *rupestris*, and *vinifera*. The rootstocks with the greatest lime tolerance are 41B and 333EM; each is the result of a *berlanderi* × *vinifera* cross and each tolerates up to 40% active lime in the soil (9). Table 1 gives the relative lime tolerance and species background of currently important rootstocks.

4.1.2. Nematode Resistance. Perold (24) was the first to report that eelworms (nematodes) caused injury and damage to grapevines. Nematodes are small roundworms which can cause damage to grapevines either by direct attack and feeding upon the roots or by serving as the vector of virus diseases. Important direct-feeding nematodes on grapevine roots include root knot (*Meloidogyne incognito*), dagger (*Xiphinema americana* and *X. index*), root lesion (*Pratylenchus vulnus*), citrus (*Tylenchulus semipenetrans*), and the ring nematode (*Criconemoides* spp.). Of these genera, Ramsdell and Bird (28) consider the dagger nematode to be of major importance as a virus vector.

Rootstocks have been selected that are resistant to nematodes (15) (Table 1). Nematode resistance is high in *V. champini* and is considered to be very important to rootstock researchers in California (15). Pongracz (27), however, takes exception to the use of any nematode-resistant stocks that are deficient in resistance to phylloxera:

> it would be sheer folly to attempt to reconstitute a vineyard in phylloxera infested regions on a nematode resistant rootstock whose resistance against the phylloxera is seriously questioned outside the English-speaking grape-growing countries (1613C, Harmony, Salt Creek, Dog Ridge, etc.).

In this, Pongracz is clearly correct. Phylloxera must be the first priority. However, soils differ in their chemical and physical characteristics and these differences influence the ease and degree of vine infestation by phylloxera. Phylloxera are in all Michigan vineyard soils, but are a much greater threat to vine health on the clay loams than they are on the sands and sandy loams. The phylloxera-resistant hybrid direct producer Baco Noir (Baco #1) is highly vigorous and produces a large vine canopy on clayloams or sandy loams in Michigan. The Kuhlman HDP, Marechal Foch, is more susceptible to phylloxera (27) than Baco Noir and is reduced in vine vigor and canopy capacity on the clay-loam soil. However on sands and sandy-loam soils, Marechal Foch grows nearly as vigorously and to as large a canopy capacity as Baco Noir. It thus appears that the value of a rootstock's high level of phylloxera resistance must be evaluated with consideration of the extent of the problem in the soil.

This is especially true in the eastern United States, where phylloxera are indigenous. As the native grape species evolved resistance in the presence of the root-feeding insect, it is equally probable that soil-borne pathogens, parasites, and predator organisms served to limit phylloxera population outbreaks in the same soils. In support of this is the observation that the *V. labruscana* cultivar 'Concord' grows to large vine capacity on their own roots in spite of a relatively low level of phylloxera resistance (27). Indeed, 'Concord' was rejected by French viticulturists because phylloxera caused vine death and resistant cultivars such as 'Noah,' 'Othello,' and 'Clinton' were selected (24). Such a vineyard condition might make the choice of a lesser phylloxera-resistant stock an acceptable choice, and if the stock also possessed salinity or nematode resistance and these were high-priority concerns, the lesser phylloxera-resistant stock could be preferred. If, however, such knowledge of soil phylloxera status is absent, one would be wise to follow the advice of Pongracz (27) and choose a rootstock with strong resistance to phylloxera.

4.1.3. Drought Resistance and Salt Tolerance. The culture of grapevines in regions where water is either lacking or in seasonal short supply makes the growth habit and rate of penetration of rootstocks into the soil in search of water a matter of considerable consequence in those geographic regions. Table 1 provides relative rankings of drought resistance. It is interesting to note that of the currently recommended rootstocks for California North Coast vineyards ('Rupestris St. George,' AXR #1), neither possesses much drought resistance; each does possess relatively high levels of salt tolerance (33).

This may be instructive. Viticultural regions that are lacking in rainfall are supplementally watered via irrigation. This is costly, and efforts continue to minimize the cost of supplemental water and its application. In almost every region where irrigation is a necessary part of viticulture, there is, sooner or later, a soil salinity problem. Even the best-quality irrigation water will have naturally dissolved salts. These will be deposited with irrigation water. It is therefore of little surprise that California, Australia, and Israel are world leaders in both vineyard irrigation technology and methods of solving vineyard soil salinity problems.

Table 1 suggests that 110R and 140 RU have superior drought tolerance and that 1103P and 41B are acceptable. Interestingly, 'Dog Ridge', 'Ramsey' (Salt Creek), 'Harmony', 'Freedom', 1202C, AXR #1, SO-4 and 1613C are all recommended stocks in California, and none has acceptable or superior drought resistance.

4.1.4. Soil-Borne Problems Potentially Resolved by Rootstock. A soil-borne fungus that devastates roots of grapevines in the U.S. southwest and in parts of Central America is the cotton root rot [*Phymototrichum omnivorum* (Shear)] Daggar (16). It is common in soils that have a high pH and low organic matter. The fungus attacks the roots of the grapevine and plugs vascular tissue (20).

Mortenson (21) did early work on the potential of native *Vitis* species for resistance to the disease and found tolerance and/or resistance in *V. candicans*, *V. berlandieri*, and *V. champini*. Further effort in plant breeding and selection is needed, however, since all presently used stocks, even the tolerant ones, do become infected by the disease (25).

Virus resistance among grape genotypes may provide a means of reducing either infection or economic losses. In preliminary work by Ramsdell (personal communication, 1985) the *V. labruscana* cultivar 'Niagara' showed great resistance to infection by peach rosette mosaic virus. Grafting studies to determine the efficacy of such resistance in protecting a scion cultivar are presently under way.

5. ROOTSTOCK–SCION RELATIONSHIPS

5.1. Influence of Rootstock on Scion Vigor

As noted in Table 2, differences in the influence of root system on the vigor of the scion cultivar have been observed. Rootstocks that have been suggested to yield high scion vigor are 'Rupestris St. George,' 99R, 140RU, 5BB, SO-4, AXR #1, 'Dog Ridge,' and 'Salt Creek' (Table 2). Rootstocks reportedly producing reduced or low vigor include 161-49C, 333EM, 'Harmony,' 41B, 1616C, 101-14 Mgt, 420A, and 'Riparia Gloire.'

The first and most obvious influence on vine vigor is the influence of the graft union (27). Stocks that tend to form poor unions will give erratic results, with the most common response being reduced vigor. However, there must be additional effects, because the relatively poor rootstock grafter 5BB is reported by Pastena (in Pongracz [27]) to impart high vigor. That has also been our Michigan observation of White Riesling and Chardonnay grafted to 5BB.

A second consideration is the growth habit and soil mass penetration of the root system of the stock. Nutrients and water are necessary for vigorous scion growth. A rootstock cultivar that possesses root characteristics resulting in deep soil penetration, a positive hydrotropism, and is efficient in uptake of water and nutrients required for growth will not limit the growth habit and physiology of the scion cultivar.

TABLE 2. Interaction of Rootstock and Scion[a]

Rootstock	Ease of Rooting	Ease of Bench Grafting	Affinity with *V. vinifera*	Scion Vigor
Riparia Gloire	3	2	2	2
Rupestris St. George	3	3	4	4
420 A	2	2	2	2
5 BB	2	2	1	4
SO-4	2	2	3	4
5 C	2	2	1	3
161-49 C	1	1	1	1
110 R	3	3	4	3
99 R	4	4	4	4
140 Ru	3	3	4	4
1103 P	3	3	4	3
3309 C	3	2	2	3
3306 C	3	2	2	3
101-14 Mgt	3	2	2	2
44-53 M	4	4	4	3
1616 C	3	2	—	2
1202 C	3	3	—	3
AXR No. 1	3	3	2	4
41 B	1	2	3	2
333 EM	1	2	3	1
1613 C	2	3	2	3
Dog Ridge	1	2	1	4
Salt Creek	1	2	1	4
Harmony	4	3	—	2
Freedom	3	3	—	3

[a]Desirability Scale: 5 = best and 1 = worst. After Galet (9); Kasamatis and Lider (15); Perold, (24); Pongracz (27); Weaver (33); and Winkler et al. (34).

5.2. Compatibility of Rootstocks with *V. vinifera* Cultivars

Perold (24) cites Teleki concerning the affinity of *V. vinifera* to native American stocks. Teleki stated that Couderc was the first to observe that some cultivars did well on one stock and poorly on another. In Perold's view, "Affinity is simply the behaviour of the European vine towards the American stock in the grafted state" (24).

In an ungrafted vine, tissues are organized and aligned so that transmission between above- and below-ground vine tissues is optimized. Once a cut is made, this organization and alignment is temporarily lost. In certain scion–stock interactions these tissues are never fully reorganized due to either anatomical or physiological differences or both. *V. riparia* differs strongly in its physiological functions from *V. vinifera*, but this is less true for other American species such as *V. rupestris* and *V. berlandieri* (27).

One concern to the reader of the viticultural literature on rootstocks is the degree to which values for a scion or rootstock characteristic are assigned on the basis of empirical observation versus critical experimentation. This problem has been approached by Rives (29). The stock–scion factors affinity and vigor were observed and Rives found that European rootstock breeders desire to produce stocks which promote moderate vigor and good affinity between the stock and scion. He expresses the frustration resulting from inadequate definitions and sets out to correct the situation.

He subdivides vigor into two aspects: (1) own-vigor—vigor related to the contribution of the scion cultivar, and (2) given vigor—vigor related to the contribution of the rootstock cultivar. Using data from several sources, an analysis was made using analysis of variance and Tukey's test for nonadditivity. This approach provided a means, short of direct experimental evidence, to evaluate the two major theories of rootstock–scion influence on vine vigor:

Theory I   Vine vigor is the sum of the own-vigor of the scion cultivar and the given vigor of the rootstock;

Theory II   Vigor results solely from an interaction between vigor contribution of both the rootstock and the scion.

If Theory I is correct, then the Tukey value for nonadditivity should not be statistically significant.

In the analysis, clear statistical differences among scion cultivars and among rootstock cultivars was shown, and the variation among rootstocks was generally smaller than that among scions.

Based on these data, Rives (29) concluded that grafted vine vigor (measured as weight of cane prunings) results from three effects: (1) an additive contribution by the scion, (2) an additive contribution of the rootstock, and (3) a nonadditive, interactive contribution specific to the scion and rootstock cultivars joined.

From this analysis, Rives gives definitions to vigor and affinity. To the additive properties of scion and rootstock cultivar he gives the terms *own-vigor* and *given vigor*, respectively. To the nonadditive property, he gives the term *affinity*. Based on this definition, the term *affinity* cannot be used as a synonym for *compatibility*, as is commonly the case in U.S. rootstock literature. In his view, *compatibility* would refer only to the relative ease of union formation and maintenance over the life of a vineyard.

In a later paper (18) the natural evolution of Rives' approach was accomplished by recognizing that the two-facet association plus an interaction are similar to the condition in quantitative genetics in which general and specific combining ability is determined. Using this approach on young grafted vines in hydroponic culture, they discovered that the genetic tool was useful in computing the quantitative contributions of the rootstock and scion cultivars as well as the interactive contribution.

5.3. Influence of Rootstock on Yield and Fruit Quality of the Scion Cultivar

Scion cultivars on different rootstocks produce different yields and varying levels of fruit quality. The prevailing wisdom was best put forth by Perold (24). He suggested that highest quality could be achieved only if the best cultivar's scions were grafted to moderately vigorous stocks and then cultivated in a manner to limit crop size. This is a theme that runs through most of European viticulture and those geographic areas culturally influenced by them today.

5.4. Influence of Rootstock on Vine Maturation and Cold Hardiness

There are reported differences among rootstocks in their vegetation cycle (22, 27). The cycle range is short to very long. Unfortunately, there is little critical data to indicate whether the cycle of the rootstock is influenced by its root system and therefore may possibly influence the scion cultivar in a similar way.

The same is true with cold hardiness. In our laboratory we have measured hardiness differences among canes and buds of rootstocks and, in fewer cases, the hardiness differences of canes and buds of scions cultivars grafted to those rootstocks (Table 3).

Table 3 give preliminary data from our laboratory which gives credence to the idea that rootstocks may influence the cold hardiness of the scion cane and bud tissues. Further studies on this topic are under evaluation in our laboratory.

TABLE 3. Ranked Relative Hardiness of Different Rootstocks and the Scion Cultivar on Those Rootstocks[a]

Rootstock	Hardiness	Scion Cultivar	Hardiness
5 A	5	W. Riesling I 239	—
3306 C	5		—
3309 C	5		5
5 C	2		1
5 BB	3		3
SO-4	1		3
Riparia Gloire	1		2
		Own-rooted	3

[a]Scale: 1–5; 5 is greatest hardiness, 1 is least hardy. Comparisons only within columns.

6. SPECIES BACKGROUND OF GRAPE ROOTSTOCKS

The following are descriptions of the important species other than *V. vinifera* that have been used in the production of grape rootstocks. The descriptions are after those published by Hedrick (12) and Munson (22):

A. Vitis riparia Michaux. Commonly called the "Riverside Grape" or "Riverbank Grape" because of the frequency with which it is found in the moist, sandy soils along the rivers and streams. Munson (22) suggests that this species is highly resistant to phylloxera, downy mildew (*Plasmopora viticola*), black rot (*Guignardia bidwelli*), and freezing stress (Table 4). In this last case, *V. riparia* is the most cold hardy of all native American grape species. *Riparia* also roots easily, preferring sandy soils with good moisture. Under such conditions, *riparia* is a strong, vigorous vine. The species is very susceptible to lime-induced chlorosis and is not resistant to drought. The root system tends to be near the soil surface, fibrous, and spreading. Munson (22) observes that soil penetration by the roots is only "fair." *Riparia* vines tend to be very early in both spring bud burst and fruit ripening.

B. Vitis rupestris Schull. Commonly called "Sand Grape" or "Sand Beach Grape." Munson (22) suggests that this species is highly resistant to phylloxera, downy mildew, and black rot and is quite hardy to freezing stress (Table 4). Cuttings are easily rooted and vines are moderately vigorous to vigorous when grown on sandy soils with moisture within 2–3 ft of the surface. While more tolerant to lime than *V. riparia*, the additional tolerance is still inadequate for high-pH soils of many vineyard soils. Similarly, *V. rupestris* is more drought tolerant than *V. riparia*, but not by very much. Roots of *V. rupestris* are not as fibrous as *V. riparia* and are more deeply penetrating of the soil mass than that species. *V. rupestris* tends to be early in both bud burst and fruit ripening, but not as early as *V. riparia*.

C. Vitis berlandieri Planchon. Commonly called "Little Mountain Grape." The species is highly resistant to phylloxera, downy mildew, and black rot and is resistant to drought stress. *V. berlandieri* has some difficulty rooting cuttings and is only moderately tolerant of freezing stress (Table 4). It is vigorous on both sandy soils and soils high in lime content. This last characteristic is very important and explains the interest in the species. The roots are less branched than *V. rupestris*, and Munson (22) observes that they are "hard, deeply penetrating" in their growth habit. This may explain why the species is very drought tolerant. Bud burst and fruit ripening are late in this species.

D. Vitis champini Planchon. Commonly called the "Adobe Land Grape." This common name relates the key characteristic of value in this species. It is very acceptable in "limy, adobe and drouthy soils and is very resistant to phylloxera." For this reason it was already being used considerably for rootstock in

TABLE 4. Adaptive Characteristics of Four Native American Species Used Extensively in Development of Grape Rootstock[a]

Species	Vigor	Soil Preference		Rooting Ease	Stress Tolerance		Insect & Disease Resistance			Seasons	
		Sandy	Lime		Drought	Freezing	Phylloxera	Mildew	Black Rot	Leaf	Ripening
Vitis berlandieri	4	4	5	2	5	3	5	5	5	Late	Late
Vitis champini	5	5	5	2	5	3	5	3	5	Early	Mid-season
Vitis riparia	4	5	1	5	1	5	5	5	5	Very Early	Very Early
Vitis rupestris	3	5	2	5	2	4	5	5	5	Early	Early

[a]Ranked 1 to 5, with 5 being greatest. After Munson (22).

California in the early parts of this century. Munson (22) says that the species is very resistant to phylloxera and black rot and moderately resistant to downy mildew. While only moderately tolerant to freezing stress, it is very drought tolerant, and grown very vigorously on either sandy or heavy, limy soils. A major weakness is the difficulty with rooting (Table 4). As in the drought-resistant *V. berlandieri* and in contrast with *V. riparia* and *V. rupestris*, the roots are "hard, wiry and deeply penetrating" into the soil mass.

7. PROPAGATION

There are several levels of propagation that are relevant to the grape rootstock question. Some are more important than others. Of lesser importance is seed propagation, since grape rootstocks are clonally propagated except for breeding efforts aimed at genetic improvement of rootstocks. Of greater importance are grafting and ease of rooting. The need for the latter is obvious. To propagate any grapevine there must be roots. Interestingly, given the data on ease of rooting grape rootstocks (Table 2), the current propagation texts make the general and erroneous suggestion that grapevines are easily rooted. Actually, a range of rooting ease exists among various hybrid direct producer (HDP) grapevines (8) and, while use of rooting-promoting hormones can facilitate the process, most rootstocks have been selected so that this is a low-priority item.

7.1. Bench Grafting

The most common method of producing grafted grapevines is via the process of bench grafting. The technique is described by Haesler and Romberger (10) and Becker and Hiller (7).

1. After leaf fall and enough cold to produce mature, hardy canes, collect both scion and rootstock canes. Cut to 12−13-in. lengths, disbud the rootstock wood, and bundle into 100−200 cane bundles. Soak the bundles in 0.5% Chinosol for 13−15 hr. Store at 1−3°C and 98% relative humidity in sealed polyethylene bags.

2. Store in this manner until the grafts are made. Storage is usually 2−4 months, so that callusing can be completed at about the time of the last hard frost.

3. The graft technique can be of several types involving both hand- and machine-produced grafts. The most common hand technique is the short whip graft, but most grapevines are machine grafted using a type of cut called the "omega" cut because of the shape. The resultant attachment is a good tight fit. This process if facilitated by a presorting of scion and stock canes by diameter.

4. Upon grafting, the union and scion portions are immediately dipped into a special grafting wax. The wax choice should melt at about 65−70°C. Special

waxes with a fungicide for botrytis is desirable. When dipping, the grafted end is briefly dipped in cold water and then very quickly in the melted wax.

5. The grafts are now ready for packing in a callusing box. The callusing box has four rigid sides, one removable side, and one open side. The box is laid down so that the removable side is up. A layer of dripping wet peat or sawdust is placed on the lower side to a depth of 2 to 3 in. The board that serves as the removable side is placed at the open end. The waxed grafts are laid horizontally on the packing and spaced so that they are evenly apart, with all scion buds oriented in the same direction and the top of the scions touching the removable side. Once a layer of grafts is in place, a small board is placed over the scion to cover the grafts just below the union. Over this is placed a layer of very wet peat or sawdust. The board over the scion and union prevents their becoming covered by the wet peat. The board is then removed and the process repeated until the box is filled. Then the removable side is replaced, the box turned upright, and a fine sifting of dry peat added to cover the buds and scions. Over this dry layer, 1 to 1.25 in. of soaked peat is evenly packed and pressed to provide a flat surface.

6. The filled boxes are then either stored at 1°C or placed in callusing rooms that are well lighted, with 96−98% relative humidity and a temperature of 30°C for three to four days and then reduced to 26°C. The light is important to produce green shoots once growth begins. To prevent botrytis infection, spray with a solution of 0.2% Benomyl at three- to five-day intervals. Callusing is completed when a complete layer of callus is formed between all points of contact between the scion and rootstock.

7. Begin lowering the temperature of the room gradually until the temperature equals the mean ambient temperature. After a week of this exposure, the vines may be removed outside. Water the vines thoroughly and protect against frosts.

8. Once the last frosts are past, plant in the nursery.

7.2. Field Grafting

Field grafting has been used in vine propagation. In the United States, it is used primarily as a means of changing cultivars. Vineyard establishment is more commonly accomplished via planting of bench grafted vines. Changes in demand for grape cultivars have resulted in increased usage of these techniques. The interested reader is referred to Twight (31), Harmon and Weinberger (11), Kimball (17), and Alley and his co-workers (1−6).

8. PHILOSOPHICAL THOUGHTS ABOUT STOCK−SCION INTERACTIONS

One of the difficulties encountered by a reader of the grape rootstock literature on the effects of stock−scion interactions is the confounded nature of the relationships. Given a complex plant characteristic such as fruit ripening, bud differentiation, cane maturation, vine productivity, or cold hardiness, the challenge for the viticultural rootstock researcher becomes one of sorting out the primary effects of the rootstock, for example, water relations, nutrient uptake, growth regulator production, from the secondary effects in which "rootstock" effect is mediated via well-understood influences on vine vigor, vine capacity, and canopy shading of both vegetative and reproductive tissues.

For example, consider the empirical observation that a given scion cultivar always ripens earlier on rootstock A than on rootstock B. However, the vine capacity of the scion cultivar is always greater on B than on A. Is this a *rootstock* effect? Occam's razor requires that we provide the simplest answer to any observed phenomenon. In this case, the multiple reports of the influence of internal vine shading on ripening is more logical an explanation than some unexplained, complex rootstock effect.

One might ask what practical difference this makes. It is of considerable practical importance. If it is indeed a canopy effect, the viticultural challenge ceases to be choice of rootstock and becomes one of effective vine canopy

TABLE 5. Differences in Cold Hardiness of Cane and Primary But Tissues of Concord Grapevines of Differing Age and Cropping Status.[a]

Treatment	Characteristics of Sample	Cane Hardiness (°C)	Primary Bud (T$_5$O) (°C)
1-Year potted vine	Dark	−27.0a	−25.0a
2-Year, nonbearing	Brown, well-exposed, small diameter (4−5 mm)	−27.5a	−27.5a
15-Year, bearing GDC,[b] balanced pruned, twice shoot positioned	Same	−26.5a	−24.5a
25-Year bearing GDC, not pruned for 3 years, no shoot positioning	Same	−27.0a	−25.0a
15-Year bearing GDC, balanced pruned, twice shoot positioned	Light brown, poorly exposed, medium diameter (6−7 mm)	−20.0b	−18.0b
25-Year bearing GDC, not pruned 3 years, no shoot positioning	Same	−20.5b	−17.5b

[a]Vine sampled on February 21, 1976.
[b]Geneva Double Curtain.

management. Further, if such an effect is attributed to the rootstock, erroneous viticultural management decisions can be the result. Finally, if there really are influences of rootstocks on the various scion characteristics mentioned earlier, then genetic improvement and selection are possible. Such improvement and selection will be severely limited unless methods which measure only primary rootstock effects are devised and employed.

Rigorous care in setting up experiments and in the use of stratified random sampling procedures (30) is required. In the case of cold hardiness, our research dictates that we sample cane tissues of comparable internode length and diameter and that have been equally exposed to sunlight and cropping stress. In our studies on vines of the same cultivar (13), we have learned that differences of up to 13°C in hardiness could exist on similar tissues on the *same* vine on the same date. Of equal importance, when *comparable tissues* were taken from nonfruiting 2-year vines, 15-year-old mature bearing vines, and from a 35-year-old abandoned vine of the same cultivar at the same location on the same date, *there was no hardiness difference* (Table 5). Using such a critical sampling procedure should establish whether rootstocks do directly influence vine maturation or cold hardiness.

A similar approach is necessary for yield and fruit quality. It is not easy to create such common conditions among rootstocks, but it can be done. Only if done can the real contribution of the rootstock to the factor being measured be determined. In addition, approaches by Rives (29) and Lefort and Legisle (18) suggest statistical tools which will be of value.

REFERENCES

1. Alley, C. J. (1981). Grapevine propagation. XVIII. Spring chip-budding of mature grapevines at high level from February through April, *Amer. Jour. Enol. Viticult.*, **32**, 26−28.

2. Alley, C. J. (1981). Grapevine propagation. XIX. Comparison of inverted with standard T-budding, *Amer. Jour. Enol. Viticult.*, **32**, 29−34.

3. Alley, C. J., and A. T. Toyama (1978). Vine bleeding delays growth of T-budded grapevines, *Cal. Agriculture*, **32**, 7.

4. Alley, C. J., R. A. Neja, and T. A. West (1979). Grapevine propagation. X. Effects of tape color, wax type, bud covering, and early vine topping on late spring chip and T-budded vines at Greenfield, Monterey County, California, 1976, *Amer. Jour. Enol. Viticult.*, **30**, 1−2.

5. Alley, C. J. (1975). Grapevine propagation. VIII. The side whip graft, an alternate method to the split graft for use on stocks 2−4 cm in diameter, *Amer. Jour. Enol. Viticult.*, **26**, 109−111.

6. Alley, C. J. (1975). Grapevine propagation. VII. The wedge graft—a modified notch graft, *Amer. Jour. Enol. Viticult.*, **26**, 105−108.

7. Becker, H., and M. H. Hiller (1977). Hygiene in modern bench-grafting, *Amer. Jour. Enol. Viticult.*, **28**, 113−118.

8. Ehrlinger, D., and G. S. Howell (1981). Differential rooting of hardwood cuttings of different grape cultivars, *The Plant Propagator*, **27**, 13−15.

9. Galet, P. (1979). *A Practical Ampelography*, trans. by L. T. Morton, Cornell Univ. Press., Ithaca, NY, 245 pp.

10. Haeseler, C. W., and G. A. Romberger (1976). The basics of bench-grafting, *Vinifera Wine Growers Jour.*, **3**, 233–239.

11. Harmon, F. N., and J. H. Weinberger (1969). *The Chip-bud Method of Propagating* Vinifera Grape Varieties on Rootstocks, USDA Leaflet No. 513, 5 pp.

12. Hedrick, U. P. (1908). *The Grapes of New York*, J. B. Lyons 15th Annual Report, State of New York, 564 pp.

13. Howell, G. S., and N. Shaulis (1980). Factors influencing within-vine variation in the cold resistance of cane and primary bud tissues, *Am J. Enol. Vitic.*, **31**, 158–161.

14. Johnson, H. (1971). *World Atlas of Wine*, Simon & Schuster, New York, 272 pp.

15. Kasamatis, A. N., and L. Lider (1980). *Grape Rootstock Varieties*, Univ. of California Extension Leaflet No. 2780, 19 pp.

16. Killough, D. T. (1916). *Grapes*, Texas Agr. Expt. Sta. Ann. Rept. 24–25.

17. Kimball, K. H. (1976). *Converting Mature Vineyards to Other Varieties*, New York Agri. Exp. Sta. Spec. Rept. No. 22, 19 pp.

18. Lefort, P. L., and N. Legisle (1977). Quantitative stock–scion relationships in vine. Preliminary investigations by the analysis of reciprocal graftings, *Vitis*, **16**, 149–161.

19. Little, V. A. (1963). *General and Applied Entomology*. Harper & Row, New York, 543 pp.

20. Lyda, S. D. (1978). Ecology of *Phymototrichum omnivorum*, *Ann. Rev. Phytopath*, **16**, 193–209.

21. Mortenson, E. (1952). *Grape Rootstocks for Southwest Texas*, Texas Agri. Expt. Sta. Progress Rept. No. 1475.

22. Munson, T. V. (1909). *Foundations of American Grape Culture*, T. V. Munson and Son, Denison, TX, 252 pp.

23. Pearis, L. M., and R. H. Davidson (1956). *Insect Pests of Farm, Garden and Orchard*, Wiley, New York, 661 pp.

24. Perold, A. I. (1927). *A Treatise on Viticulture*, Macmillan, London, 696 pp.

25. Perry, R. L., and H. M. Escamilla (1985). Rootstocks vs. cotton root rot, *Eastern Grape Grower and Winery News*, **11**, 59–61.

26. Planchon, J. E. (1883–1887). Monographie des ampelidees, *Act. C. de Condolle.*, **5**, 305–648.

27. Pongracz, D. P. (1983). *Rootstocks for Grapevines*, D. Phillip, Cape Town, S. A., 150 pp.

28. Ramsdell, D. C, and G. W. Bird (1982). *Vineyard Preparation for Nematode and Virus Disease Control*, Michigan Agri. Ext. Bull. E-806, 3 pp.

29. Rives, M. (1971). Statistical analysis of rootstock experiments as providing a definition of the terms "vigor" and "affinity" in grapes, *Vitis*, **9**, 280–290.

30. Steel, R. G. D., and J. H. Torrie (1960). *Principles and Procedures of Statistics*, McGraw-Hill, New York, 481 pp.

31. Twight, E. H. (1902). *New Methods of Grafting and Budding Vines*, Univ. of Calif. Agri. Expt. Sta. Bull. No. 146, 13 pp.

32. Wagner, P. (1972). *A Winegrowers Guide*, Knopf, New York, 234 pp.

33. Weaver, R. J. (1976). *Grape Growing*, Wiley, New York, 371 pp.
34. Winkler, A. J., J. A. Cook, W. M. Kleiwer, and L. A. Lider (1974). *General Viticulture*, Univ. of California Press, Berkeley, 710 pp.
35. Vine, R. P. (1981). *Commercial Winemaking, Processing and Controls*, AVI Publishing, Westport, CT, 493 pp.

INDEX

In this Index, each fruit type, with the exception of *Vitis* (grapes) is listed in the alphabetical sequence according to the fruit's common name, i.e., apple, pear, peach, walnut. Under this listing the salient features applicable to that particular fruit tree's rootstock are listed. A complete rootstock listing, specific to that fruit, is found under the entry "usage of rootstock." The main body of the Index lists unique or general topic entries including a full listing of genera and species. There is frequent cross-listing of some entries, for example: diseases, nematodes, viruses, etc.

473

P. syrica, 154, 157, 166, 168, 169
P. syrotina, 148
P. ussuriensis, 145, 146, 148, 154, 157, 162,
    165, 166, 168–171
P. xerophila, 154

Quiescent center, 6

Radical, 1
Replant problem, 18, 19, 60
Rhizome, 1
Rhizosphere, 24, 25
Root:
  browning, 11
  classifications:
    absorbing, 9, 10, 17
    axial, 9, 17
    conducting, 10
    fibrous, 5, 401, 465
    horizontal, 10
    intermediate, 10
    lateral, 12, 23
    primary, 9
    scaffold, 15
    secondary, 12
    vertical, 10, 15
  depth of rooting, 15, 19
  distribution, 10, 25
  exudates 17, 18, 25
  function:
    absorption, 19–22
    anchorage, 15, 19
    contribution to soil, 24, 25
    storage, 24
    synthesis, 22
  growth:
    cyclic, 14
    development, 9, 17
    extension, 15, 17, 19, 21, 23, 25
    lateral, 23, 24
    longevity, 14
    secondary growth, 10, 12, 24
    shedding, 14, 15
  hairs, 8, 19, 25
  injury to, 19, 24
  initiation of, 90
  pressure, 21
  regions, structural:
    cap, 5, 6, 15, 23
    elongation, 5, 6, 23
    maturation (differentiation), 5, 6, 18
    meristematic, 5, 6, 12
  rootstocks usage, see individual fruits

soil contact, 19, 21
tip, 22, 23

Seedling rootstocks, see individual fruits
Seedlings, 57–59
  germination, 57
  reason to use, 57
  technical requirements, 57
Self root, see Own root
Sieve tube necrosis, 85
Soil adaptation, see individual fruits
Soil solution
  ion movement, 18, 21, 22
  pH ranges, 18
Starch:
  blockage, 80
  concentration, 83, 92
  distribution, 81, 83
  reserves, 83
  storage, 93
Stele, 7, 22
Stion, 2
Stock plants, 38, 39
Stock–scion combinations, 79, 82
Stock–scion interaction, 82, 84, 113, 114,
    116, 117
Stoolbed, 34, 38
Stratification, seed, 57, 58
Suberin, 21
Suberization, 12, 16, 30
Suckers, 34, 328, 438, 440
Symplast, 21

Temperature effects:
  on absorption, 21
  freezing, 17
  high temperature, 16, 23
  low temperature, 17
  tolerance, 17
  see also individual fruits
Tissue culture, 83, 138, 353, 407
Transpiration, 20, 21
Tree harvesting, 68
Tree production, 60

Usage of rootstocks, see individual fruits

Vascular system:
  abnormal, 87
  connections, 86
  continuity, 80
  necrotic, 90
  swirling effect, 90